T0320514

Optical Switching Networks

Optical Switching Networks describes all the major switching paradigms developed for modern optical networks, discussing their operation, advantages, disadvantages, and implementation. Following a review of the evolution of optical wavelength division multiplexing (WDM) networks, an overview of the future of optical networks is set out. The latest developments of techniques applied in optical access, local, metropolitan, and wide area networks are covered, including detailed technical descriptions of generalized multiprotocol label switching, waveband switching, photonic slot routing, optical flow, burst, and packet switching. The convergence of optical and wireless access networks is also discussed, as are the IEEE 802.17 Resilient Packet Ring and IEEE 802.3ah Ethernet passive optical network standards and their WDM upgraded derivatives. The feasibility, challenges, and potential of next-generation optical networks are described in a survey of state-of-the-art optical networking testbeds. Animations showing how the key optical switching techniques work are available via the Web, as are lecture slides.

This authoritative account of the major application areas of optical networks is ideal for graduate students and researchers in electrical engineering and computer science as well as practitioners involved in the optical networking industry.

Additional resources for this title are available from www.cambridge.org/ 9780521868006.

Martin Maier is Associate Professor at the Institut National de la Recherche Scientifique (INRS), University of Quebec, Canada. He received his MSc and PhD degrees, both with distinctions (*summa cum laude*), from Technical University Berlin, Germany. He was a Postdoc Fellow at MIT and Visiting Associate Professor at Stanford University. His research interests include the design, control, and performance evaluation of next-generation optical networks and their evolutionary WDM upgrade strategies. Dr. Maier is the author of the book *Metropolitan Area WDM Networks – An AWG Based Approach*.

Optical Switching Networks

MARTIN MAIER
Université du Québec
Montréal, Canada

CAMBRIDGE
UNIVERSITY PRESS

CAMBRIDGE
UNIVERSITY PRESS

University Printing House, Cambridge CB2 8BS, United Kingdom

One Liberty Plaza, 20th Floor, New York, NY 10006, USA

477 Williamstown Road, Port Melbourne, VIC 3207, Australia

314-321, 3rd Floor, Plot 3, Splendor Forum, Jasola District Centre, New Delhi - 110025, India

79 Anson Road, #06-04/06, Singapore 079906

Cambridge University Press is part of the University of Cambridge.

It furthers the University's mission by disseminating knowledge in the pursuit of
education, learning and research at the highest international levels of excellence.

www.cambridge.org
Information on this title: www.cambridge.org/9780521868006

© Cambridge University Press 2008

First published 2008

A catalogue record for this publication is available from the British Library

Library of Congress Cataloging in Publication data
Maier, Martin.
    Optical switching networks / Martin Maier.
        p. cm.
    Includes bibliographical references and index.
    ISBN 978-0-521-86800-6 (hardback)
  1. Optical communications.   2. Telecommunication systems.   3. Computer networks.
I. Title
    TK5103.59.M36   2008
    621.382′7—dc22

                                                                    2007052216

ISBN 978-0-521-86800-6 Hardback

In love and gratitude to my wonderful wife
and our two little Canadians

Contents

Illustrations

Tables

Preface

Optical fiber is commonly recognized as an excellent transmission medium owing to its advantageous properties, such as low attenuation, huge bandwidth, and immunity against electromagnetic interference. Because of their unique properties, optical fibers have been widely deployed to realize high-speed links that may carry either a single wavelength channel or multiple wavelength channels by means of wavelength division multiplexing (WDM). The advent of Erbium doped fiber amplifiers was key to the commercial adoption of WDM links in today's network infrastructure. WDM links offer unprecedented amounts of capacity in a cost-effective manner and are clearly one of the major success stories of optical fiber communications.

Since their initial deployment as high-capacity links, optical WDM fiber links turned out to offer additional benefits apart from high-speed transmission. Most notably, the simple yet very effective concept of optical bypassing enabled network designers to let in-transit traffic remain in the optical domain without undergoing optical-electrical-optical conversion at intermediate network nodes. As a result, intermediate nodes can be optically bypassed and costly optical-electrical-optical conversions can be avoided, which typically represent one of the largest expenditures in optical fiber networks in terms of power consumption, footprint, port count, and processing overhead. More important, optical bypassing gave rise to so-called all-optical networks in which optical signals stay in the optical domain all the way from source node to destination node.

All-optical networks were quickly embraced by both academia and industry, and the research and development of novel architectures, techniques, mechanisms, algorithms, and protocols in the arena of all-optical network design took off immediately worldwide. The outcome of these global research and development efforts is the deployment of optical network technologies at all hierarchical levels of today's network infrastructure covering wide, metropolitan, access, and local areas.

The goals of this book are manifold. First, we set the stage by providing a brief historical overview of the beginnings of optical networks and the major achievements over the past few decades, thereby highlighting key enabling technologies and techniques that paved the way to current state-of-the-art optical networks. Next, we elaborate on the big picture of future optical networks and identify the major steps toward next-generation optical networks. The major contribution of this book is an up-to-date overview of the latest and most important developments in the area of optical wide, metropolitan, access, and local area networks. We pay particular attention to recently standardized and emerging high-performance switching paradigms designed for the cost-effective and bandwidth-efficient support of a variety of both legacy and new applications and services at all optical network hierarchy levels. In addition, we explain recently standardized Ethernet-based optical metro, access, and local area networks in great detail and report ongoing research on their performance enhancements. After describing the concepts and underlying techniques of the various optical switching paradigms at length, we take a comprehensive look at current testbed activities carried out around the world to better understand the implementation complexity associated with each of the described optical switching techniques, as well as to get an idea of what future optical switching networks are expected to look like. Finally, we include a chapter on the important topic of converging optical (wired) networks with their wireless counterparts.

This book was written to be used for teaching graduate students as well as to provide communications networks researchers, engineers, and professionals with a thorough overview and an in-depth understanding of state-of-the-art optical switching networks and how they support new and emerging applications and services.

Acknowledgments

I am grateful to Dr. Andreas Gladisch of Deutsche Telekom for introducing me to the exciting research area of optical networks many years ago. I also would like to thank my former advisor Prof. Adam Wolisz of the Technical University of Berlin for his guidance of my initial academic steps. In particular, I am grateful to my mentor Prof. Martin Reisslein from Arizona State University and his former PhD students Chun Fan, Hyo-Sik Yang, Michael P. McGarry, and Patrick Seeling for their immensely fruitful collaboration. I am deeply grateful to Dr. Martin Herzog for his significant contributions over the past few years and his review of parts of this book. Furthermore, I would like to acknowledge the outstanding support of Prof. Michael Scheutzow and his group members (former or current) Stefan Adams, Frank Aurzada, Matthias an der Heiden, Michel Sortais, and Henryk Zähle of the Technical University of Berlin. In addition, I am grateful to Prof. Chadi M. Assi and Ahmad Dhaini of Concordia University and Prof. Abdallah Shami of the University of Western Ontario for their excellent collaboration on performance-enhanced Ethernet PONs. I also would like to thank Prof. Eytan Modiano of the Massachusetts Institute of Technology and Prof. Leonid G. Kazovsky of Stanford University for being my hosts during my research visits and for their fruitful discussions and insightful comments.

At Cambridge University Press, I would like to thank Dr. Phil Meyler for offering me the opportunity to write this book and Anna Littlewood for making the publication process such a smooth and enjoyable experience.

Finally and most importantly, I am deeply grateful to my wife Alexie who supported and encouraged me with all her love, strength, and inspiration throughout the past year and a half while I wrote this book. This book is dedicated to my wife and our two children; it not only carries all the technical details but also the countless personal memories of our first two years in Canada.

Part I

Introduction

1 Historical overview of optical networks

Optical fiber provides an unprecedented bandwidth potential that is far in excess of any other known transmission medium. A single strand of fiber offers a total bandwidth of 25 000 GHz. To put this potential into perspective, it is worthwhile to note that the total bandwidth of radio on Earth is not more than 25 GHz (Green, 1996). Apart from its enormous bandwidth, optical fiber provides additional advantages such as low attenuation loss (Payne and Stern, 1986). Optical networks aim at exploiting the unique properties of fiber in an efficient and cost-effective manner.

1.1 Optical point-to-point links

The huge bandwidth potential of optical fiber has been long recognized. Optical fiber has been widely deployed to build high-speed optical networks using fiber links to interconnect geographically distributed network nodes. Optical networks have come a long way. In the early 1980s, optical fiber was primarily used to build and study point-to-point transmission systems (Hill, 1990). As shown in Fig. 1.1(a), an optical point-to-point link provides an optical single-hop connection between two nodes without any (electrical) intermediate node in between. Optical point-to-point links may be viewed as the beginning of optical networks. Optical point-to-point links may be used to interconnect two different sites for data transmission and reception. At the transmitting side, the electrical data is converted into an optical signal (EO conversion) and subsequently sent on the optical fiber. At the receiving side, the arriving optical signal is converted back into the electrical domain (OE conversion) for electronic processing and storage. To interconnect more than two network nodes, multiple optical single-hop point-to-point links may be used to form various network topologies (e.g., star and ring networks). Figure 1.1(b) shows how multiple optical point-to-point links can be combined by means of a star coupler to build optical single-hop star networks (Mukherjee, 1992). The star coupler is basically an optical device that combines all incoming optical signals and equally distributes them among all its output ports. In other words, the star coupler is an optical broadcast device where an optical signal arriving at any input port is forwarded to all output ports without undergoing any EO or OE conversion at the star coupler. Similar to optical point-to-point links, optical single-hop star networks make use of EO conversion at the transmitting side and OE conversion at the receiving side. Besides

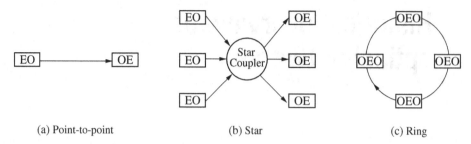

(a) Point-to-point   (b) Star   (c) Ring

Figure 1.1 Optical single-hop connections: (a) point-to-point, (b) star, and (c) ring configurations.

optical stars, optical ring networks can be realized by interconnecting each pair of adjacent ring nodes with a separate optical single-hop point-to-point fiber link, as depicted in Fig. 1.1(c). In the resultant optical ring network, each node performs OE conversion for incoming signals and EO conversion for outgoing signals. The combined OE and EO conversion is usually referred to as OEO conversion. A good example of an optical ring network with OEO conversion at each node is the fiber distributed data interface (FDDI) standard, which can be found in today's existing optical network infrastructure (Ross, 1986; Jain, 1993).

1.2   SONET/SDH

One of the most important standards for optical point-to-point links is the Synchronous Optical Network (SONET) standard and its closely related synchronous digital hierarchy (SDH) standard. The SONET standardization began during 1985 and the first standard was completed in June 1988 (Ballart and Ching, 1989). The goals of the SONET standard were to specify optical point-to-point transmission signal interfaces that allow interconnection of fiber optics transmission systems of different carriers and manufacturers, easy access to tributary signals, direct optical interfaces on terminals, and to provide new network features. SONET defines standard optical signals, a synchronous frame structure for time division multiplexed (TDM) digital traffic, and network operation procedures.

SONET is based on a digital TDM signal hierarchy where a periodically recurring time frame of 125 μs can carry payload traffic of various rates. Besides payload traffic, the SONET frame structure contains several overhead bytes to perform a wide range of important network operations such as error monitoring, network maintenance, and channel provisioning.

SONET is now globally deployed by a large number of major network operators. Typically, SONET point-to-point links are used in ring configurations to form optical ring networks with OEO conversion at each node, similar to the one depicted in Fig. 1.1(c). In SONET rings there are two main types of OEO nodes: the add-drop multiplexer (ADM) and the digital cross-connect system (DCS). The ADM usually connects to several SONET end devices and aggregates or splits SONET traffic at various speeds. The DCS is a SONET device that adds and drops individual SONET channels at any location. One major difference between an ADM and a DCS is that the DCS can be used to interconnect a larger number of links. The DCS is often used to interconnect SONET rings (Goralski, 1997).

1.3   Multiplexing: TDM, SDM, and WDM

Given the huge bandwidth of optical fiber, it is unlikely that a single client or application will require the entire bandwidth. Instead, traffic of multiple different sources may share the fiber bandwidth by means of multiplexing. Multiplexing is a technique that allows multiple traffic sources to share a common transmission medium. In the context of optical networks, three main multiplexing approaches have been deployed to share the bandwidth of optical fiber: (1) time division multiplexing (TDM), (2) space division multiplexing (SDM), and (3) wavelength division multiplexing (WDM).

- **Time division multiplexing:** We have already seen that SONET is an important example for optical networks that deploy TDM on the underlying point-to-point fiber links. Traditional TDM is a well-understood technique and has been used in many electronic network architectures throughout the more than 50-year history of digital communications (Green, 1996). In the context of high-speed optical networks, however, TDM is under pressure from the so-called "electro-optical" bottleneck. This is due to the fact that the optical TDM signal carries the aggregate traffic of multiple different clients and each TDM network node must be able to operate at the aggregate line rate rather than the subrate that corresponds to the traffic originating from or destined for a given individual node. Clearly, the aggregate line rate cannot scale to arbitrarily high values but is limited by the fastest available electronic transmitting, receiving, and processing technology. As a result, TDM faces severe problems to fully exploit the enormous bandwidth of optical fiber, as further outlined in Section 1.4.
- **Space division multiplexing:** One straightforward approach to avoid the electro-optical bottleneck is SDM, where multiple fibers are used in parallel instead of a single fiber. Each of these parallel fibers may operate at any arbitrary line rate (e.g., electronic peak rate). SDM is well suited for short-distance transmissions but becomes less practical and more costly for increasing distances due to the fact that multiple fibers need to be installed and operated.
- **Wavelength division multiplexing:** WDM appears to be the most promising approach to tap into the vast amount of fiber bandwidth while avoiding the aforementioned shortcomings of TDM and SDM. WDM can be thought of as optical frequency division multiplexing (FDM), where traffic from each client is sent on a different carrier frequency. In optical WDM networks the term wavelength is usually used instead of frequency, but the principle remains the same. As shown in Fig. 1.2, in optical WDM networks each transmitter i sends on a separate wavelength λ_i, where $1 \leq i \leq N$. At the transmitting side, a wavelength multiplexer collects all wavelengths and feeds them onto a common outgoing fiber. At the receiving side, a wavelength demultiplexer separates the wavelengths and forwards each wavelength λ_i to a different

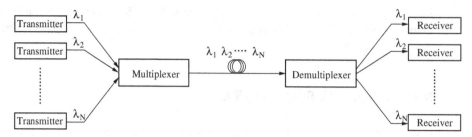

Figure 1.2 Wavelength division multiplexing (WDM).

receiver i. Unlike for TDM, each wavelength channel may operate at any arbitrary line rate well below the aggregate TDM line rate. By using multiple wavelengths the huge bandwidth potential of optical fiber can be exploited. As opposed to SDM, WDM takes full advantage of the bandwidth potential of a single fiber and does not require multiple fibers to be installed and operated in parallel, resulting in significant cost savings. Optical WDM networks have been attracting a great deal of attention by network operators, manufacturers, and research groups around the world, as discussed in Section 1.5.

It is worthwhile to note that in existing and emerging optical networks all three multiplexing techniques are used together to realize high-performance network and node architectures. By capitalizing on the respective strengths of TDM, SDM, and WDM and gaining a better understanding of their duality relationships, novel space–time–wavelength switching node structures may be found that enable future performance-enhanced optical networks (Kobayashi and Kaminow, 1996).

1.4   Optical TDM networks

Progress on the development of very short optical pulse technology enables the realization of optical time division multiplexing (OTDM) networks. OTDM networks aim at operating at an aggregate line rate of 100 Gb/s and above. At such high data rates, the transmission properties of optical fiber come into play and need to be taken care of properly. In particular, dispersion has a major impact on the achievable bandwidth-distance product of OTDM networks. Simply put, dispersion makes different parts of the optical signal travel at different speeds along a fiber link. As a result, parts of the optical signal arrive at the receiving side at different time instances, resulting in the so-called intersymbol interference (ISI), where the optical power of a given received bit interferes with that of adjacent bits. As a consequence, the optical power level of adjacent bits is changed and may lead to wrong decisions at the threshold detector of the receiver and transmission errors. This effect is exacerbated for increasing data rates and fiber lengths, translating into a decreasing bandwidth-distance product. Therefore, OTDM networks appear better suited for short-range networks where the impact of dispersion is kept small. For long-distance networks, dispersion effects can be avoided by the use of the so-called "soliton" propagation. With the soliton propagation, dispersion effects are canceled out by nonlinear effects of optical fiber, resulting in a significantly improved bandwidth-distance product (Green, 1996).

OTDM networks have been receiving considerable attention due to the progress of optical short-pulse technology. Apart from the aforementioned transmission issues in very-high-speed OTDM networks, other important topics have been addressed. Among others, research efforts have been focusing on the design of OTDM network and node architectures and advanced components (e.g., ultra-short-pulse fiber laser, soliton compression source, and optical short-pulse storage loop; Barry et al., 1996).

OTDM networks suffer from two major disadvantages: (1) due to the underlying TDM operation, nodes need to be synchronized in order to start transmission in their assigned time slot and thus avoid collisions on the channel; more important, (2) OTDM networks do not provide *transparency*. Synchronization is a fundamental requirement of OTDM networks and becomes more challenging for increasing data rates of 100 Gb/s and above. As for the missing transparency, note that OTDM network clients are required to comply with the underlying TDM frame structure. As a result, the TDM frame structure dictates the transmission and reception of client traffic and thereby destroys the transparency against arbitrary client protocols in that clients need to match their traffic and protocols to the OTDM framing format. To build optical networks that are transparent against different protocols, the optical signal must be able to remain in the optical domain until it arrives at the destination. Clearly, this can be achieved by avoiding OEO conversions at intermediate nodes. In doing so, data stays in the optical domain and is optically switched all the way from the source to the destination node, enabling end nodes to communicate with each other using their own protocol. By using optical switching components that are electronically controlled, transparent OTDM networks are getting closer to feasibility and deployment (Seo et al., 1996).

Transparent OTDM networks are an interesting type of optical network but are still in their infancy. Alternatively, optical WDM networks are a promising solution to realize transparent optical networks. In optical WDM networks, each wavelength channel may be operated separately without requiring network-wide synchronization, thus providing a transparent channel not only against protocol but also against data rate and modulation format, as opposed to OTDM networks. Compared to OTDM networks, optical WDM networks are widely considered more mature and are discussed at length in the following section.

1.5   Optical WDM networks

Optical WDM networks do not necessarily have to be transparent. Strictly speaking, optical WDM networks are networks that deploy optical WDM fiber links where each fiber link carries multiple wavelength channels rather than only a single one. Like any other optical network, optical WDM networks may consist of one or more simple point-to-point WDM links with OEO conversion at each network node, similar to the

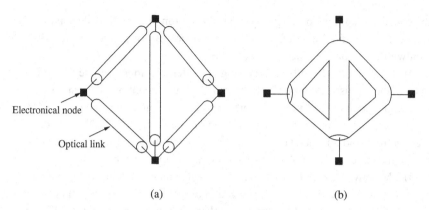

Figure 1.3 Optical WDM networks: (a) opaque and (b) transparent network architectures.

point-to-point link in Fig. 1.1(a) and ring network in Fig. 1.1(c). Optical WDM networks like that depicted in Fig. 1.1(c) are multihop networks where traffic traverses multiple intermediate nodes between any pair of source and destination nodes. Due to the fact that OEO conversion takes place at intermediate nodes, source and destination nodes are prevented from choosing their own protocol, line rate, and modulation format but have to follow the transmission requirements imposed by intermediate nodes. Thus, optical multihop networks with OEO conversion at intermediate nodes are unable to provide transparency to end nodes. In contrast, optical single-hop star networks similar to that shown in Fig. 1.1(b) inherently provide transparency for any pair of source and destination nodes. To see this, recall that the central star coupler is an optical device which does not perform any OEO conversion and leaves all in-transit traffic in the optical domain. As a result, end nodes are free to communicate with each other using their own agreed-upon protocol, data rate, and modulation format and are not hindered by any transmission requirements of intermediate nodes. The inherent transparency together with the simplicity of optical single-hop networks have led to a family of optical WDM networks known as *broadcast-and-select* networks (Mukherjee, 1992). Broadcast-and-select networks are WDM networks that are based on a central star coupler. Each transmitter is able to send on one or more different wavelengths. The star coupler broadcasts all incoming wavelengths to every receiver. Each receiver deploys an optical filter that is either fixed-tuned to a specific wavelength or tunable across multiple wavelengths. In either case, the optical filter selects a single wavelength and the destination is thus able to retrieve data sent on the selected wavelength.

Optical single-hop star WDM networks received considerable attention both from academia and industry. They are suitable for local area networks (LANs) and metropolitan area networks (MANs) where the number of nodes and distances are rather small. To build networks that are scalable in terms of nodes and coverage, optical WDM networks must be allowed to have any arbitrary topology (e.g., mesh topology). These networks can be categorized into two generations of optical WDM networks: (1) *opaque* and (2) *transparent* optical network architectures (Green, 1993). As shown in Fig. 1.3(a), in opaque optical WDM networks all wavelength channels are OEO converted at each network node, whereas in transparent optical WDM networks, as depicted in Fig. 1.3(b), intermediate nodes can be optically bypassed by dropping only a subset of wavelength channels into the electronical domain while leaving the remaining wavelength channels in the optical domain. Consequently, data sent on optically bypassing wavelengths can stay in the optical domain all the way between source and destination nodes, enabling transparent optical WDM networks. (For the sake of completeness, we note that there also exist so-called *translucent* optical networks. Translucent optical networks may be viewed as a combination of transparent and opaque optical networks where some network nodes provide optical bypassing capability while the remaining nodes perform OEO conversion of all wavelength channels. That is, translucent optical networks comprise both transparent and opaque network nodes.) Optical WDM networks with optical bypassing capability at intermediate nodes are widely referred to as *all-optical networks* (AONs) since the end-to-end path between source and destination is purely optical without undergoing any OEO conversion at intermediate nodes. AONs can be applied at any network hierarchy level. Unlike optical star networks, AONs are well suited for building not only optical WDM LANs and MANs but also optical WDM wide area networks (WANs). Due to their wide applicability, AONs have been attracting a great amount of attention by research groups and network operators worldwide.

1.5.1 All-optical networks

AONs are usually optical circuit-switched (OCS) networks, where circuits are switched by (intermediate) nodes at the granularity of a wavelength channel. Accordingly, OCS AONs are also called wavelength-routing networks. In wavelength-routing OCS networks, optical circuits are equivalent to wavelength channels. As mentioned earlier, AONs provide end-to-end optical paths by deploying all-optical node structures which allow the optical signal to stay partly in the optical domain. Such all-optical nodes are also called OOO nodes to emphasize the fact that they do not perform OEO conversion of all wavelength channels and in-transit traffic stays in the optical domain.

To understand the rationale behind the design of AONs, it is instructive to look at the similarities between AONs and SONET/SDH networks, which were discussed in Section 1.2. Note that both AONs and SONET/SDH networks are circuit-switched systems. The multiplexing, processing, and switching of TDM time slots in SONET/SDH networks are quite analogous to the multiplexing, processing, and switching of WDM wavelength channels in AONs. More precisely, in SONET/SDH networks lower-speed channels are multiplexed via byte interleaving to generate a higher-speed signal, where a SONET/SDH TDM signal can carry a mix of different traffic types and data rates. Furthermore, in SONET/SDH, ADMs and DCSs enable the manipulation and access to individual channels. Analogous functions can be found in AONs. As a matter of fact, the OOO node architectures used in AONs may be considered optical replica of the ADM and DCS node architectures of SONET/SDH, where electrical components are replaced with their optical counterparts. Accordingly, the resultant optical AON node architectures are called optical add-drop multiplexer (OADM) and optical cross-connect

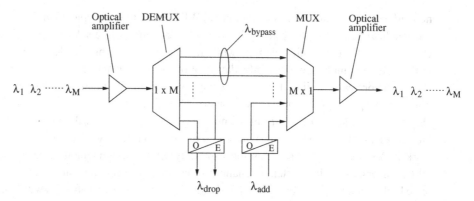

Figure 1.4 Optical add-drop multiplexer (OADM) with a single fiber link carrying M wavelengths.

(OXC), which are also known as wavelength ADM (WADM) and wavelength-selective cross-connect (WSXC), respectively (Maeda, 1998).

Figure 1.4 shows the basic schematic of an OADM with a single input/output fiber link that carries M different wavelength channels. At the input fiber the incoming optical signal comprising a total of M wavelengths $\lambda_1, \lambda_2, \ldots, \lambda_M$ is preamplified by means of an optical amplifier. A good choice for an optical amplifier is the so-called Erbium doped fiber amplifier (EDFA). A single EDFA is able to amplify multiple WDM wavelength channels simultaneously. After optical preamplification the WDM wavelength comb signal is partitioned into its M separate wavelengths by using a $1 \times M$ wavelength demultiplexer (DEMUX). In general, some bypass wavelengths λ_{bypass} remain in the optical domain and are thus able to optically bypass the local node. The remaining wavelengths λ_{drop} are dropped by means of OE conversion for electronic processing and/or storing at the local node. In doing so, the dropped wavelengths become available. The local node may use each of these freed wavelengths to insert local traffic on the available added wavelengths λ_{add}. Note that the dropped wavelengths λ_{drop} and added wavelengths λ_{add} operate at the same optical frequency but carry different traffic (locally dropped and added traffic, respectively). Subsequently, all M wavelengths are combined onto a common outgoing fiber by using an $M \times 1$ wavelength multiplexer (MUX). The composite optical WDM comb signal may be amplified by using another optical amplifier at the output fiber (e.g., EDFA).

The generic structure of an OXC with N input/output fiber links, each carrying M different wavelength channels, is shown in Fig. 1.5. An OXC is an $N \times N \times M$ component with N input fibers, N output fibers, and M wavelength channels on each fiber. A demultiplexer is attached to each input fiber (and optionally also an optical amplifier, similar to the previously discussed OADM). Each output from a demultiplexer goes into a separate wavelength layer. Each wavelength layer has a space division switch that directs each wavelength channel to a selected multiplexer. Each multiplexer collects light from M space division switches and multiplexes the wavelengths onto an output fiber. OXCs improve the flexibility and survivability of networks. They provide restoration and can reconfigure the network to accomodate traffic load changes and to compensate

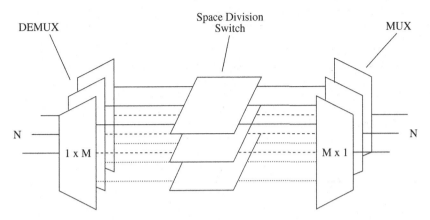

Figure 1.5 Optical cross-connect (OXC) with N fiber links, each carrying M wavelengths.

for link and/or node failures. An AON that deploys OXCs and OADMs is commonly referred to as an *optical transport network* (OTN). OTNs are able to provide substantial cost savings due to their flexibility, optical bypass capability, reconfigurability, and restoration (Sengupta et al., 2003).

AONs were examined by various research groups. Two major design goals of AONs are *scalability* and *modularity* (Brackett et al., 1993). Scalability is defined as the property that more nodes may always be added to the network, thereby permitting service to be offered to an arbitrarily large service domain. Modularity is defined as the property that only one more node needs to be added at a time. Besides scalability and modularity, AONs are intended to support a very large degree of wavelength reuse. Wavelength reuse allows wavelengths to be used many times in different locations throughout the network such that signals sent on the same wavelengths never interfere with each other. With wavelength reuse, bandwidth resources are used highly efficiently, resulting in an increased network capacity and decreased network costs. Toward the realization of scalable and modular AONs, significant progress has been made in the area of device technology, for example, acousto-optic tunable filters (AOTFs), multiwavelength lasers, multiwavelength receiver arrays, and other components (Brackett et al., 1993; Chidgey, 1994).

AONs are expected to support a number of different services and applications. For instance, provided services may comprise point-to-point as well as point-to-multipoint optical high-speed circuits. These services may be used to support applications such as voice, data, video, uncompressed high-definition TV (HDTV), medical imaging, and the interconnection of supercomputers (Alexander et al., 1993). AONs hold great promise to support all these different applications in a cost-effective fashion due to their transparency. To build large transparent AONs one must take the impact of physical transmission impairments on transparency into account, for example, signal-to-noise ratio (SNR), fiber nonlinearities, and crosstalk. For instance, the SNR poses limitations on the number of network nodes, and fiber nonlinearities constrain the number of used wavelengths and distances. As a result, transparency can be achieved only to a certain

Table 1.1. Wavelength conversion

Type	Definition
Fixed conversion	Static mapping between input wavelength λ_i and output wavelength λ_j
Limited-range conversion	Input wavelength λ_i can be mapped to a subset of available output wavelengths
Full-range conversion	Input wavelength λ_i can be mapped to all available output wavelengths
Sparse conversion	Wavelength conversion is supported only by a subset of network nodes

extent in large networks and it might be necessary to partition the network in order to guarantee transparency. In other words, large AONs might need to be split into several subnetworks, where each subnetwork is able to provide transparency. Such transparent subnetworks are often referred to as so-called *islands of transparency* (O'Mahony et al., 1995).

In AONs, the optical path between the source and destination remains entirely optical from end to end. Such optical point-to-point paths are termed *lightpaths* (Chlamtac et al., 1992). Lightpaths can be generalized in order to include multiple destination nodes. The resultant optical point-to-multipoint paths are called *light-trees*, where the source node resides at the root and the destination nodes are located at the leaf end points of the tree (Sahasrabuddhe and Mukherjee, 1999). A lightpath (as well as light-tree) may be optically amplified, keep its assigned wavelength unchanged, or, alternatively, have its wavelength altered along the path. If each lightpath has to stay at its assigned wavelength, it is said that the setup of lightpaths in the AON has to satisfy the *wavelength continuity constraint*. Clearly, this constraint makes it generally more difficult to accomodate lightpaths, leading to an increased blocking probability. To improve the blocking probability performance of AONs, OXCs may be equipped with additional wavelength converters, resulting in so-called wavelength-interchanging cross-connects (WIXCs). By using WIXCs the network becomes more flexible and powerful. The added capability of wavelength conversion helps decrease the blocking probability in AONs since the wavelength continuity constraint can be omitted.

1.5.2   Wavelength conversion

Wavelength conversion comes in several flavors. As shown in Table 1.1, wavelength conversion can be categorized into fixed, limited-range, full-range, and sparse conversion, as explained in the following. *Fixed* wavelength conversion may be considered the simplest type of wavelength conversion, where a given wavelength λ_i arriving at a node will always be converted to another fixed outgoing wavelength λ_j. This type of wavelength converter is static and does not provide any flexibility with respect to the output wavelength. The flexibility is improved by allowing the wavelength converter to map a given input wavelength λ_i to more than one output wavelength. If the output wavelengths comprise only a subset of the wavelengths that are supported on the outgoing fiber link, the type of conversion is called *limited-range* wavelength conversion. In contrast, *full-range* wavelength conversion does not impose any restrictions on the output wavelengths such that a given input wavelength λ_i can be converted to any wavelength available on the output fiber link. Wavelength converters are rather expensive devices. Clearly, the costs of AONs that deploy wavelength converters can be reduced by deploying wavelength converters only at a few well-selected network nodes rather than equipping each node with its own wavelength converter. This type of conversion is called *sparse* wavelength conversion.

Wavelength converters can be realized by OE converting the optical signal arriving on wavelength λ_i and retransmitting it on wavelength λ_j. Alternatively, wavelength converters may be realized all-optically by exploiting fiber nonlinearities (Iannone et al., 1996). In either case, wavelength converters provide several benefits. Wavelength converters help resolve wavelength conflicts on the output links of wavelength-routing network nodes and thereby reduce the blocking probability in optical circuit-switched AONs. In addition, wavelengths can be spatially reused to a larger extent, resulting in an improved bandwidth efficiency (Derr et al., 1995).

Apart from sparse wavelength conversion, another solution to reduce network costs is the sharing of wavelength converters inside a wavelength-interchanging network node. More precisely, all the wavelength converters at the node are collected in a converter bank which can be accessed by all the incoming circuits of that node. Alternatively, each outgoing fiber link owns a dedicated converter bank which can be accessed only by those circuits on that particular outbound link. The two approaches lead to converter *share-per-node* and *share-per-link* WIXCs, respectively. In general, share-per-node WIXCs outperform their share-per-link counterparts in terms of blocking probability (Lee and Li, 1993).

1.5.3   Reconfigurability

We have seen in the previous section that the use of wavelength converters helps improve the performance of wavelength-routing AONs and adds to the flexibility of the network. The flexibility of a network is the property of dynamically adjust to changing traffic and/or network conditions in order to improve the network performance. By using tunable wavelength converters (TWCs), the flexibility of AONs can be further increased. Similar to conventional wavelength converters, TWCs can have either an all-optical or a hybrid optoelectronic nature. For instance, in the latter case TWCs can be realized by using a tunable transmitter that is able to operate on several different output wavelengths instead of a transmitter that is fixed-tuned to a single output wavelength. Note that by deploying TWCs the wavelength-interchanging nodes and thereby the entire network become reconfigurable. Reconfigurability is a beneficial property of networks in that it enables the rerouting and load balancing of traffic in response to traffic load changes and/or network failures.

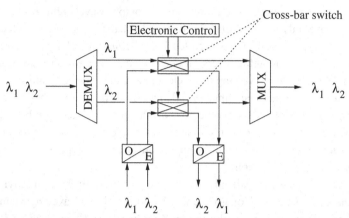

Figure 1.6 Reconfigurable optical add-drop multiplexer (ROADM) based on cross-bar switches with a single fiber link carrying two wavelengths.

The use of tunable transmitters is only one of many ways to render AONs reconfigurable and thus more flexible. In Section 1.5.1, we have briefly mentioned multiwavelength lasers. Note that a node may deploy a multiwavelength laser instead of a tunable transmitter in order to send traffic on multiple wavelengths. Alternatively, a node may deploy an array of fixed-tuned transmitters, each operating on a different wavelength. Using a multiwavelength transmitter array allows a node to send on multiple wavelengths at the same time, as opposed to a tunable transmitter which is able to send traffic on multiple wavelengths, however only on one at any given time. Similar observations can be made for multiwavelength receiver arrays which may be used instead of a single tunable receiver (e.g., AOTF).

Next, let us briefly get back to OADMs of Fig. 1.4 and discuss how they can be made reconfigurable. In conventional OADMs the optical add, drop, and in-transit paths are fixed. Hence, conventional OADMs are static and are able to add and drop only a prespecified set of wavelengths, without providing any possibility to (re)configure this set. Figure 1.6 depicts a reconfigurable optical add-drop multiplexer (ROADM) based on optical cross-bar switches. ROADMs become reconfigurable by using optical cross-bar switches on the in-transit paths between wavelength DEMUX and MUX. The cross-bar switch has two input ports and two output ports and is in either of two states at any given time: (1) *cross* or (2) *bar* state. In the bar state, the optical in-transit signal is forwarded to the opposite output port without being dropped locally. In this case, the local node is unable to add traffic. In the cross state, the optical in-transit signal is locally dropped. This allows the local node to add its own traffic which is forwarded to the wavelength multiplexer and finally onward onto the common outgoing fiber link. Thus, optical cross-bar switches enable either optical bypassing or adding/dropping operations. As shown in the figure, the state of each cross-bar switch is typically controlled electronically. In general, the cross-bar switches are controlled independently from each other such that each wavelength of the arriving WDM comb can be accessed separately by the local node.

Similar to ROADMs, OXCs can be made reconfigurable by electronically controlling the state of the space division switch of each wavelength layer (see Fig. 1.5). The resultant reconfigurable OXC (ROXC) can adapt to varying traffic loads and/or network conditions by setting its input–output connectivity pattern accordingly.

The aforementioned tunable and reconfigurable network elements together with some other previously discussed components such as the EDFA can be deployed to build high-performance multiwavelength reconfigurable AONs that are able to support a number of interesting applications, for example, video multicasting and multiparty video tele-conferencing (Chang et al., 1996).

Reconfigurable AONs can be used to realize powerful telecommunications network infrastructures but also give rise to some new problems. Given the fact that the network elements are reconfigurable, one must face the problem of how to find their optimal configuration under a given traffic scenario and to provide the best reconfiguration policies in the presence of traffic load changes, network failures, and network upgrades (Golab and Boutaba, 2004). Furthermore, the control and management of reconfigurable AONs is of utmost importance in order to guarantee their proper and efficient operation as well as to make them commercially viable (Wagner et al., 1996).

1.5.4   Control and management

For reconfigurable transparent optical networks to become commercially viable, network control and management functions need to be integrated into the optical networking architecture. Control functions are necessary to set up, modify, and tear down optical circuits such as lightpaths and light-trees across the optical network by (re)configuring tunable transmitters, receivers, wavelength converters, and reconfigurable OADMs and OXCs along the path, whereas management functions are necessary to monitor optical networks and guarantee their proper operation by isolating and diagnosing network failures, and triggering restoration mechanisms in order to achieve survivability against link and/or node failures. Survivability is considered one of the most important features of AONs apart from transparency, reconfigurability, scalability, and modularity (Maeda, 1998).

Control

In existing telecommunications networks, where each node has access to all in-transit traffic, control information might be carried in-band together with regular traffic. For instance, in SONET/SDH networks each time frame carries overhead bytes apart from payload which allow network elements to disseminate control as well as management information throughout the entire network. In transparent optical networks the situation is quite different since intermediate nodes might be optically bypassed and thereby prevented from accessing the corresponding wavelengths. Therefore, in transparent optical networks a separate wavelength channel is typically allocated to carry control and management information. This so-called *optical supervisory channel* (OSC) is OE converted, processed, and retransmitted (i.e., EO converted) at each node. As a result, the OSC can be used to distribute control and management information network-wide among all network nodes. For instance, by using the OSC the electronic controller of the ROADM of Fig. 1.6 can be instructed to set the cross-bar switch(es) as requested in order to drop or pass through the corresponding wavelength(s). Thus, the use of the OSC enables the *distributed* or *centralized* control of ROADMs as well as other reconfigurable and tunable network elements. In the distributed control approach, in principle any node is able to send control information to the electronic controller of a given reconfigurable or tunable network element and thus remotely control its state. Whereas in the centralized approach, in general only a single entity is authorized to control the state of the network elements, where the central control entity is traditionally part of the network management system (NMS).

Management

The general purpose of the NMS is to acquire and maintain a global view of the current network status by issuing queries to the network elements (OADM, OXC, etc.) and processing responses and update notifications sent by the network elements. Due to the fact that the OSC is optoelectronically dropped at each network element, a network element is able to determine and continuously update the link connectivity to its adjacent network elements and the characteristics of each of those links. Each network element stores this information in its adjacency table and sends its content to the NMS. The NMS uses the information of all network elements in order to construct and update a view of the current topology, node configuration, and link status of the whole network. Furthermore, the NMS may use this information for the set-up, modification, and tear-down of optical end-to-end connections.

In the context of reconfigurable transparent AONs much attention has been paid to the Telecommunications Management Network (TMN) framework. TMN has been jointly standardized by the International Telecommunication Union Telecommunication Standardization Sector (ITU-T) and the International Organization for Standardization (ISO). TMN encompasses a wide range of issues related to telecommunications networks, for example, planning, provisioning, installing, maintaining, operating, and administering networks. TMN incorporates a wide range of standards that cover network management issues which is often referred to as the FCAPS model. As shown in Table 1.2, the management issues of the FCAPS model covered by TMN are as follows:

- Fault Management
- Configuration Management
- Accounting Management
- Performance Management
- Security Management

Fault management consists of monitoring and detecting fault conditions, correlating internal and external failure symptoms, reporting alarms to the management system, and configuring restoration mechanisms. Examples for monitored parameters at network

Table 1.2. FCAPS model

Management issues covered by TMN

Fault management
Configuration management
Accounting management
Performance management
Security management

elements include optical signal power, SNR, and wavelength registration, which measures where the peak power of an optical signal occurs. These parameters can be used for performance management to monitor and maintain the quality of established optical circuits as well as fault management. Upon detection of fault conditions, a network element generates alarm notifications. Examples for fault conditions include transceiver card failure, environmental alarms (e.g., fire), and software failure. Accounting management, also known as billing management, provides the mechanisms to record resource usage and to charge accounts for it. Security management comprises a set of functions that protect the network from unauthorized access (e.g., cryptography).

Particular attention has been paid to the configuration management of reconfigurable transparent AONs. The configuration management provides connection set-up and tear-down capabilities. The connection management has been experimentally investigated and verified by a number of research groups. Wei et al. (1998) and Wilson et al. (2000) experimentally demonstrated a distributed implementation of a centralized TMN compliant management system. The considered connection management is able to provide connections to end users requesting transparent optical circuits or to higher layer transport facilities (e.g., SONET/SDH). Two paradigms for connection set-up and release were studied:

- **Management Provisioning:** Provisioned connection setup and release are initiated by a network administrator via a management system interface. By using global knowledge of the network, route selection and cross-connect activation at nodes along the path are coordinated. Provisioned connections have a relatively long life-span.
- **End-User Signaling:** Signaling connection set-up and release are initiated by an end user directly via a signaling interface without intervention by the management system. The signaling connection set-up and release are based on the signaling between the electronic controllers of network elements. Signaling connections are used for low-latency transport of traffic bursts.

Note that due to the transparency of AONs digital signal monitoring at the electrical level is not possible. In transparent AONs there may be failures that are hard to detect and isolate by means of optical monitoring. For instance, an OXC failure may be the placement of correct wavelengths at the correct ports, however with the incorrect digital information content. This OXC failure would go unnoticed by optical monitoring techniques. Given this shortcoming in conjunction with the aforementioned physical transmission impairments of transparent AONs, confining islands of transparency to a reasonable extent appears to be a practical solution to mitigate physical impairments and also reduce the complexity of management systems in reconfigurable transparent AONs (Maeda, 1998).

2 Optical switching networks

We have already seen that optical networks come in a large number of various flavors. Optical networks may have different topologies, may be transparent or opaque, and may deploy time, space, and/or wavelength division multiplexing (TDM, SDM, and/or WDM). They may comprise tunable devices, for example, tunable transmitters, tunable optical filters, and/or tunable wavelength converters (TWCs). Furthermore, to improve their flexibility optical networks may make use of reconfigurable optical add-drop multiplexers (ROADMs) and/or reconfigurable optical cross-connects. We will use the term *optical switching networks* to refer to all the various types of flexible and reconfigurable optical networks that use any of the aforementioned multiplexing, tuning, and switching techniques. Thus, optical switching networks are single-channel or multichannel (WDM) networks whose configuration can be changed dynamically in response to varying traffic loads and network failures by controlling the state of their tunable and/or reconfigurable network elements accordingly. Optical switching networks are widely deployed in today's wide, metropolitan, access, and local area networks and can be found at every level of the existing network infrastructure hierarchy.

2.1   End-to-end optical networks

Optical switching networks have been commonly used in backbone networks in order to cope with the ever-increasing amount of traffic originating from an increasing number of users and bandwidth-hungry applications. As shown in Fig. 2.1, optical switching networks can be found not only in wide area long-haul backbone networks but they also become increasingly the medium of choice in metro(politan), access, and local area networks (Berthelon et al., 2000). As a matter of fact, both telcos and cable providers are steadily moving the fiber-to-copper discontinuity point out toward the end users at the network periphery. Typically, phone companies deploy digital subscriber line (DSL) or any of its derivatives while cable providers deploy cable modems (Green, 2001). In both approaches, copper is used on the final network segment to connect subscribers. Clearly, this copper-based network segment forms a bottleneck between the vast amount of bandwidth available in today's optical backbone networks and the increasingly higher-speed clients at the network periphery. This bandwidth bottleneck is commonly referred

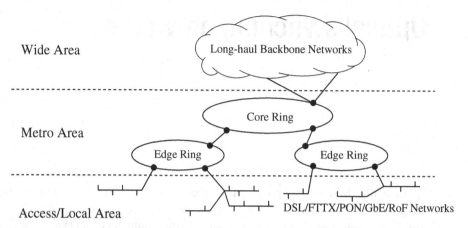

Wide Area

Metro Area

Access/Local Area

Figure 2.1 Hierarchy of optical switching networks.

to as the so-called *last mile bottleneck*, or also as the *first mile bottleneck* in order to emphasize its importance to the end users.

To mitigate or completely remove the first/last mile bottleneck, fiber is brought close or even all the way to the subscribers, who might be either business or residential users. Depending on the demarcation point of fiber, these optical access networks are called *fiber to the X* (FTTX) networks, where *X* denotes the various possible demarcation points of fiber. For instance, fiber to the building (FTTB) and fiber to the home (FTTH) networks serve a single building and home, respectively, while fiber to the curb (FTTC) networks may serve multiple buildings and homes of various business and residential subscribers.

To build cost-effective optical access networks at low capital expenditures (CAPEX) and operational expenditures (OPEX), FTTX networks are typically based on passive optical components without using amplifiers or any other powered devices. Accordingly, these FTTX networks are called *passive optical networks* (PONs). As opposed to metropolitan and wide area networks, costs are shared among a rather small number of subscribers in access networks. PONs not only help reduce the costs of access networks significantly but also simplify their operation, administration, and maintenance (OAM) as well as their management due to the completely passive nature of their underlying components. PONs come in different flavors. The transport of data traffic over PONs may be based on asynchronous transfer mode (ATM), resulting in the so-called ATM PON (APON) (van de Voorde and van der Plas, 1997). Alternatively, Ethernet frames may be transported in their native format in the so-called Ethernet PON (EPON), which aims at converging the low-cost equipment and simplicity of Ethernet and the low-cost fiber infrastructure of PONs (Kramer and Pesavento, 2002). At present, the fastest growth sector of communications networks is in the access sector. Optical access networks play a key role in providing broadband access for all, whereby cost reduction is currently considered more important than capacity and speed increase (Houghton, 2003). It is expected that currently widely used DSL access solutions will be replaced with FTTH. The question now is whether ATM- or Ethernet-based PONs are more promising to realize cost-effective broadband optical access networks. Given the low costs of Ethernet equipment and by capitalizing on the technology and products of the Ethernet LAN market, EPON appears to be in a somewhat advantageous position over ATM-based PONs. As a matter of fact, in Japan a carrier-grade EPON system has already been developed and commercially deployed in order to realize highly successful FTTH networks (Shinohara, 2005). These FTTH networks are able to provide a seamless migration from plain old telephone service (POTS) to new emerging services such as IP telephony and IP video. IP telephony and IP video are only two examples of new emerging services that will require plenty of bandwidth. To enable future bandwidth-hungry applications as well as to respond to service upgrade requests of individual customers the use of WDM on the already installed fiber infrastructure provides an evolutionary multichannel upgrade path of EPON access networks, leading to WDM EPONs (Shinohara, 2005).

Ethernet is by far the predominant technology in today's local area networks (LANs). The line rate and transmission range of Ethernet networks have been steadily increased over the last few years. The 10-Gigabit Ethernet (10GbE) provides an extended maximum transmission range of 40 kilometers, compared to Gigabit Ethernet's (GbE) 5 kilometers, and runs over optical fiber. 10GbE technology is not limited to high-speed LANs but is considered a very promising low-cost solution also for optical high-speed MANs and WANs. 10GbE equipment costs are about 80% lower than that of SONET equipment and it is expected that 10GbE services will be priced 30–60% lower than other managed network services. The use of Ethernet technology not only in today's LANs but also in future access, metropolitan, and wide area networks will potentially lead to end-to-end Ethernet optical networks (Vaughan-Nichols, 2002; Hurwitz and Feng, 2004).

Future access networks will be very likely either optical or wireless. Both optical and wireless access networks have their own merits and shortcomings. Optical access networks require fiber cabling and do not go everywhere, but provide practically unlimited bandwidth wherever they go. In contrast, wireless access networks are rather bandwidth limited but enable mobility and reachability of users. Therefore, future access networks are likely to be bimodal using optical and wireless technologies. So-called *radio-over-fiber* (RoF) networks may be viewed as the final frontier of optical networks which interface optical access networks with their wireless counterparts (Lin et al., 2004; Lin, 2005).

2.2   Applications

Many of today's applications can be roughly broken down into the following two categories (Green, 2006):

- **Latency-Critical Applications:** This type of application comprises small- to medium-size file transfers which require low latency. Examples of latency-critical applications are broadcast television, interactive video, video conferencing, security video monitoring, interactive games, telemedicine, and telecommuting.

- **Throughput-Critical Applications:** This type of application includes large-size file transfers whose latency is not so important but which require much bandwidth. Examples of throughput-critical applications are video on demand (VoD), video and still-image email attachments, backup of files, program and file sharing, and file downloading (e.g., books).

Applications have a significant impact on the traffic load and throughput-delay performance requirements of (optical) networks. Depending on the application, traffic loads and throughput-delay performance requirements may change over time. For instance, web browsing has led to rather asymmetric network traffic loads with significantly less upstream traffic than downstream traffic. Web browsing is based on the client–server paradigm, where each client sends a short request message to a server for downloading data files. In future (optical) networks, the traffic load is expected to become less asymmetric due to the fact that so-called peer-to-peer (P2P) applications become increasingly popular. Napster and its successors are good examples of P2P applications, where each client also acts as a server from which other clients may download files (e.g., photos, videos, and programs). P2P applications traffic keeps growing and may already represent the major traffic load in existing access networks (Garcia et al., 2004). The steadily growing P2P application traffic will eventually render the network traffic loads more symmetric.

The demand for more bandwidth is expected to keep increasing in order to support new emerging and future applications. Among others, high-definition television (HDTV) is expected to become increasingly popular with service operators offering much more than only a few HDTV channels, as done at present. Also, Grid computing has been receiving a great deal of attention as a means of enhancing our capabilities to model and simulate complex systems arising in scientific, engineering, and commercial applications (Darema, 2005). Grid computing provides applications with the ubiquitous on-demand access to local or remote computational resources such as storage systems and visualization systems.

2.3   Services

The aforementioned applications make use of services that are provided by the underlying (optical) network. In general, network services can be classified as follows:

- **Connection-Oriented Services:** Connection-oriented services require a connection to be established before any data can be sent by executing a specific handshake procedure between sender and destination. In doing so, sender and destination and possibly also intermediate nodes need to maintain state information for each established connection. Examples for connection-oriented network protocols are the transmission control protocol (TCP), ATM, and multiprotocol label switching (MPLS). TCP networks establish end-to-end connections where only the sender and destination of a given connection have to maintain status information. In ATM and MPLS networks, virtual circuits are established whose state information is maintained not only at the corresponding sender and destination but also at each intermediate node. By using the state information, connection-oriented services are able to recover from data loss and provide quality of service (QoS) to applications, where different services are defined by means of so-called *service-level agreements* (SLAs).

- **Connectionless Services:** Connectionless services do not require the establishment of any connection prior to sending data. The sender can simply start transmitting data to the destination according to the access control in use. Connectionless services are less reliable than their connection-oriented counterparts, but they are well suited for the transfer of best-effort traffic. A good example for connectionless network protocols is the Internet protocol (IP) which provides best-effort service to higher protocol layers and applications.

In the following, let us briefly highlight some of today's most important services that are offered to residential or business users. Cable companies offer so-called *triple-play* services to mainly residential users via their broadband access networks (e.g., FTTX networks). Triple-play offerings comprise bidirectional voice, bidirectional data, and unidirectional video services. Typically, all three services are delivered to the subscriber by the same facilities.

Besides triple-play services, *virtual private network* (VPN) services are expected to be increasingly deployed in the near to mid term. A VPN supports a closed community of authorized users, allowing them to access various network-related services and re-sources (Venkateswaran, 2001). Unlike leased private lines, a VPN is a virtual topology that is built on a physical network infrastructure whose resources may be shared by mul-tiple other VPNs. As a result, VPNs provide a more cost-effective solution to companies which want to interconnect their geographically distributed sites than costly dedicated lines. Similar to leased private lines, VPNs are able to provide privacy by isolating traffic belonging to different VPNs from each other. VPNs offer services that enable a mobile, telecommuting workforce as well as strategic partners to remotely access the intranet and/or extranet of a company. VPN services are offered to interconnect LANs located at different sites, to provide employees with dial-up access to the company's Intranet from remote locations, and to enable strategic partners, customers, or suppliers to remotely access the company's Extranet. VPNs can be realized at the link layer (L2) or network layer (L3), resulting in L2VPNs or L3VPNs, respectively (Khanvilkar and Khokhar, 2004; Knight and Lewis, 2004).

2.4   Switching granularity

The aforementioned services are offered by the underlying optical switching networks. As shown in Fig. 2.2, optical switching networks offer services to applications (e.g., file sharing). Toward this end, optical switching networks provide connectivity among various customers in a bandwidth-efficient and cost-effective manner. The connections provided by optical switching networks must be able to adapt to time-varying network

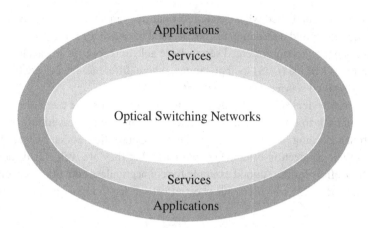

Figure 2.2 Optical switching networks offering services to applications.

conditions and traffic requirements dynamically in order to optimize the network performance and utilization of network resources. Depending on the traffic demands of various clients optical switching networks need to be able to provide different types of connections.

One way to categorize connections in optical switching networks is to look at the switching granularity (Finn and Barry, 1996). In optical switching networks, connections between source and destination may be switched at different granularities. Table 2.1 lists the major switching granularities deployed in optical switching networks, ranging from coarse granularity at the top to fine granularity at the bottom of the table.

Let us briefly explain the various switching granularities in the following, beginning at the top of the table. Fiber switching is done by switching all data arriving on an incoming fiber to another outgoing fiber without making any difference among the various wavelength channels carried on the fiber. Thus, fiber switching allows only for a very coarse switching granularity. Keeping in mind that in optical switching networks each fiber typically carries multiple wavelength channels, the next level of switching granularity is waveband switching. With waveband switching the set of wavelength channels carried on the fiber is divided into multiple adjacent wavebands, where each waveband contains two or more contiguous wavelength channels. Wavebands are switched independently from each other; that is, network nodes are able to switch individual wavebands arriving on the

Table 2.1. Switching granularity

Fiber switching
Waveband switching
Wavelength switching
Subwavelength switching
Optical circuit switching (OCS)
Optical burst switching (OBS)
Optical packet switching (OPS)

same incoming fiber to different outgoing fibers. Clearly, waveband switching provides a finer switching granularity than the aforementioned fiber switching. The switching granularity can be further improved by means of wavelength switching. Wavelength switching may be viewed as a special case of waveband switching where each waveband contains a single wavelength channel. With wavelength switching an input WDM comb signal is first demultiplexed into its individual wavelength channels and each wavelength channel is then switched independently. A given wavelength channel may carry the data of a single client at full rate or may be interleaved by means of TDM, resulting in optical TDM (OTDM). In OTDM networks, each time slot carries the data of a different client and may be switched individually. Unlike fiber, waveband, and wavelength switching, OTDM networks allow for subwavelength granularity switching.

Note that all switching techniques discussed so far are optical circuit switching (OCS) techniques. In OCS networks, circuits – fibers, wavebands, wavelengths, or time slots – are dedicated to source–destination node pairs and unused bandwidth cannot be claimed by other nodes. As a result, for bursty traffic OCS networks suffer from wasted bandwidth and they are less bandwidth-efficient than optical packet switching (OPS) networks which allow for statistical multiplexing. Statistical multiplexing does not dedicate bandwidth to nodes and is particularly useful to support bursty data traffic efficiently. OPS networks represent another example for subwavelength switching optical networks, where packets on a given wavelength channel are switched independently from each other. OPS networks are a promising solution to build efficient optical switching networks. However, they face some technological challenges due to the fact that packet-switched networks require buffers to resolve contention. At present, optical random access memory (RAM) is not feasible. Instead, so-called fiber delay lines (FDLs) are used. FDLs are fiber loops where the optical signal continues to circulate until it is forwarded. Clearly, FDLs are not suited to build very efficient OPS networks. To see this, note that the reading/writing of the information is restricted to integer multiples of the round-trip time of the loop due to its cyclic nature. Furthermore, FDLs must be designed such that the largest packet size can be stored on the loop, which leads to an increased delay for small-size packets. Given these technological constraints, current OPS networks are more suited to support fixed-size packets rather than variable-size packets. In other words, OPS networks favor cell switching, where each cell has the same fixed size.

Both OCS and OPS networks have their own merits and shortcomings. OCS networks are able to provide circuits with a high level of QoS. At the downside, OCS networks are connection oriented and typically require a two-way reservation to set up optical circuits. In contrast, OPS networks are connectionless and thus avoid any delay due to circuit set-up. But OPS networks do not necessarily meet QoS requirements. Recently, much attention has been paid to an alternative optical switching technique called optical burst switching (OBS) (Chen et al., 2004). OBS aims at combining the respective strengths of OCS and OPS while avoiding their drawbacks. More precisely, in OBS networks client data is aggregated at the network ingress and sent as bursts across the network. For each burst a reservation control packet is sent on a dedicated control wavelength channel prior to sending the burst on one of the data wavelength channels after a prespecified offset time. Hence, there is an offset time between the control packet and the corresponding burst. The control packet is used to configure intermediate OBS nodes on the fly along the path between the ingress node and egress node of the OBS network. The offset time is set such that the burst can be all-optically switched at intermediate nodes in a cut-through manner. In doing so, the burst does not need to be OEO converted, buffered, and processed at intermediate nodes, thus avoiding the need for optical RAM and FDL, as opposed to OPS networks. OBS networks deploy a one-way reservation instead of a two-way reservation, as is done in OCS networks. Moreover, OBS networks allow for statistical switching at the burst-level granularity. Also, by controlling the offset time, service differentiation and various QoS levels can be achieved.

2.5   Interlayer networking

The aforementioned switching techniques work at the so-called *data plane* of optical switching networks. The data plane encompasses all switching mechanisms and techniques required to send and forward data from source to destination. In order to work properly, the data plane needs to be controlled by means of a *control plane*. For instance, we have seen that in OBS networks a reservation control packet is sent prior to the data burst in order to configure intermediate nodes along the path from ingress to egress of the OBS network and to achieve a certain level of QoS (e.g., negligible burst loss). Generally speaking, the control plane is responsible for setting up and making the various switching techniques work in a coordinated and efficient manner throughout the optical network. The control plane can be implemented at the medium access control (MAC) layer of optical switching networks to avoid or mitigate collisions of data frames on each wavelength channel. Alternatively, the control plane can be implemented at the network layer to route and establish optical connections from source to destination.

To realize the control plane of optical switching networks there exist basically two approaches. In the first approach, new control protocols are designed for the optical switching networks under consideration, taking their respective properties into account. For instance, to set up and tear down lightpaths in wavelength-switching optical networks between hosts for large file transfers, a new signaling protocol that is implemented in hardware was proposed in Veeraraghavan et al. (2001). The second approach does not call for new control protocols to realize the control plane of optical switching networks but leverages on already existing control protocols which were successfully deployed in electronic data networks. Specifically, the adoption of IP signaling and routing protocols has been receiving a great deal of attention from both industry and academia to build an IP-centric control plane, where IP clients are able to dynamically set up, modify, and tear down point-to-point lightpaths (Rajagopalan et al., 2000). In doing so, all-optical networks (AONs) were not considered in isolation any longer. Instead, *interlayer networking* between AONs and IP clients became the next evolutionary step in designing flexible and resilient optical switching networks, leading to so-called IP/WDM networks. For such IP-over-optical networks, the following *interconnection models* have been proposed:

- **Peer Model:** In the peer model, IP networks and optical networks are considered an integrated network with a unified control plane. Hence, with respect to routing and signaling, electronic IP routers and optical nodes (e.g., OADMs and OXCs), act as peers without any difference between them. As a result, IP and optical networks can be interconnected in a seamless fashion. At the downside, the peer model implies that optical network nodes need to provide IP routers with full routing information. Therefore, the peer model may not be favored by many optical network operators for security reasons.
- **Overlay Model:** The overlay model envisions that IP networks and optical networks operate completely independently from each other. In this model, IP networks and optical networks run their own set of routing and signaling protocols, respectively. In order to interconnect geographically distributed IP routers across optical networks appropriate interfaces between both types of network need to be standardized.
- **Interdomain (Augmented) Model:** In the interdomain model, which is also known as the augmented model, both IP and optical networks have their own routing instances, but reachability information of geographically distributed IP routers is passed by the underlying optical networks onward to IP clients. Thus, the interdomain model may be viewed to be between the peer model and overlay model with the optical and electronic domains interacting to some extent.

Various standardization bodies and industry fora have been working on the standardization of a control plane for optical switching networks that enables the interoperability among multiple carrier networks and multivendor platforms (Benjamin et al., 2001). The resultant standards will allow end users and network operators to dynamically control end-to-end connections across optical switching networks. The major players involved in the standardization process are the International Telecommunication Union-Telecommunication Standardization Sector (ITU-T), Internet Engineering Task Force (IETF), Optical Internetworking Forum (OIF), and T1X1. ITU-T works on a framework referred to as *automatic switched transport network* (ASTN) for the control plane which supports arbitrary network topologies and various restoration options. The ASTN framework deals with a number of network functions (e.g., autodiscovery of network topology and resources) and a number of important interfaces such as the node-to-node interface (NNI) and optical user-network interface (O-UNI). The O-UNI allows a client to set up and tear down connections across the optical network by using *generalized multiprotocol label switching* (GMPLS) protocols, which are specified by the IETF. The O-UNI functionality is assessed in the OIF and its requirements are determined in the T1X1 and ITU-T. Note that the ASTN framework is not restricted to optical networks. In the context of optical switching networks, ASTN is also referred to as *automatic switched optical network* (ASON) (Varma et al., 2001).

Optical WDM networks using an IP-centric control plane have been studied by many research groups worldwide over the last few years. IP/WDM networks are widely considered a very promising solution to build high-performance optical switching networks at reduced costs. Many important issues in IP/WDM networks have been addressed. Among others, the routing and wavelength assignment (RWA) problem in IP/WDM networks was examined in Assi et al. (2001). The RWA problem deals with the route selection of lightpaths and the assignment of a wavelength on each WDM link along the selected route with the objective to optimize the network performance (e.g., blocking probability). To improve the bandwidth efficiency, network controllers can be used for the dynamic set-up of lightpaths (Sohraby et al., 2003). The network controller is used to trigger the set-up of lightpaths based on the current network status such as link utilization. A critical issue in IP/WDM networks is the reliability not only of the data plane but also of the control plane. The control plane must be designed so that it is able to reroute optical connections in the event of node and/or link failures in the data plane. In addition, the control plane must be designed such that failures in the control plane do not affect optical connections in the data plane (G. Li et al., 2002).

The ASON concepts and GMPLS protocols described earlier are well suited for conventional centrally managed optical switching networks of most of today's network operators. Alternatively, so-called *customer-controlled* and *customer-managed* optical networks are becoming increasingly common among large enterprise networks, research networks, and government departments (Arnaud et al., 2003). Unlike conventional centrally managed optical networks, in customer-controlled and -managed optical networks customers acquire their own dark fibers and/or point-to-point wavelengths and are fully responsible for the network control and management, including protection and restoration, without requiring any service from a central management. One way to realize customer-controlled and -managed optical networks is to build "condominium" networks, where customers purchase their own fiber infrastructure and optical network equipment in a condominium arrangement, similar to the idea of condominium apartment buildings. Buying one's own optical network may result in significant cost savings since ongoing monthly expenditures are replaced with a one-time initial CAPEX shared by the customers. Customers who control and manage their own optical networks can freely decide which control and management systems they deploy independent from any carrier. As a consequence, only customers (and not carriers) have visibility and control of the optical network.

Costumer-controlled optical networks are well suited to support data-intensive applications (e.g., Grid computing). Besides optical wavelength switching networks, user-controlled OBS networks have been proposed in Simeonidou et al. (2005). The studied user-controlled optical networks support two different optical switching granularities: wavelength switching and burst switching. Compared to wavelength-switching networks, the optical networks become more flexible and can support subwavelength traffic demands more efficiently by means of OBS.

2.6   Other issues

Many other issues need to be addressed in optical switching networks to render them viable. Apart from the aforementioned control and management, the following issues are key in the design of optical switching networks.

2.6.1   Security

Many of the well-known security mechanisms used in today's electronic networks (e.g., authentication, authorization, and encryption), can also be applied in optical networks. These security mechanisms are preferably implemented at higher electronic protocol layers. However, there are some important differences between optical and conventional networks which must be considered carefully (Médard et al., 1997). Due to the fact that optical networks operate at very-high-speed data rates on multiple WDM channels even short-lived and infrequent attacks may lead to large amounts of lost or corrupted information. Furthermore, eavesdropping by means of tapping may comprise the privacy of optical communications. More important, the desirable transparency of optical networks gives rise to severe security issues. Transparency renders optical networks particularly vulnerable since malicious optical signals are not OE converted and electronically processed at intermediate nodes. As a result, malicious signals are harder to detect in transparent optical networks than in opaque optical networks where OEO conversion takes place at each node. Also, due to the technological limitations of current optical components and devices, optical networks are susceptible to QoS degrade or even service disruption. For instance, optical Erbium doped fiber amplifiers (EDFAs), which are widely deployed in optical networks, exhibit a phenomenon which is referred to as *gain competition*, where different WDM channels share a given number of upper-state photons. Thus, a malicious high-power optical signal uses more of these upper-state photons and thereby reduces the gain of the other user signals. Also, the limited crosstalk of optical devices, for example, optical cross-connects (OXCs) and optical add-drop multiplexers (OADMs), can lead to a reduced QoS on one or more wavelength channels if the power of a malicious optical signal is sufficiently high. Note that in optical networks attacks can be easily launched from remote sites due to the fact that the propagation loss of optical fiber is very small.

Clearly, attack countermeasures in optical networks need to be able to react quickly in order to prevent significant data loss and corruption. Apart from reaction, the security of optical networks can be improved by means of preventive measures. Among others, bandlimiting optical filters may be used to discard optical signals outside the operational spectrum and thus mitigate the above mentioned gain competition of EDFAs.

2.6.2   Grooming

Most previous work on traffic grooming was done in the area of SONET/synchronous digital hierarchy (SDH) ring networks. The objective of traffic grooming was to reduce the number of required add-drop multiplexers (ADMs) and reduce the amount of electronic processing at intermediate ADMs by bypassing them. The idea of traffic grooming can be extended to optical mesh WDM networks, where low-rate circuits and data flows are assigned to wavelength channels which optically bypass OADMs (Modiano and Lin, 2001). In doing so, the number of required wavelength channels and nodal processing requirements can be reduced, resulting in cost savings and performance improvements of optical switching networks. Traffic grooming in optical switching networks is typically done electronically by the network ingress nodes at the network (IP) layer.

Apart from minimizing the number of required wavelength channels and nodal processing requirements, traffic grooming can be used to decrease the blocking probability in wavelength-switching mesh WDM optical networks that are able to dynamically set up and release lightpaths in order to efficiently support dynamic client traffic (e.g., SONET/SDH circuits, ATM virtual paths/virtual circuits, and IP packets) (Xin et al., 2004). Typically, in realistic mesh WDM optical networks traffic is dynamic (i.e., client calls dynamically arrive and depart over time).

To date, several important aspects of traffic grooming in optical WDM networks have been addressed and a significant body of research results has been created. However, many future challenges still need to be tackled in the area of traffic grooming in optical WDM networks. Among others, it would be interesting to define the required degree of opacity (number of dropped wavelength channels) of optical network nodes in order to optimize traffic grooming. Furthermore, many real-world network topologies have a special structure, for example, interconnected rings, stars, and/or trees. It would be interesting to study traffic grooming where the specific topological properties of the networks under consideration are exploited effectively. Finally, traffic grooming for more realistic traffic patterns (e.g., hot-spot and multicast traffic) appears to be an exciting research avenue for traffic grooming in optical WDM networks (Dutta and Rouskas, 2002).

3   Building blocks

In this chapter, we introduce several components which are widely used in optical switching networks. In Section 3.1, we introduce some key components and describe their respective properties. Section 3.2 outlines various transmitter and receiver types. Section 3.3 deals with the major transmission impairments encountered in optical switching networks. For an in-depth discussion and detailed information on components and transmission impairments the interested reader is referred to Mynbaev and Scheiner (2000), Ramaswami and Sivarajan (2001), and Mukherjee (2006).

3.1   Components

Let us first briefly describe the functionality of some key architectural building blocks used in many optical switching networks, which are depicted in Fig. 3.1(a)–(f):

- (a) *Combiner*: An $S \times 1$ combiner has S input ports and 1 output port, where $S \geq 2$. It collects wavelength channels from all S input ports and combines them onto the common output port. To avoid channel collisions at the output port of the combiner, the collected wavelength channels must be different. Thus, a given wavelength channel can be used only at one of the S input ports at any time.
- (b) *Splitter*: A $1 \times S$ splitter has 1 input port and S output ports, where $S \geq 2$. It equally distributes all incoming wavelength channels to all S output ports. Hence, a given wavelength channel can be received at all S output ports.
- (c) *Waveband partitioner*: A waveband partitioner Π has 1 input port and 2 output ports. It partitions an incoming set of Λ contiguous wavelength channels into two wavebands (subsets of wavelength channels) of Λ_A and Λ_B contiguous wavelength channels, where $1 \leq \Lambda_A$, $\Lambda_B \leq \Lambda - 1$, and $\Lambda = \Lambda_A + \Lambda_B$. Each waveband is routed to a different output port.
- (d) *Waveband departitioner*: A waveband departitioner Σ has two input ports and one output port. It collects two different wavebands consisting of Λ_A and Λ_B contiguous wavelength channels from the upper and lower input port, respectively. The combined set of Λ wavelength channels is launched onto the common output port, where $1 \leq \Lambda_A$, $\Lambda_B \leq \Lambda - 1$, and $\Lambda = \Lambda_A + \Lambda_B$.
- (e) *Passive star coupler (PSC)*: A $D \times D$ PSC has D input ports and D output ports, where $D \geq 2$. It works similar to a $D \times 1$ combiner and $1 \times D$ splitter interconnected

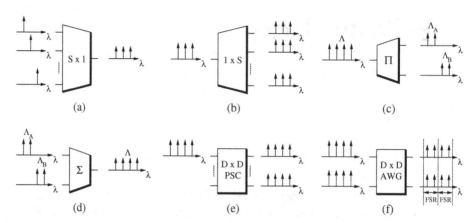

Figure 3.1 Architectural building blocks: (a) $S \times 1$ combiner, (b) $1 \times S$ splitter, (c) waveband partitioner, (d) waveband departitioner, (e) $D \times D$ passive star coupler (PSC), and (f) $D \times D$ AWG with $D = 2$.

in series. Accordingly, it collects wavelength channels from all D input ports and equally distributes them to all D output ports. Similar to the splitter, a given wavelength channel can be received at all D output ports and, similar to the combiner, to avoid channel collisions at the output ports a given wavelength channel can be used only at one of the D input ports at any time.

- (f) *Arrayed waveguide grating (AWG)*: A $D \times D$ AWG has D input ports and D output ports, where $D \geq 2$. Without loss of generality, we consider a 2×2 AWG to explain the properties of an AWG. Fig. 3.1(f) illustrates a scenario where four wavelengths are fed into both AWG input ports. Let us first consider only the upper input port. The AWG routes every second wavelength to the same output port. This period of the wavelength response is called free spectral range (FSR). In our example, there are two FSRs, each containing two wavelengths. Generally, the FSR of a $D \times D$ AWG consists of D contiguous wavelengths, i.e., the physical degree of an AWG is identical to the number of wavelengths per FSR. As depicted in Fig. 3.1(f), this holds also for the lower AWG input port. Note that the AWG routes wavelengths such that no collisions occur at the AWG output ports; that is, each wavelength can be applied at all AWG input ports simultaneously. In other words, with a $D \times D$ AWG each wavelength channel can be spatially reused D times, as opposed to the PSC. Also, note that each FSR provides one wavelength channel for communication between a given pair of AWG input and output ports. Hence, using R FSRs allows for R simultaneous transmissions between each AWG input–output port pair and the total number of wavelength channels available at each AWG port is given by $R \cdot D$, where $R \geq 1$.

Let us discuss the AWG, which is also known as phased array (PHASAR) or waveguide grating router (WGR), in greater detail and briefly explain the underlying physical concepts. The FSR is the spectral range between two successive passbands of the AWG.

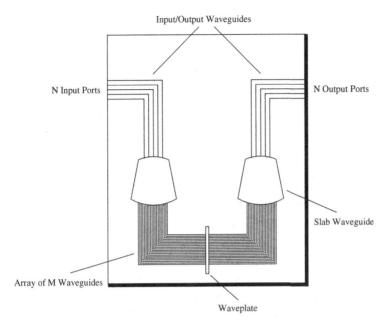

Input/Output Waveguides

N Input Ports

N Output Ports

Slab Waveguide

Array of M Waveguides

Waveplate

Figure 3.2 Schematic layout of an $N \times N$ AWG.

An $N \times N$ AWG is schematically shown in Fig. 3.2, where $N \geq 2$. It consists of N input–output waveguides, two focusing slab waveguides (free propagation regions), and an AWG, where the lengths of adjacent waveguides differ by a constant value. The waveplate at the symmetry line of the device eliminates the polarization dependence. Thus, polarization-independent AWGs can be realized. The excess loss is about 0.4 dB. Both slab waveguides work as identical $N \times M$ star couplers, where $M \gg N$, so that all the light power diffracted in the slab can be collected. If $M \gg N$ the crosstalk near the center of a passband is reduced compared with $M = N$. The signal from any of the N input ports is distributed over the M outputs of the slab waveguide to the array inputs. Each input light is diffracted in the input slab, passed through the arrayed waveguides, focused in the output slab, and coupled into the output waveguides. The arrayed waveguides introduce wavelength-dependent phase delays such that only frequencies with a phase difference of integer times 2π interfere constructively in the output slab waveguide. Thus, each output port carries periodic pass frequencies. The spacing of these periodic pass frequencies is called FSR.

In each FSR an $N \times N$ AWG accepts a total of N wavelengths from each input port and it transmits each wavelength to a particular output port. Each output port receives N wavelengths, one from each input port. There exist cyclic wavelength permutations at the output waveguides if different input waveguides are used. In Fig. 3.3 the routing connectivity of an 8×8 AWG is illustrated. Each wavelength gives routing instructions that are independent of the input port. Thus, λ_k's routing information is to exit the output port that is $(k-1)$ ports below the corresponding input port; that is, λ_1 goes from input port 1 to exit port 1 and from input port 5 to exit port 5. Similarly, λ_3

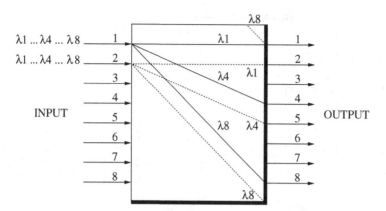

Figure 3.3 Routing connectivity of an 8×8 AWG.

incident on input port 1 is directed to output port 3, whereas if λ_3 were incident on port 5, it would be directed to output port 7. Due to the periodicity property of the AWG wavelength λ_9 (i.e., one FSR higher than λ_1) entering port 1 exits at port 1 like λ_{17} and other frequencies separated by an integral number of FSRs (Frigo, 1997). But there are also AWGs with different channel routing patterns. An $N \times N$ AWG provides full $N \times N$ interconnection. Using one FSR a total of N^2 simultaneous connections are possible. Note that an $N \times N$ PSC is capable of simultaneously carrying only N channels.

3.2   Transmitters and receivers

Besides the aforementioned components, transmitters and receivers are needed for build-ing a WDM communications network. A transmitter comprises a light source, a modu-lator, and supporting electronics. A receiver is composed of an optical filter, a photode-tector, a demodulator, and supporting electronics. In the following, we discuss different types of transmitters and receivers, including the tuning range and tuning time of the different light sources and optical filters since the tuning characteristics of transceivers play a major role in building efficient optical switching networks.

3.2.1   Broadband light sources

The light output of a broadband light source has a broad spectrum in the range of 10-100 nm. Light-emitting diodes (LEDs) are a very common and cost-effective example of a broadband light source. Due to their relatively small bandwidth-distance product LEDs are mainly applied where the data rates are low and/or distances are short. Typical output powers are of the order of -10 dBm. However, superluminescent diodes with an

Table 3.1. Transmitters: tuning ranges and tuning times

Transmitter Type	Tuning Range	Tuning Time
Mechanically tunable	500 nm	1–10 ms
Acousto-optic	~100 nm	~10 μs
Electro-optic	10–15 nm	1–10 ns
Injection current	~30 nm	15 ns

output power in single-mode fiber of 18.0 dBm and a 3-dB bandwidth of 35 nm are also commercially available.

3.2.2   Lasers

To achieve a significantly increased bandwidth-distance product lasers are deployed. Essentially, a laser is an optical amplifier enclosed within a reflective cavity that causes the light to oscillate via positive feedback. Lasers are capable of achieving high output powers, typically between 0 and 10 dBm.

Lasers can be either fixed tuned to a nominal wavelength (though it can drift with temperature and age) or tunable, where tunable lasers can be either continuously or discretely tunable. Since only wavelengths which match the period and refractive index of the laser will be constructively reinforced, a laser can be tuned by controlling the cavity length and/or the refractive index of the lasing medium. Common examples are mechanically, acousto-optically, electro-optically, and injection-current tunable lasers. Most mechanically tuned lasers use an external Fabry-Perot cavity whose length is physically adjusted. Mechanically tunable lasers exhibit a relatively wide tuning range of up to 500 nm but a relatively slow tuning time of 1–10 ms. In an acousto-optic or electro-optic laser the refractive index in the external cavity is changed by using either sound waves or electrical current, respectively. An acousto-optic laser combines a moderate tuning range of ~100 nm with a moderate tuning time of ~10 μs. Electro-optical lasers can be tuned over a range of 10–15 nm within a few nanoseconds. Injection-current-tuned lasers form a family of light sources which allow wavelength selection via a diffraction grating (e.g., distributed feedback (DFB) and distributed Bragg reflector (DBR) lasers). Tuning is achieved by changing the injected current density and thereby the refractive index. This type of laser typically consists of multiple sections in order to allow for independent control of output power and wavelength of the laser (Kobrinski et al., 1990). Fast tunable multisection transmitters which can be tuned to adjacent wavelengths within 4 ns (Fukashiro et al., 2000; Lavrova et al., 2000) and over a wide range of ~30 nm within 15 ns (Shrikhande et al., 2001) were reported. In particular, so-called SG-DBR lasers hold promise for use as fast tunable transmitters with a wide tuning range and high output power (Mason, 2000; Williams et al., 2000).

The tuning ranges and tuning times of the different transmitter types are summarized in Table 3.1. Note that instead of tunable lasers one might use an array of fixed-tuned lasers, each operating at a different wavelength, or multifrequency lasers (Zirngibl, 1998).

3.2.3   Optical filters

Optical filters are used to select a slice of a broadband signal or one wavelength out of the wavelength division multiplexing (WDM) comb. The selected wavelength is subsequently optoelectrically converted by a photodetector. Optical filters are either fixed tuned or tunable, whereby tunable filters can be either continuously or discretely tunable. Examples for fixed-tuned filters are diffraction gratings, dielectric thin-film filters, and fiber Bragg gratings (FBGs). Tunable optical filters encompass mechanically, thermally, acousto-optically, electro-optically tuned filters, and liquid-crystal Fabry-Perot filters (Sadot and Boimovich, 1998). In the following, we describe the different types of tunable optical filters and their properties in greater detail.

Mechanically tunable filters consist of one cavity (or more) formed by two parallel mirrors (facets). By mechanically adjusting the distance between the mirrors different wavelengths can be selected. This type of filter has a tuning range of $\sim$500 nm and a tuning time in the range of 1–10 ms.

The Mach-Zehnder interferometer (MZI) is an example for a thermally controlled optical filter. In an MZI a splitter splits the incoming light into two waveguides and a combiner recombines the signals at the outputs of the waveguides. A thermally adjustable delay element controls the optical path length in one of the waveguides. Due to the resulting phase difference a single desired wavelength can be selected by means of constructive interference. An MZI can be tuned over >10 nm within a few milliseconds.

In acousto-optic tunable filters (AOTFs) a sound wave periodically changes the refractive index of the filtering medium which enables the medium to act as a grating. By changing the frequency of the sound wave a single optical wavelength can be chosen to pass through while the remainder of the wavelengths interfere destructively. If more than one sound wave is applied, more than one wavelength can be filtered out. One drawback of AOTFs is that they are unable to filter out crosstalk from adjacent channels if the channels are closely spaced, thus limiting the number of channels. AOTFs can be tuned over a range of $\sim$100 nm within $\sim$10 µs.

Electro-optic tunable filters (EOTFs) use electrodes which rest in the filtering medium. Currents are applied to change the refractive index of the filtering medium which allows a desired wavelength to pass through while others interfere destructively. The tuning time is limited only by the speed of the electronics. Hence, EOTFs can be tuned on the order of 1–10 ns. However, EOTFs provide a relatively small tuning range of $\sim$15 nm.

Liquid-crystal (LC) Fabry-Perot filters appear to be an inexpensive filter technology with low power requirements. The design of an LC filter is similar to the design of a Fabry-Perot filter, but the cavity consists of an LC. The refractive index of the LC is controlled by an electrical current to filter out the corresponding wavelength. The tuning time is on the order of 0.5–10 µs and the tuning range is 30–40 nm.

Table 3.2. Receivers: tuning ranges and tuning times

Receiver Type	Tuning Range	Tuning Time
Mechanically tunable	500 nm	1–10 ms
Thermally tunable	>10 nm	1–10 ms
Acousto-optic	∼100 nm	∼10 μs
Electro-optic	10–15 nm	1–10 ns
Liquid crystal	30–40 nm	0.5–10 μs

The tuning ranges and tuning times of the different receiver types are summarized in Table 3.2. Note that, alternatively to tunable optical filters, arrays of fixed-tuned receivers or multiwavelength receivers can be deployed (Tong, 1998; Ohyama et al., 2001).

3.3   Transmission impairments

To build communications systems the previously described components are connected by fibers. In such a system, a light signal which propagates from the transmitter to the receiver undergoes a number of impairments, which are discussed next.

3.3.1   Attenuation

Beside the optical power loss caused by the components, the fiber further reduces the signal power. Figure 3.4 shows the attenuation loss of a fiber as a function of wavelength. The peak in loss in the 1400-nm region is due to hydroxyl ion (OH^-) impurities in the fiber. However, in Lucent's AllWave fiber this peak is reduced significantly. In today's optical communications systems three wavelength bands are used: 0.85, 1.3, and 1.55 μm, where the latter band provides the smallest attenuation of ∼0.25 dB/km.

3.3.2   Dispersion

Dispersion is the name given to any effect wherein different components of the trans-mitted signal travel at different velocities in the fiber, arriving at different times at the receiver. As a result, the pulse widens and causes intersymbol interference (ISI). Thus, dispersion limits the minimum bit spacing (i.e., the maximum transmission rate). The amount of accumulated dispersion depends on the length of the link. The important forms of dispersion are modal dispersion, chromatic (material) dispersion, waveguide dispersion, and polarization mode dispersion (PMD).

Modal dispersion
Modal dispersion arises only in multimode fiber where different modes travel at different velocities. Clearly, in single-mode fibers (SMFs) modal dispersion is not a problem.

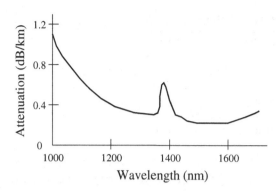

Figure 3.4 Attenuation of an optical fiber.

Waveguide dispersion

Waveguide dispersion is caused because the propagation of different wavelengths depends on waveguide characteristics such as indices and shape of the fiber core and cladding. After entering an SMF, an information-carrying light pulse is distributed between the core and the cladding. Its major portion travels within the core, the rest within the cladding. Both portions propagate at different velocities since the core and the cladding have different refractive indices.

Chromatic dispersion

Chromatic or material dispersion arises because different frequency components of a pulse (and also signals at different wavelengths) travel at different velocities due to the fact that the refractive index of the fiber is a function of the wavelength. It is typically measured in units of ps/(nm·km), where ps refers to the time spread of the pulse, nm is the spectral width of the pulse, and km corresponds to the link length. Typically, standard (SMFs) have a chromatic dispersion of 17 ps/(nm·km) at 1550 nm.

Recently, so-called nonzero dispersion shifted fibers (NZ-DSFs) are installed more often. By controlling the waveguide dispersion accordingly, NZ-DSFs have a chromatic dispersion between 1 and 8 ps/(nm·km), or between −1 and −8 ps/(nm·km) at 1550 nm. For example, Alcatel's TeraLight Metro Fiber has a dispersion of 8 ps/(nm·km). Such a low-dispersion fiber is targeted toward 10 Gb/s operation over 80–200 km without requiring dispersion compensation. Another example is Corning's MetroCor fiber. Its low negative dispersion enables the use of low-cost directly modulated DFB lasers. Both fibers are mainly devised for metro WDM networks in order to reduce network complexity and costs.

Polarization mode dispersion (PMD)

PMD arises because the fiber core is not perfectly circular, particularly in older installations. Thus, different polarizations of the signal travel at different velocities. PMD is proving to be a serious impediment in very-high-speed systems operating at 10 Gb/s and beyond.

3.3.3   Nonlinearities

As long as the optical power within an optical fiber is small, the fiber can be treated as a linear medium (i.e., the loss and refractive index of the fiber are independent of the signal power). However, when the power levels get fairly high in the system the nonlinearities can place significant limitations on high-speed systems as well as WDM systems. Nonlinearities can be classified into two categories. The first set of effects occurs owing to the dependence of refractive index on the optical power. This category includes self-phase modulation (SPM), cross-phase modulation (XPM), and four-wave mixing (FWM). The second set of effects occurs owing to scattering effects in the fiber medium due to the interaction of light waves with phonons (molecular vibrations) in the silica medium. The two main effects in this category are stimulated Raman scattering (SRS) and stimulated Brillouin scattering (SBS).

Self-phase modulation
SPM is caused by variations in the power of an optical signal and results in variations in the phase of the signal. SPM leads to the spectral broadening of pulses. Instantaneous variations in a signal's phase caused by changes in the signal's intensity will result in instantaneous variations of the frequency around the signal's central frequency. For very short pulses, the additional frequency components generated by SPM combined with the effects of material dispersion also lead to spreading or compression of the pulse in the time domain, affecting the maximum bit rate and the bit error rate (BER).

Cross-phase modulation
XPM is a shift in the phase of a signal caused by the change in intensity of a signal propagating at a different wavelength. XPM can lead to asymmetric spectral broadening, and combined with SPM and dispersion may also affect the pulse shape in the time domain.

Four-wave mixing
FWM occurs when two wavelengths, operating at frequencies f_1 and f_2, respectively, mix to cause signals at frequencies such as $2f_1 - f_2$ and $2f_2 - f_1$. These extra signals can cause interference if they overlap with frequencies used for data transmission. Similarly, mixing can occur between combinations of three and more wavelengths.

Stimulated Raman scattering
SRS is caused by the interaction of light with molecular vibrations. Light incident on the molecules creates scattered light at a longer wavelength than that of the incident light. A portion of the light traveling at each frequency is downshifted across a region of lower frequencies. The light generated at the lower frequencies is called the *Stokes wave*. The fraction of power transferred to the Stokes wave grows rapidly as the power of the input signal is increased. In multiwavelength systems, the shorter-wavelength channels will lose some power to the longer-wavelength channels. To reduce the amount of loss, the power on each channel needs to be below a certain level.

Stimulated Brillouin scattering

SBS is similar to SRS, except that the frequency shift is caused by sound waves rather than molecular vibrations. Other characteristics of SBS are that the Stokes wave propagates in the opposite direction of the input light. The intensity of the scattered light is much greater in SBS than in SRS, but the frequency range of SBS is much lower than that of SRS. To counter the effects of SBS, one must ensure that the input power is below a certain threshold. Also, in multiwavelength systems, SBS may induce crosstalk between channels. Crosstalk occurs when two counterpropagating channels differ in frequency by the Brillouin shift, which is around 11 GHz for wavelengths at 1550 nm.

3.3.4   Crosstalk

Crosstalk decreases the signal-to-noise ratio (SNR) leading to an increased BER. Crosstalk may either be caused by signals on different wavelengths (interchannel crosstalk) or by signals on the same wavelength on another fiber (intrachannel crosstalk) due to imperfect transmission characteristics of components (e.g., AWG). Interchannel crosstalk must be considered when determining channel spacing. In some cases, interchannel crosstalk may be removed through the use of appropriate narrowband filters. Intrachannel crosstalk usually occurs in switching/routing nodes where multiple signals on the same wavelength are being switched/routed from different inputs to different outputs. This form of crosstalk is more of a concern than interchannel crosstalk because intrachannel crosstalk cannot be removed through filtering.

3.3.5   Noise

The SNR is deteriorated by different noise terms. In particular, we consider amplified spontaneous emission (ASE) of optical Erbium doped fiber amplifiers (EDFAs), shot noise of photodetectors, and thermal noise of electrical amplifiers.

Amplified spontaneous emission

An optical EDFA amplifies an incoming light signal by means of stimulated emission. Besides stimulated emission spontaneous emission also takes place which has a deleterious effect on the system. The amplifier treats spontaneous emission radiation as another input signal and the spontaneous emission is amplified in addition to the incident light signal. The resulting ASE appears as noise at the output of the EDFA.

Shot noise

A photodetector converts the optical signal into an electrical photocurrent. The main complication in recovering the transmitted bit is that in addition to the photocurrent there is a shot noise current. Shot noise current occurs due to the random distribution of the electrons generated by the photodetection process even when the input light intensity is constant. (Note that the shot noise current is not added to the generated photocurrent but is merely a convenient representation of the variability in the generated photocurrent as a separate component.)

Thermal noise

Since the photocurrent is rather small it is subsequently amplified by an electrical amplifier. This electrical amplifier introduces an additional thermal noise current due to the random motion of electrons that is always present at typical temperatures.

4   Summary

The ultimate goal of the Internet and communications networks in general is to provide access to information when we need it, where we need it, and in whatever format we need it (Mukherjee, 2000). To achieve this goal wireless and optical technologies play a key role in future communications networks. Wireless and optical networks can be thought of as quite complementary. Optical fiber does not go everywhere, but where it does go, it provides a huge amount of available bandwidth. Wireless networks, on the other hand, potentially go almost everywhere and are thus able to support mobility and reachability, but they provide a highly bandwidth-constrained transmission channel, susceptible to a variety of impairments (Ramaswami, 2002). As opposed to the wireless channel, optical fiber exhibits a number of advantageous transmission properties such as low attenuation, large bandwidth, and immunity from electromagnetic interference. Future communications networks will be bimodal, capitalizing on the respective strengths of wireless and optical networks.

4.1   Historical review

Optical networks have been long recognized to have many beneficial properties. Among others, optical fiber is well suited to satisfy the growing demand for bandwidth, transparency, reliability, and simplified operation and management (Green, 1996). In this part, we have first reviewed the historical evolution of optical networks from point-to-point links to reconfigurable all-optical WDM networks of arbitrary topology. In our review, we introduced the basic concepts and techniques of optical networking, highlighted key optical network elements (e.g., reconfigurable OADM and OXC), elaborated on the rationale behind the design of all-optical networks, and outlined their similarities to SONET/SDH networks. Furthermore, we identified and explained the most important features of optical networks, namely, transparency, reconfigurability, survivability, scalability, and modularity. At the end of our historical overview, we focused on the management and control of reconfigurable optical networks. Both management and control are of utmost importance to make optical networks commercially viable. We briefly discussed the TMN framework, which encompasses a wide range of standards that cover network management issues which are commonly known as the FCAPS model. We also elaborated on the need and importance of the control plane which is responsible to control the data plane by making sure that the various deployed data forwarding techniques are put in place and operate properly in an efficient and coordinated fashion in each single optical network as well as guaranteeing interoperability among multiple optical networks of different vendors and operators.

4.2   Big picture

After our historical overview of optical networks, we provided the big picture of current optical networks. We introduced the term *optical switching networks* to denote all the various types of flexible, resilient, and reconfigurable optical networks which use any of the described multiplexing, tuning, and switching techniques and whose configuration can be changed dynamically in response to varying traffic loads and network failures. Optical switching networks can be found at each level of today's network infrastructure, including wide, metropolitan, access, and local areas. These end-to-end optical switching networks offer connection-oriented and connectionless services to existing as well as new emerging latency-critical and throughput-critical applications (e.g., interactive games, videoconferences, telecommuting, file sharing, video on demand, and HDTV). To offer the services in a bandwidth-efficient manner, switching in optical networks may be done at many different switching granularities, for example, at the fiber, waveband, wavelength, and subwavelength (time slot, burst, packet, cell, frame) granularity. Depending on the applied switching paradigm, optical switching can be roughly categorized into optical circuit, burst, and packet switching networks. All these switching techniques operate at the data plane of optical switching networks and are used to forward data from source to destination nodes. The switching nodes need to perform certain control functions to ensure the data forwarding is done properly. The control plane of optical switching networks typically operates at the data link or network layer. In particular, IP-centric control planes have been receiving a great deal of attention, where extended routing and signaling protocols are used to set up, modify, and release lightpaths dynamically. The resultant interlayer networking between electrical and optical layers is also useful to address other issues in optical switching networks such as security and grooming, which are best done in the electrical domain.

Apart from interlayer networking, future optical switching networks are very likely to contain both optical and electronical components (Medina, 2002). Due to the limitations of current optical technologies as well as the fact that the coverage of all-optical networks is limited to islands of transparency one interesting research avenue is to quantify the trade-offs between optical and electronic technologies and exploit their unique strengths to further reduce the costs and improve the performance of optical switching networks.

Next-generation optical switching networks will provide a transparent networking platform with enhanced control and management functionalities. After solving most of the technical problems, the focus of research efforts is expected to shift to other issues that will become increasingly important in next-generation optical switching networks. One of the key issues will be the value creation and revenue growth by developing new services and applications. An adaptive service shell will be designed around transparent optical networks which helps realize novel business strategies (Hirosaki et al., 2003).

4.3   Further reading

4.3.1   Books

There exist a number of excellent books that either provide a general overview of optical networks or focus on one or more selected topics. In the following, we list selected books in chronologically descending order for each of the two categories and briefly describe their scope and content.

Books with general overview

- **Optical Switching**, *by T. S. El-Bawab (Editor), Springer, 2006:*
  This book surveys the state of the art of a wide range of optical switching technologies for optical circuit, packet, and burst switching networks, for example, acousto-optic, thermo-optic, liquid-crystal, and microelectromechanical systems (MEMS) based optical switching technologies. In addition, the book provides a detailed description of several optical switch fabrics and architectures.
- **Optical WDM Networks**, *by B. Mukherjee, Springer, 2006:*
  This book provides an excellent overview of problems encountered in the design of optical WDM networks and describes a number of proposed architectures, protocols, and algorithms for optical local, access, metropolitan, and wide area networks. Among others, the book covers single-hop and multihop local area WDM networks and elaborates on the routing and wavelength assignment problem and design of virtual topologies. Furthermore, performance-enhancing techniques such as wavelength conversion, traffic grooming, and optical impairment-aware routing are discussed in great detail. The book also addresses the survivability as well as control and management of optical WDM networks. At the end of the book, optical packet switching (OPS) and optical burst switching (OBS) are explained.
- **IP over WDM: Building the Next Generation Optical Internet**, *by S. S. Dixit (Editor), Wiley, 2003:*
  This book addresses the convergence of the IP and optical layers, including the interworking between IP routing and signaling protocols, higher-layer protocols such as TCP or UDP and the protocols at the optical WDM layer. After highlighting the salient features of network and tranport layer protocols and introducing the key optical enabling technologies (e.g., wavelength conversion), the book explores quality of service (QoS) support, traffic engineering, routing, and signaling in IP-over-WDM networks.
- **Optical Networks: A Practical Perspective (Second Edition)**, *by R. Ramaswami and K. N. Sivarajan, Morgan Kaufmann, 2002:*
  This book is an excellent bottom-up reference for optical networks. It starts with a technically detailed description of the physical layer of optical networks, components,

modulation/demodulation, and optical network elements, for example, optical add-drop multiplexer (OADM) and optical cross-connect (OXC). It then elaborates on the design, control, and management of optical WDM networks and their deployment. Besides wavelength routing, the book also covers OPS/OBS to some extent.

- **Optical Networks: Architecture and Survivability**, *by H. T. Mouftah and P.-H. Ho, Springer, 2002:*

  This book focuses on the control and management of optical wavelength-routing WDM networks. It provides a good overview of GMPLS and an in-depth discussion of dynamic routing and wavelength assignment (RWA) algorithms for optical networks with multigranularity optical cross-connects (MG-OXCs). The book also addresses the interesting topics of spare capacity allocation and survivable routing under dynamic traffic. The book mainly deals with optical circuit-switching networks, but briefly touches on OBS toward the end.

- **Next Generation Optical Networks: The Convergence of IP Intelligence and Optical Technologies**, *by P. Tomsu and C. Schmutzer, Prentice Hall, 2001:*

  After briefly describing the main features of SONET/SDH and ATM networks, the authors discuss various possible layer stacks for next-generation optical WDM networks. The book provides a nice overview of recent optical networking standardization activities for optical local, access, metro, and wide area circuit- and packet-switched networks. Several control plane architectures are presented in detail, including MPLS. Of particular interest is the discussion of various application scenarios and case examples for next-generation optical networks.

- **Optical WDM Networks: Principles and Practice**, *by K. M. Sivalingam and S. Subramaniam (Editors), Springer, 2000:*

  This book briefly overviews widely used optical network components and then concentrates on a number of problems encountered in the design and performance evaluation of optical wavelength-routing wide area networks and broadcast-and-select local area networks. It also elaborates on optical access networks and the future of optical WDM networks.

- **Multiwavelength Optical Networks: A Layered Approach**, *by T. E. Stern and K. Bala, Prentice Hall, 1999:*

  This book offers a nice introduction to the key building blocks, techniques, and mechanisms to be used in optical WDM networks, with an emphasis on wavelength-routing wide area networks. Moreover, the book discusses the pros and cons of various logical topologies embedded on optical physical topologies (e.g., Hypernets). One of the most interesting parts of the book is dedicated to the description of so-called linear lightwave networks (LLNs) that support routing not only of individual wavelengths but also wavebands that consist of multiple contiguous wavelengths.

- **Wavelength Division Multiple Access Optical Networks**, *by A. Borella, F. Chiaraluce, and G. Cancellieri, Artech House, 1998:*

  This book nicely surveys a wide range of experimental optical broadcast-and-select single-hop networks and explains widely studied optical multihop networks which have received a great deal of attention in the context of designing logical network topologies (e.g., Manhattan Street, Shuffle, and De Bruijn topologies).

Books on selected topics

- **Fiber to the Home: The New Empowerment**, *by P. E. Green, Wiley, 2006:*

  This book is an excellent reference for the latest developments in the area of passive optical networks (PONs) and fiber-to-the-home (FTTH) networks. It compares ATM- and Ethernet-based PONs and provides an update of their current worldwide deployments. It also explains the underlying technologies and installation of PONs.

- **Path Routing in Mesh Optical Networks**, *by E. Bouillet, G. Ellinas, J.-F. Labourdette, and R. Ramamurthy, Wiley, 2006:*

  This book provides detailed information on new routing/switching technologies and combines theoretical and practical aspects of routing and dimensioning for mesh optical networks.

- **Survivability and Traffic Grooming in WDM Optical Networks**, *by A. K. Somani, Cambridge University Press, 2006:*

  This book covers the latest protection and restoration techniques for optical WDM networks. It covers emerging resilience techniques such as p-cycles that aim at combining the benefits of protection and restoration. The discussion of survivability also takes multiple link failures into account. The state of the art of both static and dynamic grooming techniques for optical WDM networks is described at length, including their deployment in so-called light-trail-based optical WDM networks.

- **GMPLS: Architecture and Applications**, *by A. Farrel and I. Bryskin, Morgan-Kaufmann, 2005:*

  This book provides an in-depth description of MPLS and its extension to GMPLS. GMPLS signaling, routing, and link management protocols are explained in great detail. Among others, the book elaborates on the service restoration and traffic engineering capabilities of GMPLS and elaborates on its path computation with and without constraints. Finally, the management of GMPLS networks is addressed.

- **Ethernet Passive Optical Networks**, *by G. Kramer, McGraw-Hill, 2005:*

  This book describes the technical details of Ethernet passive optical networks (EPONs) and evaluates their performance. It explains the architecture of EPONs and the format and function of the various multipoint control protocol (MPCP) messages. In addition, more involved EPON related topics such as forward error correction and encryption are addressed.

- **Optical Burst Switched Networks**, *by J. P. Jue and V. M. Vokkarane, Springer, 2004:*

  After an introductory description of enabling technologies, this book explores the functional building blocks of OBS networks in detail. Apart from burst assembly, signaling, contention resolution, scheduling, and QoS support, the authors also discuss other interesting topics of OBS networks (e.g., multicasting and TCP over OBS) and provide a brief overview of OBS testbeds.

- **Network Recovery: Protection and Restoration of Optical, SONET-SDH, IP, and MPLS**, *by J.-P. Vasseur, M. Pickavet, and P. Demeester, Morgan Kaufmann, 2004:*

  This book introduces various key recovery mechanisms, including multilayer recovery approaches. After highlighting the resilience features of SONET/SDH networks, the authors outline recovery schemes for optical ring and mesh networks. Furthermore, they discuss how IP and MPLS protocols can be used for load balancing, traffic engineering, and routing in the event of network failures, and how they can be combined with recovery mechanisms at the optical layer to allow for efficient multilayer fault recovery.

- **Metropolitan Area WDM Networks: An AWG-Based Approach**, *by M. Maier, Springer, 2003:*

  This book provides a comprehensive and technically detailed overview of metropolitan area WDM network experimental systems, architectures, and access protocols. Apart from optical ring networks, the book focuses on the design and performance evaluation of wavelength-routing metro WDM star networks based on an arrayed waveguide grating (AWG).

- **Optical Network Control: Architecture, Protocols, and Standards**, *by G. Bernstein, B. Rajagopalan, and D. Saha, Addison-Wesley, 2003:*

  This book focuses on the control plane architecture of SONET/SDH, ATM, MPLS, and GMPLS networks. It provides a detailed description of neighbor discovery, signaling for connection provisioning and fault recovery as well as intradomain and interdomain routing protocols.

- **WDM Mesh Networks: Management and Survivability**, *by H. Zang, Springer, 2002:*

  This book deals with various types of static and dynamic RWA algorithms for optical wavelength-routing WDM mesh networks with or without wavelength converters. The presented RWA algorithms can be used either for dedicated or shared protection of paths and links. In addition, the book provides some insight in slot assignment algorithms and contention resolution schemes for photonic slot routing (PSR) networks.

- **Gigabit Ethernet for Metro Area Networks**, *by P. Bedell, McGraw-Hill, 2002:*

  After reviewing the origins of Ethernet, this book provides a detailed description of Gigabit and 10-Gigabit Ethernet networks. The authors explain the role and deployment of Gigabit Ethernet in both access and metropolitan areas and compare it to competing technologies such as SONET/SDH, Fibre Channel, Frame Relay, and ATM. The book also briefly discusses Resilient Packet Ring (RPR) and other emerging metro network solutions.

- **Photonic Slot Routing in Optical Transport Networks**, *by G. Wedzinga, Springer, 2002:*

  This book describes the basic architecture and enabling technologies of PSR networks as well as their performance evaluation and possible upgrade approaches.

- **Radio over Fiber Technologies for Mobile Communications Networks**, *by H. Al-Raweshidy and S. Komaki (Editors), Artech House, 2002:*

  This book provides insight in the various devices and components used in radio-over-fiber (RoF) networks and discusses a wide range of important topics. Beginning with a description of the microwave properties of optical links the book elaborates on the physical transmission issues in RoF systems. It then overviews several applications of RoF technology and outlines its future potential in mobile wireless networks.

- **SONET (Second Edition)**, *by W. J. Goralski, McGraw-Hill, 2000:*

  This book provides an in-depth description of SONET networks. After reviewing the beginnings of SONET, the book explains in great detail the SONET architecture, protocols, equipment, and lists key equipment providers. The deployment and service support as well as the future and issues of SONET networks are described at length.

- **Understanding SONET/SDH and ATM: Communications Networks for the Next Millenium,** *by S. V. Kartalopoulos, IEEE Press, 1999:*

    This book explains the architecture and operation of SONET/SDH and ATM networks in great detail. It also discusses interworking issues between these two layers as well as the IP and optical WDM layers.

4.3.2   Journals and magazines

Technically profound information about research and development of enabling technologies, devices, components, systems, architectures, techniques, protocols, services, and applications for next-generation optical switching networks can be found in the following journals and magazines.

Journals
- IEEE/OSA Journal of Lightwave Technology
- IEEE Journal on Selected Areas in Communications – Optical Communications and Networking Series
- OSA Journal of Optical Networking
- Optical Switching and Networking
- Photonic Network Communications
- OSA Journal of Optical Technology
- IEEE/ACM Transactions on Networking
- IEEE Transactions on Communications
- IEICE Transactions on Communications
- IET Communications
- IET Optoelectronics
- IEEE Photonics Technology Letters
- OSA Applied Optics
- OSA Optics Letters
- OSA Optics Express
- IET Electronics Letters
- IEEE Communication Letters

Magazines
- IEEE Communications Magazine – Optical Communications Supplement
- IEEE Network
- IEEE Communications Surveys & Tutorials

4.3.3   Web links

Up-to-date information on emerging applications, services, technologies, architectures, protocols, standards, products, and their commercial adoption can be found at:

- `http://www.lightreading.com/`
- `http://www.fiberopticsonline.com/`
- `http://fibers.org/`
- `http://optics.org/`
- `http://www.eetimes.com/`
- `http://www.convergedigest.com/`
- `http://www.electronics-manufacturers.com/`
- `http://www.allbusiness.com/`
- `http://www.freshpatents.com/`
- `http://www.wired.com/`
- `http://citeseer.ist.psu.edu/`
- `http://www.iec.org/`

Part II

Optical wide area networks

Overview

In this part, we discuss and describe in great detail various switching techniques for optical wide area networks (WANs). A number of different optical switching techniques have been proposed for backbone wavelength division multiplexing (WDM) networks over the last few years. Our overview will focus on the major optical switching techniques that can be found in today's operational long-haul WDM networks or are expected to be likely deployed in future optical WANs. In our overview we do not claim to provide a comprehensive description of all proposed switching techniques. Instead, we try to focus on the major optical switching techniques and describe their underlying principles and operation at length. We believe that our overview of carefully selected optical switching techniques fully covers the different types of switching techniques available for optical WANs and helps the reader gain sufficient knowledge to anticipate and understand any of the unmentioned optical switching techniques that in most cases might be viewed as extensions or hybrids of the optical switching techniques discussed. For instance, a so-called *light-trail* is a generalization of a conventional point-to-point lightpath in which data can be dropped and added at any node along the path, as opposed to a light-path where data can be added only by the source and dropped only by the destination node, respectively (Gumaste and Zheng, 2005). Another good example is *fractional lambda switching* (FλS) (Baldi and Ofek, 2002). FλS uses the globally available coordinated universal time (UTC) as a common time reference to synchronize all optical switches throughout the FλS network. FλS might be viewed as a subwavelength circuit switching technique where periodically recurring time slots are all-optically switched at intermediate nodes without requiring optical processing and optical buffering due to the network-wide synchronization of optical switches via UTC.

Wide area networks were one of the first network segments where optical technologies were widely deployed in order to provide sufficient capacity in support of heavy long-haul traffic. By means of WDM today's optical WANs offer huge bandwidth pipes where a single fiber may carry tens or even hundreds of wavelength channels, each operating at 10 Gb/s or higher. Given these vast amounts of bandwidth, one of the major design criteria of today's backbone networks is not to maximize the utilization of bandwidth resources but to simplify network operation and reduce costs. Toward this end, most of today's operational optical WANs deploy circuit switching at the wavelength granularity, which supports bursty data traffic only inefficiently. Due to ever-increasing higher-speed access technologies, for example, fiber to the home (FTTH), optical backbone networks are likely to run out of capacity and therefore may have to resort to more efficient switching techniques at the subwavelength granularity in the near to mid term. In the following chapters, we explain some of the most promising optical switching techniques that provide different switching granularities, ranging from switching entire fibers to switching individual packets on each wavelength channel.

Chapter 5

In Chapter 5, we elaborate on the generalized multiprotocol label switching (GMPLS) protocol suite which encompasses a number of standardized signaling and routing protocols. GMPLS can be used to control a wide range of different devices that perform not only packet switching but also switch time slots, wavelengths, wavebands, or fibers. GMPLS builds on the well-known multiprotocol label switching (MPLS). GMPLS differs from MPLS in that it supports other types of switching apart from switching only packets of fixed or variable size. GMPLS is a powerful tool to form a common control plane for disparate types of optical networks (e.g., packet switched, wavelength switched, or waveband switched), and to offload the management plane of large optical networks.

Chapter 6

Ordinary optical cross-connects (OXCs) perform only wavelength switching. Given the fact that optical fibers may carry hundreds of wavelengths, the complexity and costs of wavelength-switching OXCs become increasingly prohibitive. Waveband switching (WBS) is a practical solution to reduce the size and complexity of photonic and optical cross-connects by grouping several wavelengths together as a waveband and switching wavebands as an entity without demultiplexing the arriving WDM comb signal into its individual wavelengths. Chapter 6 describes WBS at length, including various waveband grouping algorithms and the architecture of so-called multigranularity optical cross-connects (MG-OXCs) which are able to perform switching at different switching granularities.

Chapter 7

To improve the bandwidth utilization of optical networks under bursty nonregular data traffic, a subwavelength switching technique, called photonic slot routing (PSR), was proposed. PSR networks perform switching of fixed-size time slots that span all wavelengths, referred to as photonic slots. In PSR, each photonic slot may carry multiple fixed-size packets on distinct wavelength channels that are switched together as a single entity all-optically. Similar to conventional wavelength-switching networks, PSR networks provide optical transparency. Unlike lightpath-based optical networks, however, PSR allows traffic aggregation to be done optically at intermediate nodes in order to use the bandwidth of each wavelength channel more efficiently without requiring electronic traffic grooming at source nodes, as discussed in greater detail in Chapter 7.

Chapter 8

In Chapter 8, we introduce optical flow switching (OFS). OFS may be viewed as a fast circuit switching technique that dynamically sets up and releases wavelength channels in order to switch large transactions and/or long-duration flows at the optical layer, thereby offloading electronic routers. OFS helps reduce the computation load and processing delay at electronic routers, but it is mainly suitable for switching large-size flows which justify the set-up of a lightpath that takes at least one round-trip time between source and destination routers.

Chapter 9

To avoid the round-trip delay enountered in setting up a lightpath dynamically or statically via two-way reservation, a source node may start sending data without waiting for any reservation acknowledgment sent back by the destination, as done in optical burst switching (OBS). In OBS, a source node aggregates multiple data packets into a data burst. A source node transmits a control packet on a separate control wavelength channel to announce the properties of the burst, which is sent on any of the available data wavelengths after a certain offset time. Control packets are OEO converted and processed electronically by intermediate OBS nodes while the corresponding data bursts are switched in the optical domain. As explained at length in Chapter 9, OBS aims at combining the merits of optical circuit switching and optical packet switching.

Chapter 10

Finally, Chapter 10 describes the state of the art of optical packet switching (OPS) networks. Unlike OBS, OPS requires neither any burst assembly and disassembly at source and destination nodes, respectively, nor any offset between packet header and payload. OPS networks are inherently bandwidth efficient due to their statistical multiplexing capability. At present, however, they face technological challenges and feasibility issues not only due to the lack of optical RAM but also the fact that current all-optical header processing techniques allow only for rather simple logical functions.

5 Generalized multiprotocol label switching

We have briefly introduced the automatic switched optical network (ASON) frame-work for the control plane of optical networks in Section 2.5. The ASON framework facilitates the set-up, modification, reconfiguration, and release of both *switched* and *soft-permanent* optical connections. Switched connections are controlled by clients as opposed to soft-permanent connections whose set-up and tear-down are initiated by the network management system. An ASON consists of one or more domains, where each domain may belong to a different network operator, administrator, or vendor plat-form. In the ASON framework, the points of interaction between different domains are called *reference points*. Figure 5.1 depicts the ASON reference points between various optical networks and client networks (e.g., IP, asynchronous transfer mode [ATM], or Synchronous Optical Network/synchronous digital hierarchy [SONET/SDH] networks), which are connected via lightpaths. Specifically, the reference point between a client net-work and an administrative domain of an optical network is called user–network interface (UNI). The reference point between the administrative domains of two different optical networks is called external network–network interface (E-NNI). The reference point between two domains (e.g., routing areas), within the same administrative domain of an optical network is called internal network–network interface (I-NNI).

5.1 Multiprotocol label switching

The ASON framework may be viewed as a reference architecture for the control plane of optical switching networks. It is important to note that the framework addresses the ASON requirements but does not specify any control plane protocol. In transparent opti-cal networks, such as ASON, intermediate optical add-drop multiplexers (OADMs) and optical cross-connects (OXCs) may be optically bypassed and thereby prevented from accessing the corresponding wavelength channels. Due to this fact, in-band signaling techniques are ruled out in favor of out-of-band control techniques for optical switching networks. *Multiprotocol label switching* (MPLS) provides a promising foundation for the control plane of optical switching networks since MPLS decouples the control and data planes (Nadeau and Rakotoranto, 2005).

The MPLS architecture was standardized by the Internet Engineering Task Force (IETF) (Rosen et al., 2001). It reuses and extends existing IP routing and signaling

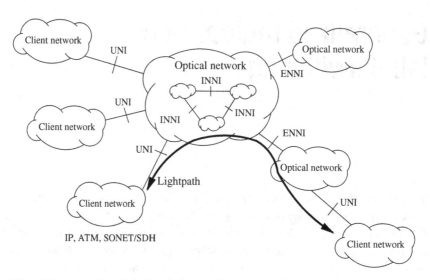

Figure 5.1 Automatic switched optical network (ASON) reference points. After Zhang et al. (2001). © 2001 IEEE.

protocols and thus avoids the need to reinvent the wheel, minimizes the risk related to protocol development, and reduces the time to market. MPLS introduces the connection-oriented model in the connectionless IP context. MPLS requires the encapsulation of IP packets into labeled packets. The realization of the label depends on the link technology in use. For instance, in ATM networks the virtual channel identifier (VCI) and virtual path identifier (VPI) may be naturally used as labels. Alternatively, an MPLS shim header may be added to the IP packet and used as label. The labeled packets are forwarded along a virtual connection called *label switched path* (LSP), similar to virtual circuits and virtual paths in ATM networks. MPLS routers are called *label switched routers* (LSRs). LSRs at the edge of an MPLS domain, which are also referred to as *label edge routers* (LERs), are able to set up, modify, reroute, and tear down LSPs by using the aforementioned signaling and routing protocols with the appropriate extensions. Intermediate LSRs within an MPLS domain do not examine the IP header during forwarding. Instead, they forward labeled IP packets according to the *label swapping* paradigm. With label swapping, each intermediate LSR maps a particular input label and port to an output label and port. Edge LSRs (LERs) establish LSPs by configuring each intermediate LSR to perform label swapping properly by using the input port and label of the arriving labeled IP packet to determine the output port and outgoing label.

MPLS provides a number of advantageous features that help network operators build converged multiservice networks and eliminate redundant network layers by incorporating some of the functions provided by ATM and SONET/SDH to the IP/MPLS control plane. MPLS supports the reservation of network resources as well as the possibility of explicit and constraint-based routing. Constraint-based routing can be used for traffic engineering (TE) and fast reroute (FRR). In doing so, IP/MPLS can replace ATM for the purpose of TE and SONET/SDH for protection and restoration. Moreover, MPLS

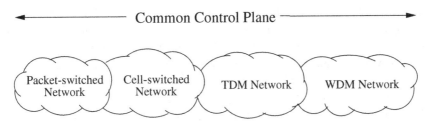

Figure 5.2 Common control plane for disparate types of optical switching networks.

provides the possibility of *stacking labels*. As a result, in MPLS networks a labeled IP packet can have one, two, or more labels, as opposed to only two labels (VCI, VPI) in ATM networks. The label stacking capability allows to build arbitrary *LSP hierarchies* in MPLS networks, as discussed in greater detail later in the chapter.

At the downside, MPLS suffers from a few shortcomings. In the MPLS architecture, LSPs are basically unidirectional. To set up a bidirectional LSP two separate counterdirectional LSPs must be set up independently. Thus, the IP/MPLS control plane is unable to establish bidirectional connections in a single request, resulting in an increased control overhead and set-up delay. Furthermore, protection bandwidth cannot be used by lower-priority traffic during failure-free network operation, which would be pre-empted in the event of a network failure in order to carry the protected higher-priority traffic. As a result, in MPLS networks protection bandwidth goes unused during failure-free operation.

5.2   Generalized MPLS (GMPLS)

The MPLS architecture is designed to support only devices that perform packet switching. Recall from Chapter 2, however, that optical switching networks deploy a wide range of different switching and multiplexing techniques. As shown in Fig. 5.2, end-to-end optical switching networks may comprise (variable-size) packet-switched, (fixed-size) cell-switched, time division multiplexing (TDM), and wavelength division multiplexing (WDM) networks. Clearly, to support not only packet-switching devices, MPLS needs to be extended in order to also support devices that switch time slots, wavelengths, wavebands, or fibers apart from packets. In other words, MPLS must be generalized to encompass non-packet-switching devices, leading to GMPLS (Mannie, 2004). GMPLS differs from MPLS in that it supports multiple types of switching; besides packet switching GMPLS also supports TDM, lambda, and fiber (port) switching. To deal with the widening scope of MPLS into the time and optical domains, several new forms of label are required. These new forms of label are collectively referred to as a *generalized label*. A generalized label contains enough information to allow the receiving node to program its cross-connect, regardless of the type of this cross-connect. The generalized label extends the traditional label (e.g, VCI, VPI, or shim header), by allowing the representation of not only labels which travel in-band with associated data packets, but also labels which are identical to time slots, wavelengths, or fibers (ports). Since the nodes sending and receiving the generalized label know what kinds of link they are using, the nodes know from the context what type of label to expect (Berger, 2003a).

The GMPLS architecture builds on MPLS. Many MPLS concepts can be found in the GMPLS architecture in conjunction with additional enhancements, as discussed at length in the following. The rationale behind GMPLS is to define a common control plane for disparate types of network technologies (e.g., IP, ATM, SONET/SDH, TDM, and WDM), as depicted in Fig. 5.2. Traditionally, each specific technology has its own set of control protocols which are layered one on top of the other, resulting in rather complex overlay networks that consist of multiple layers. GMPLS aims at achieving seamless interconnection across disparate network technologies and performing end-to-end connection set-up and release across heterogeneous networks (Nadeau and Rakotoranto, 2005). Toward this end, GMPLS adds the required intelligence to the control plane of optical networks, leading to so-called *intelligent optical networks* (IONs) (Puype et al., 2005).

5.2.1   Interface switching capability

GMPLS is a multipurpose control plane which supports not only packet-switching devices but also devices that perform switching in the time, wavelength, and space domains. The GMPLS control plane is able to operate over a wide range of heterogeneous network devices (e.g., IP/MPLS routers, SONET/SDH network elements, ATM switches, as well as optical network elements such as OXCs and OADMs). All these heterogeneous network devices represent LSRs that perform different types of switching. The different types of LSRs encountered in GMPLS networks can be categorized according to their *interface switching capability* (ISC). The interfaces of a given LSR can be subdivided into the following classes (Mannie, 2004):

- **Packet Switch Capable (PSC) Interfaces:** Interfaces that recognize packet boundaries and can forward data based on the content of the packet header. Examples include interfaces on routers that forward data based on the content of the IP header and interfaces on routers that switch data based on the content of the MPLS shim header.
- **Layer-2 Switch Capable (L2SC) Interfaces:** Interfaces that recognize frame/cell boundaries and can switch data based on the content of the frame/cell header. Examples include interfaces on Ethernet bridges that switch data based on the content of the medium access control (MAC) header and interfaces on ATM-LSRs that forward data based on the ATM VPI/VCI.
- **Time-Division Multiplex Capable (TDM) Interfaces:** Interfaces that switch data based on the data's time slot in a repeating cycle. Examples of such an interface is that of a SONET/SDH digital cross-connect system (DCS) or add-drop multiplexer (ADM).
- **Lambda Switch Capable (LSC) Interfaces:** Interfaces that switch data based on the wavelength on which the data is received. An example of such an interface is that of an OXC that can operate at the level of an individual wavelength. Note that the class

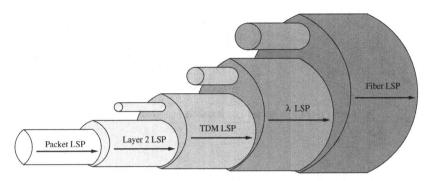

Figure 5.3 Hierarchy of GMPLS label switched paths (LSPs). After Iovanna et al. (2003). © 2003 IEEE.

of LSC interfaces also includes interfaces that switch data at the level of a group of wavelengths (waveband), resulting in waveband switching.

• **Fiber Switch Capable (FSC) Interfaces:** Interfaces that switch data based on a position of the data in the physical space. An example of such an interface is that of an OXC that can operate at the level of a single or multiple fibers.

Note that an interface of a given LSR may support a single ISC or multiple ISCs. For instance, consider a fiber link carrying a set of lambdas (wavelengths) that terminates on a given LSR interface that could either cross-connect one of these lambdas to some other outgoing optical channel, or could terminate the lambda and extract (demultiplex) data from that lambda using TDM and then cross-connect these TDM channels to some outgoing TDM channels (Kompella and Rekhter, 2005c).

Also note that, in GMPLS networks, an LSP can be established only between and through interfaces of the same type. That is, LSPs always have to start and terminate on network elements that support the same ISC. LSPs established between pairs of network elements with different ISCs, however, can be nested inside each other, giving rise to a hierarchy of LSPs.

5.2.2 LSP hierarchy

Recall from earlier that MPLS provides the possibility of label stacking. Label stacking allows MPLS LSP hierarchies to be realized by letting intermediate LSRs add labels to the header of the packet, resulting in a label stack. In doing so, LSPs can be nested inside other LSPs, giving rise to a hierarchy of LSPs.

The notion of LSP hierarchy can be extended to GMPLS networks that use generalized labels. Similar to MPLS, a forwarding hierarchy of LSPs can be built between generalized LSRs with the same ISC. For example, in the case of TDM interfaces a lower-order SONET/SDH LSP (e.g., OC12) can be nested inside a higher-order SONET/SDH LSP (e.g., OC48). Unlike MPLS, in GMPLS networks the nesting of LSPs can also be done between different types of interfaces. As shown in Fig. 5.3, the GMPLS hierarchy of LSPs is based on the different switching capabilities of the LSR interfaces (Iovanna

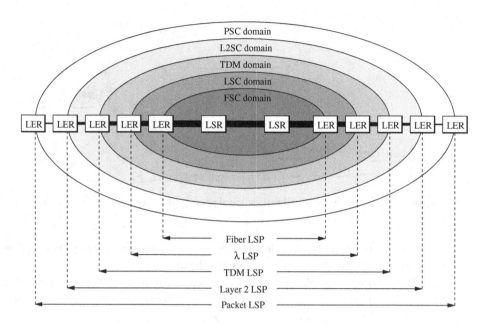

Figure 5.4 GMPLS label switched path (LSP) tunnels. After Banerjee et al. (2001a). © 2001 IEEE.

et al., 2003). Specifically, an LSP that starts on a PSC interface, thus forming a packet LSP, can be nested inside a layer 2 LSP. The layer 2 LSP can be nested together with other layer 2 LSPs inside a TDM LSP which starts and ends at two LSRs whose interfaces are both TDM capable. As depicted in the figure, this nesting procedure can be continued toward higher-order LSPs. The TDM LSP can be nested together with other TDM LSPs inside a λ (lambda) LSP, which in turn can be nested together with other λ LSPs inside a fiber LSP. At the top of the resultant LSP hierarchy are FSC interfaces, followed by LSC interfaces, followed by TDM interfaces, followed by L2SC interfaces, followed by PSC interfaces. It is important to note that each type of LSP starts and ends at LSRs whose interfaces have the same switching capability. Thus, an LSP that starts and ends on a PSC interface can be nested into an LSP that starts and ends on an L2SC interface. This LSP, in turn, can be nested into an LSP that starts and ends on a TDM interface. In turn, this LSP can be nested into an LSP that starts and ends on an LSC interface. And finally, this LSP, in turn, can be nested into an LSP that starts and ends on an FSC interface. Figure 5.4 illustrates the nesting of lower-order LSPs inside higher-order LSPs, where each LSP of a given order is established between a pair of LERs whose interfaces support the same switching capability. Recall from earlier that in MPLS networks, an LER is an LSR located at the edge of an MPLS domain. In GMPLS networks, an LER is an LSR located at the ingress or egress of a domain that comprises LSRs whose interfaces support the same type of switching capability. For instance, in Fig. 5.4, the FSC domain comprises two fiber-switching LSRs within the domain and two fiber-switching LERs at the edge of the domain. At the ingress fiber-switching LER, multiple λ LSPs are multiplexed on a common fiber LSP and subsequently demultiplexed at the egress LER after traversing the two intermediate fiber-switching LSRs. In doing so, the λ LSPs are tunneled inside the fiber LSP. This tunneling principle is valid for all GMPLS domains, where higher-order LSPs may be viewed as *LSP tunnels* for nested lower-order LSPs. LSP tunnels are formed by LSRs at the border of two GMPLS domains that differ from each other with respect to their ISC.

5.2.3   LSP control

We have seen in the previous section that lower-order LSPs are tunneled inside an existing higher-order LSP, provided the latter one has sufficient spare capacity to support the lower-order LSPs. Note that so far we assumed that the higher-order LSP was already established to serve as a tunnel to carry lower-order LSPs. However, if a new lower-order LSP does not find any appropriate existing higher-order LSP, the lower-order LSP will trigger the set-up of higher-order LSPs (Banerjee et al., 2001a).

The space-time diagram in Fig. 5.5 illustrates the basic principle of how the various GMPLS LSP tunnels of Fig. 5.4 are set up. Let us consider the following scenario. The PSC LER at the left-hand side wants to set up a packet LSP to the other PSC LER at the right-hand side, without any pre-existing LSPs being established between them. Toward this end, the PSC LER at the left-hand side generates a packet LSP set-up request and sends it to the neighboring L2SC LER. The arrival of the set-up request at the L2SC LER triggers the set-up of a layer 2 LSP between the L2SC LER at the left-hand side and its counterpart, the L2SC LER at the right-hand side. Hence, the left-hand-side L2SC LER generates a layer 2 LSP set-up request and sends it to the neighboring TDM LER. At the TDM LER, the arrival of the layer 2 LSP set-up request triggers the set-up of a TDM LSP between both TDM LERs. Accordingly, the TDM LER at the left-hand side generates a TDM LSP set-up request and sends it to the other TDM LER at the right-hand side. This triggering procedure continues for the set-up request of a λ LSP and fiber LSP. Once the right-hand-side FSC LER receives the fiber LSP set-up request generated and sent by the left-hand FSC LER, the right-hand-side FSC LER returns a fiber LSP set-up acknowledgement to its counterpart, the FSC LER at the left-hand side. After receiving the acknowledgement, the fiber LSP is established between both FSC LERs. Following the successful set-up of the fiber LSP, the fiber LSP serves as a link across the two intermediate FSC LSRs along the λ LSP. The λ LSP set-up request is tunneled through the established fiber LSP toward the LSC LER at the right-hand side. The latter one sends a λ LSP set-up acknowledgement back to the left-hand-side LSC LER whose reception completes the set-up of the λ LSP. Next, the established λ LSP is used to tunnel the TDM LSP request. The procedure of setting up higher-order LSPs and subsequently tunneling lower-order LSP set-up requests inside them is repeated until the TDM as well as layer 2 and packet LSPs are established. The established GMPLS LSPs form a forwarding hierarchy of LSPs that can be used for data transmission.

In the remainder of this section, we are going to look at the control of GMPLS LSPs in greater detail, including routing and TE, and explain the underlying concepts and mechanisms at length.

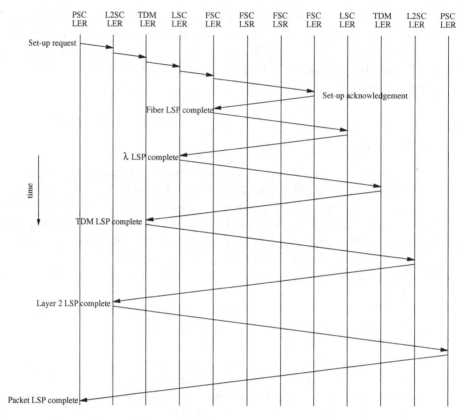

Figure 5.5 Setup of GMPLS label switched path (LSP) tunnels. After Banerjee et al. (2001a). © 2001 IEEE.

TE link and forwarding adjacency

In order to facilitate not only legacy shortest path first (SPF) routing but also constraint-based SPF routing of LSPs, LSRs that perform LSP establishment need more information about the links in the network than standard interior gateway protocols (IGPs) (e.g., open shortest path first (OSPF) and intermediate system to intermediate system (IS-IS) routing protocols) provide by means of flooding (flooding is a technique that disseminates link state information to all nodes within a single routing domain). This additional link information is provided by the so-called *TE attributes* (Mannie, 2004). TE attributes describe the properties of each link. TE attributes associated with a link capture its characteristics such as ISC, unreserved bandwidth, the maximum reservable bandwidth, protection/restoration type, and *shared risk link group* (SRLG). The SRLG represents a group of links that share the same fate in the event of failures. In other words, all links belonging to the same SRLG are affected by the same link and/or node failure. A link together with its associated TE attributes is called a *TE link*. The IGP floods the link state information about TE links just as it floods the link state information about any other conventional link. Note that the LSRs connected by a TE link are adjacent devices.

TE links can be extended to nonadjacent devices by using the concept of *forwarding adjacency* (FA). More specifically, an LSR can advertise an LSP as a TE link into a single routing domain. Such a link is called an FA and the corresponding LSP is called an FA-LSP. Note that FAs provide an abstraction of the underlying physical topology of the network in which FA-LSPs have been established. As a consequence, FAs provide a virtual (logical) topology to upper layers. In the resultant virtual topology, two FAs may be identical (i.e., they interconnect the same LSRs) even though the paths of the corresponding FA-LSPs in the underlying physical topology may differ. Information about FAs are flooded by the IGP like that of TE links (Kompella and Rekhter, 2005a; Vigoureux et al., 2005).

To reduce the amount of link state information flooded by the IGP and thereby improve the scalability of GMPLS networks, TE links and FAs can be *bundled* and/or *unnumbered*, as explained in the following:

- **Link Bundling:** Link bundling denotes the process of aggregating the attributes of several TE links and FAs of similar characteristics and assigning these aggregated attributes to a single bundled link, which may consist of a mix of TE links and FAs. The IGP then needs to flood the link state information only of the bundled link rather than of all the component TE links and FAs. As a result, the number of links whose state information is flooded by the IGP can be reduced significantly, resulting in an improved scalability of the GMPLS network (Banerjee et al., 2001b). However, link bundling is subject to certain constraints. Link bundling must meet the following restrictions. All component TE links and FAs in a bundled link must have the same link type (i.e., point-to-point or multi-access), the same TE metric, and must begin and end on the same pair of LSRs (Kompella et al., 2005).
- **Unnumbered Links:** Typically, in GMPLS networks all links are assigned IP addresses (either IPv4 or IPv6 addresses). Clearly, the fact that each link must have an IP address may lead to scalability, control, and management issues in GMPLS networks. In case there are not enough IP addresses available or the overhead of managing them becomes too high, unnumbered links can be used to solve this problem (Banerjee et al., 2001b). Unnumbered links are links that do not have any IP addresses. However, without having an IP address, an alternative must be found to uniquely identify each link in a GMPLS network. To do so, each LSR in GMPLS networks is assigned a unique IP address and each LSR numbers its links locally. The tuple [LSR IP address, local link number] can be used to uniquely identify each link in GMPLS networks. Both TE links and FAs can be unnumbered links (Kompella and Rekhter, 2005a).

Link management

Recall that in GMPLS networks data and control planes are decoupled. Consequently, the control channels between two neighboring nodes are no longer required to use the same physical medium as the data channels between those nodes. For instance, a control channel could use a separate virtual circuit, wavelength, fiber, Ethernet link, or IP network. The control channels used to exchange the GMPLS control-plane messages exist independently of the TE links they manage. The so-called *link management proto-col* (LMP) was specified to establish and maintain such out-of-band control channels between neighboring nodes and to manage the data TE links between them (Lang, 2005). LMP is designed to accomplish four tasks: (1) control channel management, (2) link property correlation, (3) link connectivity verification, and (4) fault management. In GMPLS networks, control channel management and link property correlation are mandatory per LMP, while link connectivity verification and fault management are optional (Mannie, 2004). The four tasks are described in greater detail in the following:

- **Control Channel Management:** A control channel between two neighboring nodes is a pair of mutually reachable interfaces that are used to enable communication between nodes not only for link management but also for routing and signaling (routing and signaling will be addressed later). In LMP, one or more bidirectional control channels must be activated. The exact implementation of the control channel(s) is not specified in Lang (2005). For instance, a control channel can be a separate wavelength or fiber, an Ethernet link, an IP tunnel through a separate management network, or the overhead bytes of a data link protocol. Each node assigns a local control channel identifier to each control channel. Note that this identifier is taken from the same space as the unnumbered links, discussed earlier. To establish a control channel, the destination IP address on the remote end of the control channel must be known to the source node on the local end of the control channel. This knowledge may be explicitly configured or automatically discovered. Currently, LMP assumes that control channels are explicitly configured while their configuration can be dynamically negotiated (Mannie, 2004). More precisely, control channels can be established in different ways. For example, one might be implemented in-fiber while another one might be implemented out-of-fiber. Therefore, control channel parameters must be individually negotiated for each control channel. LMP consists of two phases: a *parameter negotiation* phase and a subsequent *keep-alive* phase. During the first phase, several negotiable parameters are negotiated and nonnegotiable parameters are announced. Among others, the HelloInterval and HelloDeadInterval parameters must be agreed upon by both neighboring nodes prior to sending any keep-alive messages. Those parameters are used in the Hello protocol during the subsequent keep-alive phase. The Hello protocol can be used to maintain control channel connectivity between the neighboring nodes and to detect control channel failures, whereby the HelloInterval parameter denotes the time period between two subsequently transmitted Hello messages, and the HelloDeadInterval parameter time indicates how long a device should wait to receive a Hello message before declaring a control channel dead. Clearly, the HelloDeadInterval parameter must be greater than the HelloInterval parameter. Both parameters are typically measured in milliseconds. Note that alternatively to using the Hello protocol, it may be possible to detect control channel failures by using lower-layer protocols (e.g., SONET/SDH overhead bytes), if available, since each control channel is electrically terminated at either node.
- **Link Property Correlation:** The link property correlation makes use of a message exchange between two neighboring nodes and is defined for TE links to ensure that

both local and remote ends of a given TE link is of the same type (i.e., IPv4, IPv6, or unnumbered). Furthermore, the link property correlation allows change in a link's TE attributes (e.g., minimum/maximum reservable bandwidth) and to form and modify link bundles (e.g., addition of component links to a link bundle). The link property correlation should be done before the link is brought up and may be done any time a link is up and not in the verification process, as discussed next.

- **Link Connectivity Verification (Optional):** In all-optical networks (AONs), transparent network nodes pose a challenge for verifying the physical connectivity of data TE links between two neighboring nodes. To enable connectivity verification of transparent data TE links, it is necessary that they must be electrically terminated at both ends (i.e., become opaque) until they carry user data traffic. Formally, there is no requirement that all data TE links be electrically terminated all at once. Instead, data TE links can be verified one by one with respect to their connectivity between two neighboring nodes. The verification procedure consists of sending test messages in-band over the data TE links. The link connectivity verification should be done when establishing a TE link, and subsequently, on a periodic basis for all unallocated data TE links.

- **Fault Management (Optional):** The fault management enables the network to survive node and link failures. The fault management includes the following three steps: *fault detection, fault notification,* and *fault localization*. Fault detection should be handled at the layer closest to the failure. In AONs, this is the optical layer, where, for instance, fault detection can be achieved by means of loss-of-light measurements. After detecting a failure, a network operator needs to know where exactly the failure occurred (fault localization) and a network node needs to be notified in order to initiate fault recovery actions (fault notification). In LMP, to localize a failed TE link between two neighboring nodes, the downstream node that has detected the fault informs its neighboring node about the fault by sending a control message upstream (fault notification). When the upstream node receives the fault notification it can correlate the fault with the corresponding interfaces to determine whether the fault is between the two neighboring nodes (fault localization). Once the failure has been localized, the signaling protocols (discussed in the following subsections) may be used to initiate LSP protection and restoration procedures (LSP protection and restoration will be explained in greater detail in Section 5.2.5).

Routing

To facilitate the establishment of LSPs, LSRs need more information about the links in the network than standard IGPs provide. Toward this end, TE routing extensions to the widely used link state routing protocols OSPF and IS-IS in support of carrying TE link state information for GMPLS were defined in Kompella and Rekhter (2005b–c), whereby manufacturers need to support either one or both of them. The TE routing extensions allow not only conventional topology discovery but also resource discovery throughout the routing domain by exploiting the inherent *link state advertisement* (LSA) mechanism of OSPF/IS-IS routing protocols. The link state routing protocols OSPF/IS-IS flood LSAs to distribute the topology and resource information among all

LSRs belonging to the same routing domain. Each LSR disseminates in its LSAs the resource information of its local TE links across the control channel(s). Recall from earlier that both the TE link information and control channel(s) are provided by the LMP. Apart from TE resource information, LSRs may also advertise optical resource information. Optical resource information can include wavelength value (frequency), physical layer impairments, such as polarization mode dispersion, amplified sponta- neous emission (ASE), nonlinear effects, or crosstalk (Zhang et al., 2001). However, routing information about the optical layer increases the amount of information needed to be distributed in LSAs, leading to increased distribution and settling times. One way to avoid this scalability problem is to run the link state routing protocol only in all-optical networks of limited geographic size, so-called *islands of transparency*. In each island of transparency, all paths have sufficiently adequate optical signal quality and therefore the advertising of optical resource information can be neglected. The LSAs enable all LSRs in a given routing domain to dynamically acquire and update a coherent picture of the network. This picture of the network is referred to as the *link state database*. The link state database consists of all LSRs and TE attributes of all links in a given routing domain (Banerjee et al., 2001b). Note that OSPF/IS-IS routing protocols with TE extensions flood the information about FAs just as they flood the information about TE links and any other links (Mannie, 2004). Consequently, an LSR has in its link state database the information about not just conventional links and TE links but also FAs. This information is used by an LSR to perform path computation, as explained next.

Path computation

While routing (and signaling) protocols with TE extensions are standardized for GM- PLS networks, path computation is typically proprietary and thus allows manufacturers and vendors to pursue diverse strategies and differentiate their products. In optical wavelength-switching GMPLS networks, where LSPs are identical to lightpaths, path computation faces several important issues and challenges (Zhang et al., 2001). The computation of lightpaths is commonly referred to as the routing and wavelength as- signment (RWA) problem. The RWA problem is usually decomposed into two separate subproblems: (1) route selection and (2) wavelength assignment. The path-selecting al- gorithms can be categorized into *fixed*, *fixed-alternate*, and *adaptive (dynamic)* routing algorithms. The fixed routing algorithm selects a single fixed path for each pair of source and destination nodes. The fixed-alternate routing algorithm selects one path out of mul- tiple alternative fixed paths, whereas the adaptive routing algorithm, which is also known as a dynamic routing algorithm, dynamically selects a path depending on the current network status and traffic conditions. Hence, the adaptive routing algorithm takes the current network status and traffic loads into account for computing the route of each lightpath request, as opposed to the fixed and fixed-alternate routing algorithms which use predetermined paths that in general are suboptimal. For the wavelength assignment subproblem, a wide variety of heuristics has been examined (e.g., first-fit or least-loaded). It is important to note that decomposing the RWA problem into two separate subprob- lems is well suited for lightpath computation in optical wavelength-switching networks that deploy wavelength converters. If wavelength converters are not available, the computation of lightpaths becomes slightly more involved due to the so-called *wavelength continuity constraint*. The wavelength continuity constraint imposes that a given lightpath must be established using the same wavelength on all links along the selected path. Due to this constraint, the lightpath computation problem cannot be decomposed into the route selection and wavelength assignment subproblems. To see this, consider the following example. Between a given pair of source and destination nodes a path could exist on which each link has unused wavelengths, but no common wavelength could be found for all these links. Consequently, no path satisfying the wavelength continuity constraint could be set up, despite the fact that both subproblems of route selection and wavelength assignment could be solved independently. One approach to solve the RWA problem subject to the wavelength continuity constraint is to decompose the optical wavelength-switching network into separate wavelength layers, each representing the (logical) network topology for a different wavelength. For each layer the lightpath computation is done separately. If a lightpath is found in any wavelength layer the wavelength continuity constraint is satisfied in that the lightpath can be established on a single wavelength between source and destination nodes. Such a lighpath that uses the same wavelength along the entire path is called a *wavelength path*. Clearly, this approach does not scale well. It may be used for a small to medium number of wavelength layers, but it becomes impractical for large numbers of wavelengths.

Apart from the special case of lightpath computation in optical wavelength-switching networks, paths need to be computed for GMPLS networks of any ISC. In general, path computation is achieved by running an SPF routing algorithm over a weighted graph. The weighted graph is built by using the TE link information present in the link state database of each LSR. More precisely, the link state database is used to assign a cost to all links of the network and thereby construct a weighted graph that satisfies the requirements of a given connection set-up request, leading to constrained shortest path first (CSPF) routing. For instance, TE links that do not have sufficient unreserved bandwidth to meet the bandwidth requirements of the new connection could be pruned from the TE database and thus do not appear in the weighted graph (Vigoureux et al., 2005).

Path computation needs to support different classes of service (CoS) in order to differentiate high-priority traffic from low-priority traffic. To fulfill the respective quality of service (QoS) requirements of different classes of service, various path computation approaches have been proposed. A hybrid routing approach for GMPLS networks based on both offline and online methods was described in Iovanna et al. (2003). In the proposed approach, an offline routing procedure is used to compute fixed paths for high-priority LSPs. An online routing procedure is invoked to allow prompt reaction to time-varying traffic loads by dynamically accommodating new LSP requests (adaptive routing) or rerouting or pre-empting low-priority LSPs in order to prevent congestion. The proposed hybrid offline–online routing procedure aims at efficiently supporting unpredictable Internet traffic while satisfying the QoS requirements of high-priority LSPs at any time, independent of current traffic demands. Another hybrid offline–online routing procedure for GMPLS networks was investigated in Elwalid et al. (2003). The proposed approach makes use of available traffic demand information based on customer prescriptions, traffic measurements, and traffic projections to build an approximate traffic demand matrix for path computation. It is worthwhile to note that the proposed path computation approach also takes path protection into account, whereby a set of disjoint working and backup paths are computed for each node pair.

Signaling

After computation of an appropriate path, signaling is used to establish the LSP. Similar to the GMPLS routing protocols, TE extensions have been defined for widely used and well understood signaling protocols to avoid the reinventing of the wheel. Specifically, the Resource Reservation Protocol with Traffic Engineering (RSVP-TE) (Berger, 2003b) and Constraint-Based Routing Label Distribution Protocol (CR-LDP) (Ashwood-Smith and Berger, 2003) have been standardized for signaling in GMPLS networks. The RSVP-TE and CR-LDP signaling protocols enable the set-up of LSPs, as previously discussed and depicted earlier in Fig. 5.5. Besides the set-up of LSPs, the GMPLS signaling protocols can be used to modify and release LSPs.

GMPLS signaling provides a number of advantageous features. Among others, GM-PLS signaling allows a label to be suggested by an upstream LSR, albeit the *suggested label* may be overwritten by a downstream LSR. The suggested label can be used in optical wavelength-switching networks with limited wavelength conversion capability where the wavelength assignment can be performed by the source LSR to minimize blocking probability. Furthermore, in RSVP-TE, but not CR-LDP, the *Notify* message has been defined in order to inform nonadjacent LSRs of LSP-related failures. The Notify message does not replace existing RSVP error messages, but it differs from them in that it can be targeted to any LSR other than the immediate upstream or downstream LSR. As a consequence, intermediate LSRs do not need to process Notify messages, resulting in a decreased failure notification delay and an improved failure recovery time (Banerjee et al., 2001a).

If applied in ASONs, GMPLS signaling should support *crankback* (Papadimitriou et al., 2005). Crankback allows an LSP set-up request to be retried on an alternate path that detours around a link or node which has insufficient resources to satisfy the LSP constraints. Crankback signaling comprises the following steps. The blocking resource (link or node) must be identified and returned in an error message to the repair node. The repair node is the upstream node that intercepts and processes the error message. The repair node computes an alternate path around the blocking resource that satisfies the LSP constraints. After path computation, the repair node reinitiates the LSP set-up request. To prevent an endless repetition of LSP set-up attempts, the number of retries should be limited. When the number of retries at a particular repair node is exceeded, the current repair node reports the error message upstream to the next repair node, where further rerouting attempts may be performed. When the maximum number of retries for a specific LSP has been exceeded, the current repair node should send an error message upstream back to the ingress node with no further rerouting attempts. The ingress node may choose to retry the LSP set-up according to local policy.

5.2.4   Bidirectional LSP

In the basic MPLS architecture, LSPs are unidirectional. Thus, in order to establish a bidirectional LSP, two unidirectional LSPs in opposite directions must be established independently (Banerjee et al., 2001a). Let the term initiator denote the LSR that starts the LSP establishment and the term terminator denote the LSR that is the target of the LSP. In traditional MPLS networks, the establishment of a bidirectional LSP requires two pairs of initiator and terminator, one for each direction. This approach has the following disadvantages (Mannie, 2004):

- The latency to establish the bidirectional LSP is equal to one round-trip signaling time plus one initiator–terminator transit delay.
- The control overhead is twice that of a unidirectional LSP. This is because separate control messages must be generated for both segments of the bidirectional LSP.
- Since the resources are established in separate segments, route selection is complicated.
- It is more difficult to provide a clean interface for SONET/SDH equipment that may rely on bidirectional paths for protection switching.

The need for bidirectional LSPs comes from non-PSC applications. For instance, bidirectional optical LSPs (lightpaths) are seen as a requirement for many optical networking service providers, where both directions of such LSPs have the same QoS requirements (e.g., latency, jitter, protection, and restoration). With bidirectional LSPs, there is only one pair of initiator and terminator and both the downstream and upstream data paths (i.e., from initiator to terminator and terminator to initiator) are established using a single set of signaling messages. This reduces the set-up latency to essentially one initiator–terminator round-trip time plus processing time and limits the control overhead to the same number of messages as a unidirectional LSP.

For bidirectional LSPs, two labels must be allocated, one downstream label and one upstream label. Bidirectional LSP set-up is indicated by the presence of an upstream label in the signaling message. It may occur that two bidirectional LSP set-up requests traveling in opposite directions allocate the same labels at effectively the same time, leading to a contention for labels. This contention may be resolved by imposing a policy at each initiator, for example, the initiator with the higher ID will win the contention (Berger, 2003a).

5.2.5   LSP protection and restoration

LSPs may be affected by link and/or LSR failures. To build fault-tolerant GMPLS networks, several fault recovery techniques can be deployed to support LSP protection and restoration (Banerjee et al., 2001a). Fault recovery typically takes place in the following four steps:

- Fault detection,
- Fault localization,

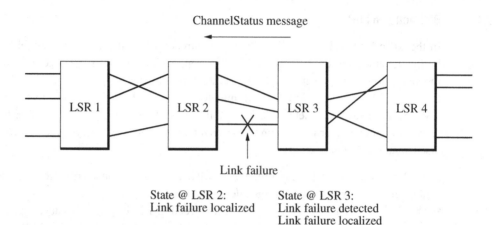

ChannelStatus message

State @ LSR 2:
Link failure localized

State @ LSR 3:
Link failure detected
Link failure localized

Figure 5.6 Fault localization using the LMP fault management procedure. After Lang and Drake (2002). © 2002 IEEE.

- Fault notification, and
- Fault mitigation.

It is recommended that fault detection takes place at the layer closest to the failure, which is the physical layer for optical networks. There exist several techniques to perform fault detection at the physical layer. For instance, a fault can be detected by detecting loss of light (LOL) or measuring the optical signal-to-noise ratio (OSNR), dispersion, crosstalk, or attenuation. Fault localization is achieved through communication between nodes to determine where the failure has occurred. In GMPLS networks, the fault management procedure of the LMP can be used to localize failures (Lang and Drake, 2002). Specifically, in the LMP fault management procedure a *ChannelStatus* message exchange between two neighboring LSRs is used for fault localization. The *ChannelStatus* message, defined in LMP, can be sent unsolicited to a neighboring LSR to indicate the current link status. Such a *ChannelStatus* message may be *SignalOkay*, *SignalDegrade*, or *SignalFail*. An example for fault localization using the LMP fault management procedure is shown in Fig. 5.6. In the figure, we assume that a fiber cut has occurred between upstream LSR 2 and downstream LSR 3. After detecting LOL, LSR 3 sends a *ChannelStatus* message to its upstream neighbor LSR 2, indicating that a failure has occurred. LSR 2 receives the message and then correlates the failure to see if the failure is also detected locally. If the failure is detected locally, LSR 2 has successfully localized the failure. Once the failure is localized, the next step of fault notification is initiated using GMPLS signaling. Fault notification is achieved by sending RSVP-TE or CR-LDP error messages to the source LSR or an intermediate LSR. Upon receipt of the fault notification, fault mitigation is initiated. A number of fault mitigation techniques are available, as discussed in greater detail in the remainder of this section.

Fault mitigation techniques used in GMPLS networks can be categorized into *protection* and *restoration* (Mannie, 2004). Let us first clarify the fundamental difference between protection and restoration. In protection, resources between the protection end points are established *before* failure and connectivity after failure is achieved simply by switching performed at the protection end points. Thus, protection may be viewed as a proactive technique. In contrast, restoration uses path computation and signaling *after* failure to dynamically allocate resources along the recovery path. Accordingly, restoration may be viewed as a reactive technique. Protection aims at achieving fast recovery time, at the expense of redundancy. Restoration takes more time than protection but is able to provide fault mitigation in a more bandwidth-efficient manner. Hence, protection and restoration provide a trade-off between recovery time and resource redundancy.

Both protection and restoration can be applied at various levels throughout the network (Lang and Drake, 2002). At the link (span) level, protection and restoration can be used to protect a pair of neighboring LSRs against a single link or channel failure by switching traffic to an alternate link or channel connecting the two LSRs. At the segment level, protection and restoration can be used to protect a connection segment against one or more link or node failures by switching traffic to an alternate segment that is routed around the failure. At the path level, protection and restoration can be used to protect the entire path between source and destination LSRs against one or more link or node failures by switching to an alternate path around the failure. Correspondingly, the three different types of protection and restoration techniques are referred to as *line switching*, *segment switching*, and *path switching*, respectively.

For any of these switching techniques, several protection schemes exist in GMPLS networks (Mannie, 2004):

- **1 + 1 Protection**: Two pre-provisioned disjoint resources (link, segment, path) are used in parallel to transmit data simultaneously. Both resources should be disjoint with respect to links and nodes, as well as SRLG. The receiving LSR uses a selector to choose the best signal. Note that although the switchover itself does not require signaling, it is often done to inform the source LSR that a switchover at the receiving LSR has taken place.
- **1:1 Protection**: One working resource and one protecting resource are pre-provisioned, but data is not replicated onto the protecting resource. If the working resource fails, the data is switched to the protecting resource.
- **1:N Protection**: N working resources and one protecting resource are pre-provisioned. If a working resource fails, the data is switched to the protecting resource. At this time, the remaining $(N-1)$ working resources are no longer protected.
- **M:N Protection**: N working resources and M protecting resources are pre-provisioned, where $1 \leq M \leq N$. If a working resource fails, the data are switched to one of the protecting resources. At this time, the remaining $(N-1)$ working resources are protected by $(M-1)$ protecting resources.

We note that 1:1 protection and 1:N protection are special cases of M:N protection. In both 1:N and M:N protection, the protecting resources are shared by N working resources. Correspondingly, this type of protection is called *shared* protection. An important characteristic of shared M:N protection is the ability to revert back to the working resource once the failure has been cleared. This *reversion* of traffic frees up the protecting resources for the remaining working resources. Conversely, both 1:1 and 1 + 1 protection are called *dedicated* protection, since the protecting resource is dedicated to a single working resource and is not shared by other working resources. However, there exists an important difference between dedicated 1:1 and 1 + 1 protection. In 1 + 1 protection, the protecting resource is used to carry replicated data. In contrast, even though the protecting resource is pre-provisioned in 1:1 protection, lower-priority extra traffic may use the protecting resource. In the event of a failed working resource, however, the extra traffic must be pre-empted such that the protecting resource can be used to carry the protected traffic. Note that the aforementioned GMPLS routing TE extensions also include the so-called *link protection type* which is flooded in each LSA (Berger, 2003a). The link protection type indicates the protection capabilities of a link (e.g., dedicated 1 + 1, dedicated 1:1, or shared 1:N protection). Path computation algorithms may take this information into account when computing paths for establishing LSPs with protection requirements.

Similar to protection, several restoration schemes exist in GMPLS networks for any of the line, segment, and path switching techniques (Mannie, 2004):

- **Restoration with Reprovisioning:** A restoration path is established after failure. The restoration path may be dynamically calculated after failure or precalculated before failure. It is important to note that the restoration path is only precalculated but no bandwidth is reserved for it via signaling.
- **Restoration with Presignaled Recovery Bandwidth Reservation and No Label Preselection:** A restoration path is precalculated before failure and a signaling message is sent along this preselected path to reserve bandwidth, but labels are not selected. The resources reserved on each link of a restoration path may be shared by different working LSPs that are not expected to fail simultaneously. Upon failure detection, signaling is initiated along the restoration path to select labels.
- **Restoration with Presignaled Recovery Bandwidth Reservation and Label Preselection:** A restoration path is precalculated before failure and a signaling message is sent along this preselected path to reserve bandwidth and select labels. The resources reserved on each link of a restoration path may be shared by different working LSPs that are not expected to fail simultaneously.

The decision of which of the aforementioned protection and restoration schemes will be actually used in an operational GMPLS network depends on a number of criteria such as robustness, recovery time, and resource sharing. Robustness is an important criterion so that GMPLS networks also become tolerant against multiple failures (Park, 2004). The decision will be made by GMPLS network operators according to given preferences, cost constraints, and service requirements.

To avoid simultaneous fault recovery actions at multiple GMPLS layers with different interface switching granularities (e.g., packet flows, lightpaths, and fibers), it is necessary to coordinate the various involved protection and restoration schemes in order to improve resource utilization and guarantee stable network operation. To achieve this, so-called *escalation strategies* must be deployed which are able to coordinate fault recovery across multiple GMPLS layers (Puype et al., 2005). In GMPLS networks, there are basically two types of escalation strategies: bottom-up and top-down (Mannie, 2004). The bottom-up escalation strategy assumes that lower-level recovery schemes are more expedient, whereby recovery starts at the lowest layer and then escalates upward for all affected traffic that cannot be restored at lower layers. In doing so, coarse switching granularities such as fibers and wavebands are handled first, followed by fine switching granularities such as wavelengths, times slots, frames, and packets. The bottom-up escalation strategy can be realized by using a hold-off timer that is set increasingly higher as one moves up in the GMPLS layer stack. Conversely, the top-down escalation strategy attempts recovery at the higher GMPLS layers before invoking lower-level protection and restoration schemes, whereby lower-level recovery is only activated if higher GMPLS layers cannot restore all traffic. The benefit of the top-down escalation strategy is given by the fact that it is service selective and permits per-CoS or per-LSP rerouting by differentiating between high-priority and low-priority traffic.

5.3   Implementation

Recently, there have been several experimental studies on implementing the various functions of a GMPLS-based control plane for optical switching networks. In Xin et al. (2001), a control plane was implemented for a mesh optical network consisting of wavelength-switching OXCs, used to interconnect high-speed IP/MPLS client routers. OXCs perform wavelength (lambda) switching; that is, OXCs have lambda switch capable (LSC) interfaces. GMPLS networks that support wavelength (lambda) switching but no other switching granularities are also referred to as multiprotocol lambda switching (MPλS) networks. The control plane may be integrated in the same box as the OXC or a separate router used to control the OXC. Between two neighboring OXCs a dedicated out-of-band wavelength is preconfigured for IP connectivity and the Transmission Control Protocol (TCP) is used to set up a reliable control channel for transmitting control messages. The control plane and proposed extensions implement key GMPLS functions such as routing, signaling, protection, and restoration.

An MPλS LSR that also supports IP packet switching was demonstrated in Sato et al. (2002). This so-called Hikari router has both PSC interfaces and LSC interfaces. In addition, the Hikari router offers 3R regeneration functions of the optical signal (reamplifying, reshaping, retiming) and wavelength conversion. At the protocol level, proprietary extensions to the GMPLS routing and signaling protocols were implemented. The path computation selects the path with the least number of wavelength converters because wavelength conversion is an expensive operation in all-optical networks. Furthermore, all packet-switched paths are monitored by IP routers. When the IP traffic becomes heavy, an optical bypass lightpath is set up by using the RSVP-TE extension. Based on IP traffic measurements, the lightpaths are dynamically reconfigured to match current IP traffic, resulting in a dynamic virtual network topology that is adaptive to current IP traffic demands (Shiomoto et al., 2003). The proposed multilayer traffic engineering heuristic approach finds the optimum establishment of lightpaths in response to IP traffic fluctuations, so as to utilize network resources in the most cost-effective manner. It was estimated that dynamic reconfiguration of lightpaths in response to IP traffic fluctuations leads to a cost reduction of more than 50%. Besides reconfiguration, grooming was implemented to further cut costs. When the IP traffic demand between a given pair of source and destination routers is much less than the capacity of a lightpath, the bypass wavelength is not fully utilized, resulting in wasted bandwidth resources. In this case, grooming is used to merge several IP traffic flows to better utilize the bypass wavelength. In Oki et al. (2005), two routing policies were implemented in the Hikari routers. Both routing policies first try to allocate a newly requested packet LSP to an existing lambda LSP that directly interconnects the source and destination nodes in a single hop. If such a lambda LSP does not exist, the two routing policies follow different directions. Policy 1 tries to find a series of existing lambda LSPs with two or more hops and sufficient resources that interconnect source and destination nodes. Policy 2 tries to set up a new single-hop lambda LSP between source and destination nodes. The obtained results show that policy 1 outperforms policy 2 in terms of admissible traffic volume when the number of PSC interfaces in the Hikari router is large, while policy 2 outperforms policy 1 when the number of PSC interfaces is small, where the admissible traffic volume is defined as the maximum admissible traffic volume under the condition that the blocking probability of packet LSP set-up requests is below a certain value.

5.4   Application

GMPLS has a great potential to reduce network costs significantly. The impact of GMPLS on the operational expenditures (OPEX) in an operational network was quantitatively analyzed in Pasqualini et al. (2005). In the quantitative analysis, several cost factors were taken into account (e.g., continuous cost of infrastructure, routine operations, reparation, and operational network planning). For the majority of considered operational models it was shown that OPEX can be reduced on the order of 50% from traditional operations when introducing GMPLS. Interestingly, the presented results hold not only for incumbent operators but also for new entrants.

GMPLS-based connection-oriented optical networks represent a good candidate for Grid computing. Besides the huge capacity of optical networks, GMPLS-based networks as connection-oriented networks are better suited to deliver rate- and delay-guaranteed services than the existing connectionless best-effort Internet. Furthermore, GMPLS is able to meet the adaptability, scalability, and heterogeneity goals of a Grid (Veeraraghavan et al., 2006).

6 Waveband switching

We have seen in Chapter 5 that generalized multiprotocol label switching (GMPLS) networks are able to support various switching granularities, covering fiber, waveband, wavelength, and subwavelength switching. To realize GMPLS networks, the underlying network nodes need to support multiple switching granularities rather than only one. Hence, ordinary optical cross-connects (OXCs) that perform only wavelength switching, such as the one shown in Fig. 1.5, must be upgraded in order to support multiple switching granularities, leading to so-called *multigranularity optical cross-connects* (MG-OXCs). Compared to ordinary OXCs, MG-OXCs hold great promise to reduce the complexity and costs of OXCs significantly by switching fibers and wavebands as an entity without demultiplexing the arriving WDM comb signal into its individual wavelengths, giving rise to *waveband switching* (WBS).

Recently, WBS has been receiving considerable attention for its practical importance in reducing the size and complexity of photonic and optical cross-connects. Due to the rapid development and worldwide deployment of dense wavelength division multiplexing (DWDM) technologies, current fibers are able to carry hundreds of wavelengths. Using ordinary wavelength-switching cross-connects would require a large number of ports. WBS comes into play here with the promise to reduce the port count, control complexity, and reduce the cost of photonic and optical cross-connects. The rationale behind WBS is to group several wavelengths together as a waveband and switch the waveband optically using a single input and a single output port instead of multiple input/output ports, one for each of the individual wavelengths of the waveband. As a result, the size of ordinary cross-connects that traditionally switch at the wavelength granularity can be reduced, including the associated control complexity and cost (Cao et al., 2003b).

6.1 Multigranularity optical cross-connect

Let us first elaborate on the necessary extensions of ordinary wavelength-switching (i.e., single-granularity) OXCs in order to support WBS. Clearly, apart from traditional wavelength switching, the upgraded OXCs must be able to switch incoming traffic at the waveband level and best also at the subwavelength level, resulting in MG-OXCs.

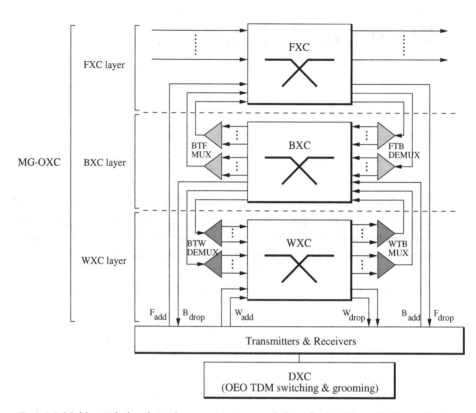

Figure 6.1 Multigranularity photonic cross-connect consisting of a three-layer multigranularity optical cross-connect (MG-OXC) and a digital cross-connect (DXC). After Cao et al. (2003b). © 2003 IEEE.

Figure 6.1 depicts a typical multigranularity photonic cross-connect consisting of a three-layer MG-OXC and a digital cross-connect (DXC) (Cao et al., 2003b). An MG-OXC not only switches traffic at multiple levels of granularities such as fiber, waveband, and wavelength levels but also allows to add and drop traffic at multiple granularities by deploying a bank of transmitters and receivers. Traffic can be shifted from one granularity level to another by using appropriate multiplexers and demultiplexers, as described in greater detail shortly. Note that for subwavelength switching (e.g., TDM switching and grooming), the MG-OXC is equipped with an additional DXC that performs OEO conversion. More specifically, the MG-OXC of Fig. 6.1 consists of the fiber cross-connect (FXC), band cross-connect (BXC), and wavelength cross-connect (WXC) layers. All three layers use add and drop ports to locally add and drop traffic at the corresponding granularity level, respectively. Unlike the FXC layer, both the BXC and WXC layers deploy multiplexers (MUXs) and demultiplexers (DEMUXs). The WXC layer comprises a WXC switch fabric that is used to switch wavelengths, i.e., lightpaths, and one or more W_{add}/W_{drop} ports to add/drop wavelengths locally. Furthermore, the WXC layer makes use of one or more band-to-wavelength (BTW) DEMUXs to demultiplex wavebands into wavelengths and one or more wavelength-to-band (WTB) MUXs to multiplex wavelengths into wavebands. Similarly, the BXC layer comprises a BXC switch fabric for waveband switching and one or more B_{add}/B_{drop} ports to add/drop wavebands locally. In addition, the BXC layer comprises one or more fiber-to-band (FTB) DEMUXs to demultiplex fibers into wavebands and one or more band-to-fiber (BTF) MUXs to multiplex wavebands into fibers. The various MUXs and DEMUXs are used for transporting traffic from one layer to another. Finally, the FXC layer consists of an FXC switch fabric for space switching of fibers.

In contrast to an ordinary single-granularity OXC that switches each wavelength individually using a separate port, the MG-OXC is able to switch a fiber using one port if none of the corresponding wavebands or wavelengths need to be dropped or added. Otherwise, the fiber will be demultiplexed into wavebands. Only those wavebands are further demultiplexed whose wavelengths need to be added or dropped. Each of the remaining wavebands will be switched as an entire waveband using one port at the BXC layer. In doing so, fibers and wavebands that carry bypass traffic do not need to undergo demultiplexing and multiplexing and can be switched as an entity, resulting in a reduced number of required ports compared to ordinary wavelength-switching OXCs. Thus, relatively small-scale modular switch fabrics are sufficient to build scalable MG-OXCs.

At the downside, MG-OXCs suffer from a few drawbacks. First, besides the switch fabrics each MG-OXC requires additional (DE)MUXs to shift traffic between the different layers. Second, the optical signal quality may deteriorate significantly within each MG-OXC due to the fact that lightpaths may need to go through multiple layers. Both shortcomings can be mitigated by using single-layer MG-OXCs instead of multilayer MG-OXCs. In single-layer MG-OXCs, all lightpaths traverse only a single switch fabric. Besides complexity, cost, and signal quality issues, it is important to consider the given traffic loads in making a choice between single-layer and multilayer MG-OXCs. It was shown in Cao et al. (2004a) that for static traffic single-layer MG-OXCs provide a greater reduction in size than multilayer MG-OXCs, and vice versa for dynamic traffic.

Several optimization approaches for design and dimensioning of MG-OXCs have been proposed recently. Among others, a heuristic approach to minimize the number of extra fibers required to extend an ordinary single-granularity wavelength-switching OXC to an MG-OXC that supports not only wavelength-switching but also waveband and fiber switching was proposed in Ho et al. (2003a,b). A heuristic algorithm for reducing the number of ports used and improving the blocking probability of an MG-OXC under dynamic traffic was studied in Cao et al. (2003a).

6.2 Waveband grouping

We have seen above that MG-OXCs enable WBS and help reduce the size and complexity of optical cross-connects. While it is somewhat obvious to recognize the merits of switching several wavelengths as a single waveband entity, it is more involved to find out how many and which wavelengths need to be grouped together into a single waveband in order to meet certain performance requirements. Toward this end, waveband grouping strategies for WBS networks were developed that satisfy different performance metrics. The existing waveband grouping strategies can be classified into the following two categories: (1) *end-to-end* waveband grouping and (2) *intermediate* waveband grouping. In general, intermediate waveband grouping strategies outperform end-to-end grouping strategies in terms of cost savings. To see this, consider the following illustrative example. Let there be three geographically distributed source nodes, each using a separate wavelength to send traffic to the same destination node D. Through intermediate waveband grouping, the three wavelengths can be grouped into a single waveband at any common intermediate node X, resulting in a reduced number of required ports at MG-OXCs between intermediate node X and destination node D. On the contrary, end-to-end waveband grouping strategies are unable to group the three wavelengths into a single waveband because the sources are not collocated.

Both waveband grouping strategies were investigated and compared in greater detail for mesh WDM networks that support wavelength switching and waveband switching. In Li et al. (2005b), an end-to-end waveband grouping strategy that groups wavelengths with the same source-destination pair into a waveband was compared with an intermediate waveband grouping strategy that groups wavelengths with the same destination at an intermediate node. In the former case, a waveband route is generated from the source node and terminated at the destination node. In the latter case, a waveband route is generated at an intermediate node, routed along the following common links, and terminated at the destination node. Under the assumption of shortest path routing, random waveband/wavelength assignment, and waveband/wavelength continuity constraint along the selected route, both waveband grouping strategies were evaluated in terms of cost savings and blocking probability. The obtained results confirm that the intermediate waveband grouping strategy outperforms its end-to-end counterpart in terms of cost savings. In terms of blocking probability, however, the end-to-end waveband grouping strategy slightly outperforms the intermediate waveband grouping strategy. To improve the blocking probability performance of the proposed intermediate waveband grouping strategy, an algorithm which defines a cost (weight) for each candidate route of a given connection request was introduced in Li et al. (2005a). Under the assumption of dynamic traffic it was shown that a good balance between minimizing the switching cost and minimizing the blocking probability can be achieved by selecting the route with the smallest weighted cost.

6.3   Routing and wavelength assignment

The routing and wavelength assignment (RWA) problem was introduced in the context of traditional optical wavelength-switching WDM networks in Section 2.5. We have seen that the RWA problem deals with the route selection of lightpaths and the assignment of a wavelength on each WDM link along the selected route. Apart from the special case of lightpath computation in optical wavelength-switching networks, paths of finer and coarser switching granularity must be computed in GMPLS networks that deploy MG-OXCs, where paths may be fiber, waveband, wavelength (lambda), and subwavelength (TDM, packet) label switched paths, as discussed in greater detail in Section 5.2.3. The RWA problem in waveband-switching networks that use MG-OXCs is in general more involved than that in wavelength-switching networks with ordinary single-granularity OXCs since further constraints must be taken into account apart from the wavelength continuity constraint. As a consequence, several new RWA-related problems in waveband-switching networks arise which need to be identified and solved in order to optimize the performance of WBS networks.

Let us first concentrate on WBS networks without wavelength conversion. Apart from routing and wavelength assignment, tunnel assignment is another important problem encountered in WBS networks, leading to the so-called *routing and wavelength/tunnel assignment* (RWTA) problem. The RWTA problem was investigated in Ho and Mouftah (2001). A tunnel is defined as a group of consecutive wavelength channels grouped and switched together. A tunnel can be either a waveband or fiber tunnel. A waveband tunnel contains multiple consecutive wavelengths. A fiber tunnel consists of multiple waveband tunnels. The RWTA problem deals with the bundling and switching of consecutive wavelengths or wavebands as a waveband tunnel or fiber tunnel, respectively, and routing lightpaths through them. Rules for allocating tunnels in mesh WBS networks were provided and studied in Ho and Mouftah (2001). Those rules recommend to use existing fiber tunnels and waveband tunnels for lightpath set-up. If no appropriate tunnels exist, it is recommended to give priority to creating a new fiber tunnel over creating new waveband tunnels. If neither fiber tunnels nor waveband tunnels can be newly established, the requested lightpath is set up without any tunnels by solving the conventional RWA problem.

An efficient approach to routing and wavelength assignment in mesh WBS networks with fiber, waveband, and wavelength switching capabilities was presented in Lee et al. (2004). The authors refer to this problem as the *RWA+* problem. The RWA+ problem is formulated as a combinatorial optimization problem with the objective to minimize the bottleneck link utilization of mesh WBS networks. The obtained results show that the proposed optimization approach outperforms conventional linear programming approaches in both accuracy and computational time complexity particularly for larger WBS networks.

The so-called *routing, wavelength assignment, and waveband assignment* (RWWBA) problem was investigated in Li and Ramamurthy (2006). The RWWBA problem addresses the optimal routing and wavelength/waveband assignment in mesh WDM networks that deploy both wavelength switching and waveband switching. The RWWBA problem aims at maximizing cost savings and minimizing blocking probability. To solve the RWWBA problem, the online integrated intermediate waveband switching algorithm was proposed which controls the creation of new waveband routes and determines the waveband grouping node and waveband disaggregating node along the selected route.

Several other routing and wavelength assignment heuristics to minimize the ports needed at MG-OXCs in WBS networks for a given set of lightpath requests were proposed in Cao et al. (2003c). The obtained results show that WBS is particularly beneficial in multifiber networks (i.e., networks where a link consists of multiple fibers in parallel). It was shown that using MG-OXCs can save up to 50% of ports in single-fiber networks and up to 70% of ports in multifiber networks compared with using ordinary single-granularity OXCs that perform conventional wavelength switching.

Next, let us consider the routing and wavelength/waveband assignment in WBS networks that make use of wavelength conversion. The aforementioned RWTA problem was re-visited in Ho and Mouftah (2002) under the assumption that ordinary OXCs are able to perform full-range wavelength conversion. Due to the added wavelength converting capability, WBS networks become more flexible and able to adapt better to given traffic demands, resulting in an increased utilization of waveband and fiber tunnels.

Apart from full-range wavelength conversion, the impact of various other types of wavelength conversion on the blocking probability of WBS networks was examined in Cao et al. (2004b). Besides full-range and limited-range wavelength conversion, the so-called *intraband wavelength conversion* was studied which can be applied in practical WBS networks. With intraband wavelength conversion, a wavelength can only be converted to any other wavelength within the same waveband. For example, let us assume that waveband b_1 contains wavelengths w_1, w_2, w_3, and waveband b_2 contains wavelengths w_4, w_5, w_6. Then wavelength w_3 can only be converted to wavelengths w_1 or w_2 but not to wavelengths w_4, w_5, w_6. Note that intraband wavelength conversion differs from limited-range conversion in that it introduces an additional constraint by allowing wavelength conversion take place only within a given waveband. The proposed wavelength assignment heuristic is able to reduce the blocking probability by efficiently grouping wavelengths into wavebands and reducing the number of used wavelength converters when satisfying a new lightpath request. The special case of limited-range wavelength conversion was addressed in greater detail in Cao et al. (2005).

6.4   TDM switching and grooming

So far, our discussion of WBS networks focused on MG-OXCs that deploy fiber, waveband, and wavelength switching. As shown in Fig. 6.1, MG-OXCs can be equipped with an additional DXC in order to perform TDM switching and grooming in the electrical domain by means of OEO conversion of wavelengths and wavebands.

A hybrid optoelectrical switch architecture which integrates all-optical fiber and waveband switching and electrical TDM switching was investigated in Yao et al. (2003b). The proposed switch architecture does not support wavelength switching, but it combines the scalability and cost savings of WBS with the flexibility of subwavelength TDM switching in the electrical domain. In addition, the electrical TDM switch can be used to perform wavelength conversion and multicasting. Based on this hybrid optoelectrical switch, several heuristics were proposed to optimize the overall switch cost of mesh WBS networks by minimizing the cost of waveband switch ports and OEO TDM switch ports, including associated transponders.

A similar hybrid optoelectrical switch architecture that integrates all-optical waveband switching and electrical traffic grooming was investigated in Yao and Mukherjee (2003).

The presented design study of WBS networks based on such hybrid WBS-OEO grooming switch architectures also takes physical transmission impairments on all-optical waveband switching paths into account (e.g., fiber nonlinearities, amplified spontaneous emission (ASE), polarization mode dispersion (PMD), etc.). Those impairments limit the maximum distance and number of nodes the optical signal can traverse without undergoing OEO conversion. The network design aims at minimizing the network cost, finding virtual topologies consisting of waveband paths, and routing connections along the waveband paths. Several routing strategies were proposed and evaluated in terms of achieved cost savings.

6.5   Implementation

The practical implementation and feasibility issues of waveband switching in existing transparent optical networks were demonstrated in Toliver et al. (2003). The reported waveband switching experiments utilized paths through transparent wavelength-selective cross-connects (WSXCs). The transmission of a waveband consisting of four 25-GHz spaced contiguous wavelengths in a 200-GHz passband through a transparent reconfigurable optical network was demonstrated and it was shown that physical transmission impairments have a tolerable impact on the bit error rate (BER). More importantly, the presented experiments successfully demonstrated that WBS is a technique that can be applied to existing transparent wavelength-switching networks within the given passband of the underlying optical network elements. The passband that originally carried only a single wavelength can be used to carry a waveband of multiple adjacent wavelengths by reducing the channel spacing. Thus, WBS networks can be realized by using wavebanding techniques at the edge of wavelength-switching networks without requiring changes to existing optical networks and fiber infrastructure.

7   Photonic slot routing

In our introductory discussion of all-optical networks (AONs) in Section 1.5.1 we have seen that the concept of lightpath plays a key role in wavelength-routing optical networks. A lightpath is an optical point-to-point path of light that interconnects a pair of source and destination nodes, where intermediate nodes along the lightpath route the signal all-optically without undergoing OEO conversion. As each lightpath requires one wavelength on every traversed link and the number of both wavelengths and links in AONs is limited for cost and efficiency reasons, it is impossible to interconnect every pair of nodes by a dedicated lightpath. Nodes that cannot be directly connected via a lightpath may use multiple different lightpaths to exchange data. In the resultant multihop optical network, each intermediate node terminating a lightpath performs OEO conversion. As a consequence, such opaque multihop optical networks are unable to provide transparency. Also, note that the transmission capacity between node pairs connected via a lightpath is equal to the bandwidth of an entire wavelength channel. This transmission capacity is dedicated and cannot be shared by other nodes, leading to wasted bandwidth under bursty nonregular traffic. To improve the bandwidth utilization of lightpaths, electronic traffic grooming becomes necessary at each source node.

To avoid the loss of transparency and the need for electronic traffic grooming of lightpath-based optical networks, a novel solution for the design of transparent mesh wavelength division multiplexing (WDM) wide area networks was proposed in Chlamtac et al. (1999b). This new design approach, called *photonic slot routing* (PSR), allows entire slots, each carrying multiple packets on distinct wavelength channels, to be switched all-optically and individually. Unlike lightpath-based optical networks, PSR enables all nodes to communicate all-optically with each other (i.e., without OEO conversion at intermediate nodes) and allows traffic aggregation to be done optically in order to use the bandwidth of each wavelength channel efficiently (i.e., without electronic traffic grooming), as explained in greater detail in the following.

7.1   Photonic slot

In PSR networks, time is divided into fixed-size slots. Each slot spans all wavelengths in the PSR network with both slot boundaries aligned across all wavelengths. Correspondingly, the resultant multiwavelength slot is called *photonic slot*. Each wavelength in the

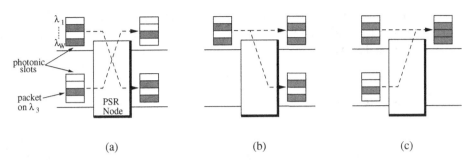

(a)   (b)   (c)

Figure 7.1 Photonic slot routing (PSR) functions: (a) photonic slot switching, (b) photonic slot copying, and (c) photonic slot merging. After Elek et al. (2001). © 2001 IEEE.

photonic slot may contain a single data packet. Thus, given W wavelength channels, the photonic slot may carry up to W data packets. Furthermore, all packets in a given photonic slot are required to be destined for the same node, but each photonic slot may be destined for a different node. Due to the fact that all packets must have the same destination, the photonic slot can be routed as a single entity by intermediate nodes, thereby avoiding the need for demultiplexing the individual wavelengths and routing them individually. By requiring that photonic slots are routed as a single entity, PSR networks exhibit a number of advantages. First, no wavelength-sensitive components are required at intermediate nodes. Instead, wavelength-insensitive components can be used, resulting in lower network costs and avoidance of interchannel switching crosstalk in adjacent wavelength channels since wavelengths are not separated at intermediate nodes. Second, PSR reduces the complexity of the switching operation and its electronic control at intermediate nodes by a factor of W. Third, PSR nodes can be built in a cost-effective fashion by using relatively simple optical components based on proven technologies.

As shown in Fig. 7.1, PSR network nodes may perform the following three functions on a per photonic slot basis (Elek et al., 2001):

- **Photonic Slot Switching:** Photonic slots arriving on any input port of a given PSR node can be switched individually to any output port. Photonic slots that are switched to the same output port simultaneously cause contention at the output port. Several approaches exist to resolve contention in PSR networks, as discussed in Section 7.4.
- **Photonic Slot Copying:** A photonic slot arriving on an input port of a given PSR node is duplicated and switched to two or more output ports. Photonic slot copying can be used to realize multicasting of photonic slots (i.e., transmitting a photonic slot to multiple destination nodes). Note that this function may be exploited by a source node to transmit packets in the same photonic slot even though the packets are not intended for the same destination.
- **Photonic Slot Merging:** Photonic slots concurrently arriving on multiple input ports of a given PSR node are switched to the same output port, thus overlapping with one another to form a new single photonic slot that departs from the PSR node. Clearly, the photonic slot merging function is only permissible if the merging slots are compatible

in the sense that they do not carry packets on the same wavelengths. This function allows packets coming from different input ports to be forwarded together toward a common destination.

7.2   Synchronization

The merits of PSR come at the expense of achieving and maintaining slot synchronization network wide. To achieve the aforementioned functions of photonic slot switching, copying, and merging, PSR requires that photonic slots arrive synchronized at PSR nodes. One way to achieve slot synchronization is to require the length of each fiber link to be an integer multiple of the photonic slot size. Toward this end, fiber lengths can be adjusted by inserting fiber delay lines (FDLs) at the input ports of PSR nodes in order to delay arriving photonic slots (Elek et al., 2001). The FDLs act as optical synchronizers. To avoid the use of optical synchronizers at the input ports of PSR nodes, photonic slots must be made to arrive aligned at the input ports of each PSR node by means of network-wide photonic slot synchronization. One possibility to achieve network-wide photonic slot synchronization was described in Chlamtac et al. (1997a) for a two-layer PSR network which consists of a central unidirectional ring network that interconnects multiple folded bus networks. The proposed photonic slot synchronization works as follows. Photonic slots circulating within the central unidirectional ring network are synchronized first, followed by the synchronization of the photonic slots transmitted in the attached folded bus networks. For photonic slot synchronization on the ring, the length of the photonic slot is designed to be equal to an integer fraction of the length of the round-trip path in the ring. Such a condition can be practically obtained by nominating one of the PSR ring nodes as master. Photonic slot synchronization on the ring is achieved through the master PSR ring node by using a ping technique based on the transmission of a subcarrier signal (the ping) at the boundary of each photonic slot and modifying the length of the photonic slot until the transmitted pings, after a ring round-trip time, overlap with the newly generated pings. In doing so, the master PSR ring node is able to achieve photonic slot synchronization. The remaining PSR ring nodes achieve photonic slot synchronization by monitoring the pings on the ring sent by the master PSR ring node. The pings on the PSR ring are copied onto each attached folded bus PSR network. Photonic slot synchronization in each folded bus is achieved in a similar way. Within each folded bus, every PSR node transmits a comb of pings assuming a nominal photonic slot length. After a certain propagation delay on the folded bus, these pings return to the PSR node together with a copy of the pings of the central PSR ring network. By comparing the two received ping combs and modifying its ping frequency, the PSR node in the folded bus reaches the required photonic slot synchronization when the two ping combs overlap.

Furthermore, in PSR networks dispersion needs to be taken into account since packets in every photonic slot must be aligned within an acceptable limit. The alignment of photonic slots may deteriorate due to dispersion when propagating over long fiber spans. The detrimental effects of dispersion can be mitigated by inserting a time guard

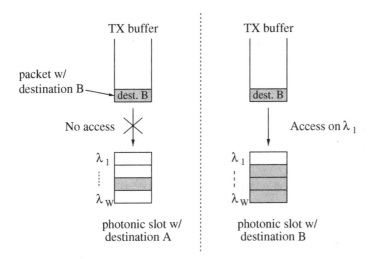

Figure 7.2 Access control in PSR networks based on destination of photonic slot. After Chlamtac et al. (1997a). With permission from IOS Press.

between adjacent photonic slots at the expense of a decreased bandwidth utilization. Alternatively, a transmission system with sufficiently low dispersion can be created by deploying an additional dispersion-compensating fiber in order to increase the coverage of PSR networks.

7.3   Sorting access protocol

Besides synchronization, PSR requires additional processing at the periphery of the network to place packets in the appropriate photonic slots destined for the corresponding node. Recall that in PSR each photonic slot is assigned a destination. In principle, the destination address might identify a single node, a group of multiple nodes (multicast group), or a network segment (e.g., a folded bus in the two-layer network discussed in the previous section). A source node is allowed to send a packet on any of the free wavelengths of an arriving photonic slot if the destination address of the photonic slot matches the packet's destination address. The access control in PSR networks is illustrated in Fig. 7.2. In our example, a packet destined for node B is stored in the transmission (TX) buffer of the source node. As shown on the left-hand side of Fig. 7.2, the current photonic slot is intended for another destination node A. Therefore, the packet transmission is deferred by one or more slots until a photonic slot with destination B appears at the source node. When such a photonic slot arrives at the source node, the packet can be transmitted on any free wavelength in the photonic slot. The right-hand side of Fig. 7.2 shows a photonic slot with destination B whose wavelength λ_1 is unused. The source node is allowed to transmit its packet in this photonic slot on λ_1 since the destination addresses of the locally generated packet and photonic slot are identical and at least one wavelength (λ_1) is free.

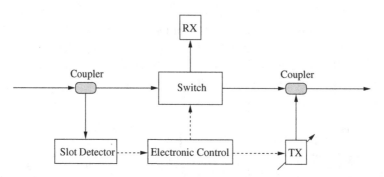

Figure 7.3 Architecture of a PSR node. After Chlamtac et al. (1999a). © 1999 IEEE.

To coordinate the packet transmission at each source node, the so-called *sorting access protocol* is used (Chlamtac et al., 1999a). The sorting access protocol ensures collision-free wavelength channel access at each source node and organizes the transmission of packets in photonic slots according to the destination address of both photonic slots and packets. A given photonic slot may propagate through the PSR network either with or without assigned destination address. The destination address of a given photonic slot is determined by the destination address of the first packet transmitted in the photonic slot. At each source node, packets are selected for transmission based on a destination-dependent queueing approach. To achieve this, packets are stored in separate transmission buffers according to their destination address. Note that the buffer management complexity is proportional to the number of buffers. Depending on the type of destination address, buffers might be required for each individual destination node, multicast group, or network segment. Clearly, buffering based on network segments requires less complex buffer management than using dedicated buffers, one for each individual destination node. If the arriving photonic slot is already assigned a destination, the packet selected for transmission is the head-of-the-line packet of the buffer associated with that destination, provided not all wavelengths are used. If the arriving photonic slot is not assigned a destination, the packet selected for transmission is the oldest packet (i.e., the packet with the longest queueing delay) among the head-of-the-line packets that do not cause any wavelength collision in the photonic slot.

Figure 7.3 depicts a possible architecture of a PSR node that enables not only the aforementioned access control but also the reception of photonic slots. Specifically, a coupler that taps a portion of the incoming optical signal off the input fiber link is used in conjunction with a slot detector to find out (1) whether an arriving photonic slot is destined for the local node and (2) which wavelengths in the arriving photonic slot carry packets. Based on this information, the photonic switch is set such that the local node extracts the packet(s) of the corresponding photonic slot from the network if the photonic slot is intended for the local node, or lets the packet(s) remain in the network otherwise. Another coupler is deployed to insert locally generated packets into photonic slots by tuning the local transmitter to one of the free wavelengths in the photonic slot whose destination address matches that of the packet to send.

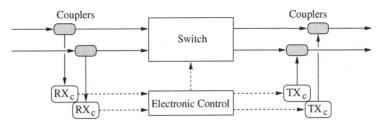

Figure 7.4 Architecture of a PSR bridge. After Chlamtac et al. (1999a). © 1999 IEEE.

To interconnect different PSR network segments one might use a PSR bridge. The architecture of a possible PSR bridge is shown in Fig. 7.4. We assume that the bridge is used to interconnect two network segments (the bridge can be easily extended to provide interconnection of three and more network segments). As shown in the figure, two pairs of control receiver RX_c and control transmitter TX_c are used, one for each of the two network segments. The receivers inform the bridge control unit of the destination of photonic slots arriving on each network segment. The electronic control of the bridge uses this information to configure the 2×2 photonic space switch accordingly, whose cross-bar configuration is determined on a photonic slot basis. In bar state, the 2×2 switch lets arriving photonic slots stay in their respective network segments (i.e., there is no exchange of photonic slots between the two network segments), whereas in cross-state, a connection between the two network segments is established and photonic slots are switched from one network segment to the other one.

In PSR networks consisting of PSR nodes and PSR bridges, photonic slots optically propagate from the source PSR node to the destination PSR node. On their way toward the destination PSR node, photonic slots may experience contention at intermediate PSR bridges. Contention at PSR bridges occurs when more than one photonic slot among those simultaneously reaching the PSR bridge need to be switched to the same output port. Clearly, one way to deal with contention is to switch only one photonic slot while the remaining ones are dropped by the PSR bridge and retransmitted by the source PSR node (Chlamtac et al., 1999a). Several approaches to resolve contention at PSR bridges were proposed and investigated, as discussed next.

7.4   Contention resolution

One way to completely avoid contention at PSR bridges network wide is the use of time division multiplexing (TDM). Recall from Section 7.2 that in PSR networks all nodes (and bridges) are synchronized. Based on the available network-wide synchronization, the use of TDM to allocate bandwidth to the contention-free transmission of photonic slots was examined in Chlamtac et al. (1999b). With TDM, transmission on each link takes place in predefined periodically recurring time frames which comprise an equal number of time slots. In each time frame, connections each with a transmission capacity of one packet per frame can be established between any pair of source and destination

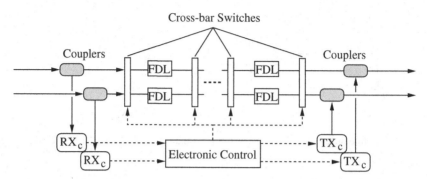

Figure 7.5 Architecture of an SDL bridge. After Chlamtac et al. (1997a). © 1997 IEEE.

PSR nodes. Given a PSR network topology and a set of connection requests, the problem of constructing optimal TDM frames was investigated in Chlamtac et al. (1999b). A heuristic algorithm was proposed that constructs TDM frames with the objective to maximize the aggregate throughput of the PSR network.

Note that contention resolution in PSR networks is somewhat tricky since conventional PSR bridges do not use any form of optical buffers for resolving contention. The contention can be mitigated by upgrading PSR bridges and adding a so-called *switched delay line* (SDL) to them, resulting in SDL bridges (Chlamtac et al., 1997b). The SDL consists of a combination of concatenated FDLs and photonic cross-bar switches that are controlled on a photonic slot basis, with the goal of delaying photonic slots until contentions at the SDL bridge are resolved. SDL bridges help reduce contentions of photonic slots significantly without compromising the merits of PSR networks, namely, all-optical end-to-end communication and optical traffic aggregation. Figure 7.5 depicts the architecture of an SDL bridge. In general, the SDL contains i FDLs and $i + 1$ photonic 2×2 cross-bar switches, where $i \geq 0$. Note that for $i = 0$ the SDL disappears and the SDL bridge degenerates into the conventional PSR bridge of Fig. 7.4. Similar to the conventional PSR bridge, the configuration of the SDL bridge is determined by the electronic control on a photonic slot basis based on the information provided by the local receivers RX_c as well as on the photonic slots already stored in the FDLs.

Figure 7.6 shows a generalized architecture of a PSR node with multiple input and multiple output ports that can be used as a source PSR node as well as an SDL bridge in mesh WDM networks (Zang et al., 1999, 2000). The PSR node consists of a wavelength-insensitive optical packet switch, SDLs to temporarily store photonic slots at intermediate PSR nodes, and electronic buffers to hold locally generated packets at source PSR nodes, one buffer for each destination. The authors proposed to use an additional control wavelength in each photonic slot that carries the header of the photonic slot. Among others, the header contains information such as the destination of the photonic slot and which wavelengths in the photonic slot are occupied by packets. At each input fiber of the photonic slot, a demultiplexer is used to extract the header of each photonic slot. Since it takes some time to process the header and to configure the optical packet switch,

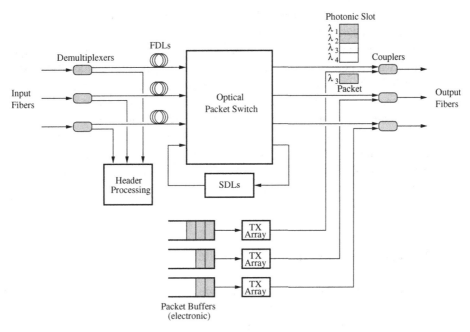

Figure 7.6 PSR node with multiple input/output ports. After Zang et al. (2000). © 2000 IEEE.

FDLs are used to delay the arriving photonic slots. On a given output fiber, the PSR node may insert packets into photonic slots which are destined for the appropriate destination by using couplers. For instance, in Fig. 7.6 a departing photonic slot carries packets on wavelengths λ_1 and λ_2. The PSR node may insert a packet into the photonic slot on wavelength λ_3. In Zang et al. (1999, 2000), several schemes to resolve contention were examined. Apart from retransmitting dropped photonic slots by source PSR nodes and buffering contending photonic slots in SDLs at intermediate PSR nodes, the benefit of *deflection routing* was investigated to resolve contention in mesh PSR networks. With deflection routing, in the event of contention a photonic slot is routed to some nonbusy link other than the link specified by the routing algorithm in use (e.g., shortest path first [SPF]). Thus, if two photonic slots are contending, one slot is routed through the output fiber link specified by the routing algorithm whereas the other slot is routed through any of the remaining free output fiber links. To prevent photonic slots from deflecting too often and thereby consuming too many network resources, photonic slots will be dropped (and need to be retransmitted) if they have been deflected up to a predefined number of times. Toward this end, a deflection counter is placed in the header of each photonic slot. Each time a photonic slot is deflected the counter is decremented. Under the assumption of shortest path routing, the obtained results indicate that under low load the throughput-delay performance of the two contention resolution techniques, buffering and deflection routing, are similar. This is because at low load there is only little contention and therefore no buffering or deflection routing is necessary. With increasing load, deflection routing may outperform buffering. This is due to the fact that shortest path routing leads to unbalanced link loads. Deflection routing is able to balance the load on the links by using alternate paths to each destination beside the shortest path.

It is important to note that the benefits of deflection routing heavily depend on the network topology and the applied routing algorithm. It might occur that deflection routing actually performs worse than buffering due to the excess consumption of network resources by deflected photonic slots. Generally, deflection routing becomes beneficial if there are one or more alternate paths to the destination which are less loaded than the path specified by the selected routing algorithm. By applying adaptive (dynamic) routing algorithms instead of fixed shortest path routing the load can be better balanced across network links, resulting in reduced contention at intermediate PSR nodes and an improved throughput-delay performance.

7.5   Evolution toward optical packet switching

In PSR, photonic slots are switched as a single entity throughout the network without requiring individual wavelengths to be demultiplexed and multiplexed at PSR nodes. As a result, PSR networks do not require wavelength-sensitive devices and cross-connects. Instead, PSR networks are able to carry optical multiwavelength packets using inexpensive wavelength-insensitive optical packet switches based on proven technologies. In doing so, costly technological problems of packet switching on a per-wavelength basis are replaced with the easier problem of efficiently forming photonic slots by using the aforementioned sorting access protocol.

At the downside, PSR suffers from some inefficiency. Due to the fact that in PSR all wavelengths in a given photonic slot are forced to be destined for the same node, only PSR nodes ready to send packets to this destination node may use the photonic slot. Other PSR nodes are prevented from using the photonic slot, even though most of the wavelengths may be free. Clearly, this inefficiency can be avoided by breaking up the photonic slot and neglecting the requirement that all wavelengths need to be destined for the same node. As a consequence, each individual wavelength can be accessed independently from each other, giving rise to *individual wavelength switching* (IWS) (Elek et al., 2001). As optical technology advances, the cost of optical components and switches is expected to decrease significantly in the long run. It will then become feasible to develop wavelength-sensitive optical packet switches in a cost-effective manner. Such wavelength-sensitive optical packet switches are able to switch packets on each individual wavelength, independent from the packet flows on the other wavelengths. The resultant IWS switch might be viewed as wavelength-selective cross-connect (WSXC) whose configuration can be altered at every time slot. It was shown in Elek et al. (2001) that the network capacity can be increased significantly by carefully replacing a relatively small percentage of conventional PSR switches with IWS switches. Thus, by incrementally replacing PSR switches with IWS switches, the network capacity can be gradually increased according to given traffic demands and/or cost constraints. Note that such an incremental upgrade affects only a subset of PSR nodes while all remaining PSR switches are left unchanged,

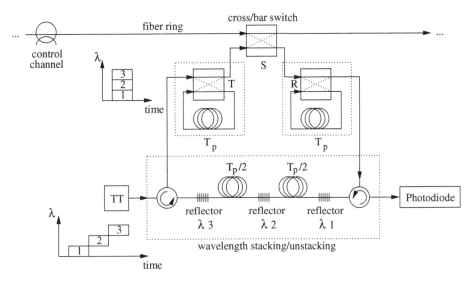

Figure 7.7 Node architecture for wavelength stacking. After Smiljanic et al. (2002). © 2002 IEEE.

enabling cautious upgrade and smooth migration paths from PSR networks to *optical packet switching* (OPS) networks. In OPS networks, packets are switched on each wavelength individually rather than jointly across all wavelengths.

It is important to note that IWS networks, similar to PSR networks, require network-wide synchronization and operate based on fixed-size time slots, each able to carry one packet. Therefore, IWS networks support the optical switching of fixed-size packets. OPS networks that do not require network-wide synchronization and are also able to support variable-size packets will be described in Chapter 10.

7.6   Implementation

PSR networks were experimentally investigated only to a limited extent so far. An interesting packet transmission technique that allows a node to simultaneously send/receive packets on different wavelengths in the same photonic slot despite the fact that the node has only one transceiver was experimentally demonstrated in Boroditsky et al. (2001a, b, 2003). This technique, referred to as *wavelength stacking*, enables a node to add/drop packets on multiple wavelengths of a photonic slot (Smiljanic et al., 2001, 2002). To realize wavelength stacking a node is equipped with one fast tunable transmitter (TT) and one photodiode, as depicted in Fig. 7.7. Time is divided into slots of duration T_p. The length of the fixed-size packets equals one time slot. The TT at a given node starts transmitting packets W time slots before its scheduled time slot, where W denotes the number of wavelengths in the PSR network. As illustrated in Fig. 7.7 for $W = 3$, in each following time slot the node transmits a packet on each different wavelength. The optical signal passes through an array of fiber grating based wavelength reflectors separated by FDLs such that the W packets transmitted on different wavelengths are aligned in time, thereby forming the photonic slot. The photonic slot is then sent to the PSR ring by setting switch S to the cross state. On the receiver side, the reverse procedure takes place. A photonic slot is received when switch S is in the cross state. The received photonic slot is subsequently unstacked by passing through the same array of reflectors and FDLs. Note that a single broadband wavelength-insensitive photodiode without optical filter is sufficient for reception of the individual packets since at most one wavelength needs to be converted from the optical to the electrical domain at any given time. The photodiode converts the optical signal into an electrical signal irrespective of the wavelength.

Since wavelength stacking takes W time slots, a node needs to decide in advance when to access the PSR ring. Toward this end, a separate wavelength is used as a control channel for reservation. Time slots are grouped into cycles of length W slots. The switches T and R in Fig. 7.7 synchronize wavelength stacking and unstacking. Wavelength stacking is completed in the last time slot of a given cycle and the photonic slot is stored in the FDL by setting switch T in the cross state. The photonic slot is stored as long as switch T is in the bar state. The photonic slot is put on the PSR ring by setting switches T and S in the cross state exactly $2W$ time slots after the reservation. Whenever a node recognizes its address on the control channel, it stores the photonic slot in the FDL by setting switches S and R in the cross state $2W$ time slots after the address notification. The node starts unstacking the photonic slot at the beginning of the next cycle by setting switch R in the cross state.

The node architecture of Fig. 7.7 enables PSR nodes to send/receive multiple packets in each photonic slot using only one tunable transceiver. However, the quality of the optical signal may suffer from passing numerous FDLs and space switches in a node.

8   Optical flow switching

In the previous section, we have seen that photonic slot routing (PSR) can be transformed into individual wavelength switching (IWS) and used to realize synchronous optical packet switching (OPS) networks with the restriction that packets need to be of fixed size. Unlike electronic IP packet switching networks, these OPS networks require network-wide synchronization and are able to transport only fixed-size packets. In contrast, IP networks do not require network-wide synchronization and support variable-size IP packets. In addition, contention resolution can be done more easily and more efficiently in electronic networks than in optical networks by using electronic random access memory (RAM). Packets contending for the same router output port can be stored in electronic RAM and sent sequentially through the same port without collision. In optical networks, RAM is not feasible with current technology. Instead, bulky switched delay lines (SDLs) and/or inefficient deflection routing need to be deployed in order to resolve contention in OPS networks. Clearly, electronic packet-switched networks are able to resolve contention more efficiently by using electronic RAM. Given the steadily growing line rates and amount of traffic, however, electronic routers may become the bottleneck in high-speed communications networks that use electronic routers for storing and routing and optical fiber links for transmitting packets of variable size. This bottleneck is commonly referred to as the *electro-optical bottleneck*.

One of the main bottlenecks in today's Internet is (electronic) routing at the IP layer. Several methods have been proposed to alleviate the routing bottleneck by switching long-duration flows at lower layers (e.g., GMPLS; see Chapter 5). In doing so, routers arc offloaded and the electro-optical bottleneck is alleviated. This concept of lower-layer switching can be extended to switching large transactions and/or long-duration flows at the optical layer, giving rise to *optical flow switching* (OFS) (Chan et al., 1998). A flow denotes a unidirectional sequence of IP packets between a given pair of source and destination IP routers. Both source and destination IP addresses, possibly together with additional IP header information such as port numbers and/or type of service (ToS), are used to identify a flow. In OFS, a lightpath is established for the transfer of large data files or long-duration and high-bandwidth streams across the network. The simplest form of OFS is to use an entire wavelength for a single transaction. Alternatively, flows with similar characteristics may be aggregated and switched together by means of grooming techniques in order to improve the utilization of the set-up lightpath. OFS helps reduce the computation load and processing delay at IP routers, but it also raises several issues that need to be addressed. For instance, how can the start and end of flows be recognized.

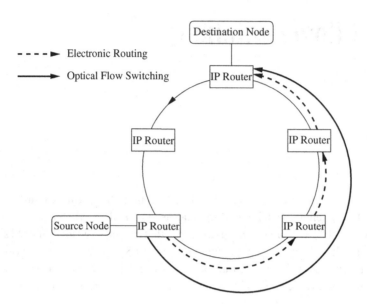

Figure 8.1 Optical flow switching (OFS) versus conventional electronic routing. After Froberg et al. (2000). © 2000 IEEE.

Or, since in OFS the set-up of a lightpath takes at least one round-trip time between source and destination IP routers, the size of a flow should be in the order of the product of round-trip propagation delay and line rate at which the lightpath operates.

8.1   Optical flow switching

Figure 8.1 illustrates the difference between OFS and conventional electronic routing by IP routers (Froberg et al., 2000). In OFS, a large set of data is routed all-optically in order to bypass and thereby offload intermediate IP routers in the data path and their associated queues. Toward this end, a dedicated lightpath is established from the source node to the destination node, allowing the source node to transmit its large data set directly to the destination node. The set-up lightpath eliminates the need for packet processing at intermediate routers (e.g., buffering, routing, etc.). It is important to note that OFS can be end-user initiated or IP-router initiated. In the first case, the end user has a large amount of data to send and requests a lightpath to the corresponding destination node. In the second case, an IP router detects large-volume flows and diverts them to a lightpath. OFS is practical if the flow lasts at least as long as it takes to set up the lightpath. With current technology and taking typical propagation delays into account, the duration of a flow must be in the order of a few up to tens of milliseconds in order to justify the round-trip time to set up the lightpath.

OFS provides several advantages. Apart from mitigating the aforementioned electro-optical bottleneck by optically bypassing and thus offloading electronic IP routers, OFS represents the highest-grade quality of service (QoS) since the established lightpath provides a dedicated connection that is not impaired by the presence of other users. At the downside, OFS must carefully determine when to set up a lightpath since wavelengths are typically a scarce resource that cannot be arbitrarily assigned (Modiano, 1999). Without the use of wavelength converters, it is necessary to assign the same wavelength to a lightpath along its entire path. This wavelength continuity constraint further restricts the number of available wavelengths.

8.2   Integrated OFS approaches

By setting up lightpaths dynamically, OFS allows wavelengths to be shared among multiple users. To enable dynamic lightpath set-up in OFS, several important issues must be addressed. These issues include flow routing, wavelength assignment, and connection set-up. Routing involves selecting an appropriate path, wavelength assignment involves assigning an appropriate wavelength on each link of the selected path, and connection set-up involves signaling to create the lightpath, given a route and a wavelength on each link along the route. In Ganguly and Modiano (2000), two integrated OFS approaches were proposed that address flow routing, wavelength assignment, and connection set-up together. Under the assumption that a reliable control network for signaling is available and flow requests are generated by end users, the following two integrated OFS approaches were proposed.

8.2.1   Tell-and-go reservation

The first integrated OFS approach, called *Tell-and-Go* (TG), is a distributed algorithm with no wavelength conversion. TG is based on link state updates. Received updates are processed at each network node to acquire and maintain the global network state. In TG, the global network state may contain inaccurate information due to the latency involved in sending updates. Given the network state, TG uses a combined routing and wavelength assignment strategy. For any routing decision, the K shortest paths are considered. Routing proceeds by using first-fit wavelength assignment over the K paths. First-fit wavelength assignment numbers all wavelengths in sequential order and routes an optical flow over the path with the lowest-numbered wavelength available at each hop. If no route with an available wavelength can be found, the optical flow is dropped. Connection set-up is achieved using the tell-and-go principle, whereby a control packet precedes the optical flow along the chosen route. The control packet establishes a lightpath for the trailing optical flow. Note that TG represents a one-way reservation, where the control packet travels only in the direction from source to destination and not backward again to the source, as is done in a two-way reservation. If at any intermediate node sufficient resources are not available, the control packet and the corresponding optical flow are terminated.

8.2.2   Reverse reservation

In the second integrated OFS approach, termed *Reverse Reservation* (RR), the initiator of an optical flow sends information-gathering packets, so-called info-packets, to the destination node on the K shortest paths. These info-packets record the link state information at each hop. Upon arrival at the destination, the info-packets contain information about the link state of all links along the K routes. Routing and wavelength assignment is then performed by the destination node, once all K info-packets from the sender have arrived. The calculation is done using the first-fit wavelength assignment, as described earlier. Similar to TG, if no route with an available wavelength can be found, the optical flow is dropped. Once a route has been selected by the destination, a reservation control packet is sent along the chosen route in reverse to establish the connection. This reservation control packet configures intermediate switches along the selected route for the chosen wavelength and finally informs the initiator that the lightpath has been set up successfully for sending the optical flow. Otherwise, if the reservation control packet does not find sufficient available resources along the selected route, the reservation is terminated and all resources held by the reservation in progress are released by sending additional control packets. Note that unlike TG, RR does not require (periodic or event-driven) updates to acquire and maintain global network state.

8.3   Implementation

OFS was experimentally investigated in the Next Generation Internet Optical Network for Regional Access using Multiwavelength Protocols (NGI ONRAMP) testbed (Froberg et al., 2000; Kuznetsov et al., 2000). The mission of the ONRAMP testbed is to implement and demonstrate apart from OFS other features such as protection, MAC protocols, control, and management. Figure 8.2 depicts the network architecture of ONRAMP. It consists of a bidirectional feeder wavelength division multiplexing (WDM) ring network which connects multiple access nodes (ANs) with one another and with the backbone network. The bidirectional feeder WDM ring carries 8 wavelength channels in each direction, each wavelength operating at a data rate of up to 10 Gb/s (OC-192). ONRAMP is envisioned to accomodate 10 to 20 ANs and 20 to 100 end users on each attached distribution network. Distribution networks carry both the feeder wavelength channels and additional local distribution wavelength channels. Each AN serves as a gateway to distribution networks of variable topologies on which the end users reside. Each AN consists of an IP router and a reconfigurable optical add-drop multiplexer (ROADM), which routes data flows between the feeder ring, the associated IP router, and the attached distribution network. The role of the AN is to route optical wavelength channels and individual IP data packets inside wavelength channels toward their destinations.

In ONRAMP, two basic categories of services are offered: an "IP" service and an "optical" service. The IP service involves conventional electronic routing of IP packets

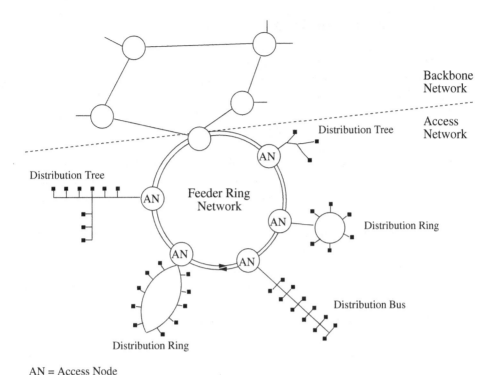

AN = Access Node

Figure 8.2 NGI ONRAMP architecture. After Kuznetsov et al. (2000).

(i.e., IP packets undergo OEO conversion and electronic buffering and processing at each intermediate IP router). The optical service represents OFS that provides the end user with an all-optical connection from source to destination in order to allow the user to send large optical flows directly to its destination. In ONRAMP, OFS is accomplished with dual-attached workstations having two interfaces: a standard IP-packet interface for low-speed data and control, and a high-speed interface to optically flow-switch bursts of extended duration. When a workstation has a large burst to send, it sends control messages over the low-speed interface to negotiate with the ONRAMP network control and management system for an all-optical end-to-end connection. ONRAMP is able to set up not only unidirectional but also bidirectional lightpaths between workstations.

An OFS scheduling approach to dynamically set up lightpaths on demand in ONRAMP was demonstrated in Ganguly and Chan (2002). In this scheduling approach, time is divided into fixed-length time slots and transmissions are scheduled across the required number of time slots. This approach was shown to work efficiently, but it requires the synchronization of network control processes. In the reported implementation, network-wide synchronization of IP routers was achieved by using the Network Time Protocol (NTP), which is a widely used Internet protocol for synchronizing the clocks of routers throughout the network.

8.4   Comparison between OFS and OBS

One of the most critical issues in OFS is the detection of flows which triggers the dynamic set-up of lightpaths. There are several possibilities to detect the beginning and end of a flow. For instance, a node may detect the beginning and end of a flow by using an x/y classifier, where x denotes the number of passing packets belonging to a given flow during a prespecified period of time y. Depending on whether the value of the classifier is above or below a predefined threshold, the flow is considered active or inactive, respectively. That is, if the number of passing packets per time period exceeds a given threshold the node detects the beginning of a flow, whereas if the number of passing packets per time period falls below the threshold the node assumes that the flow ends.

In Xin and Qiao (2001), the performance of OFS was evaluated by means of simulation using the following flow detection method. When a packet arrives, the ingress router checks if there is an existing flow to which the packet belongs. If so, the packet is either sent immediately over the lightpath already established for that flow, or is buffered if such a lightpath has been requested and is being set up. If not, the packet is considered to be the first packet of a new flow and is buffered. A lightpath set-up request is then sent to establish a lightpath to the egress router for this packet and following packets belonging to the same flow. Buffered packets of a flow will be discarded when a negative acknowledgment (NAK) arrives at the ingress router, informing that a lightpath for the flow cannot be established due to the lack of wavelength resources, or sent one by one after the lightpath is established after receiving an acknowledgment (ACK). An ingress router considers that a flow ends if there is no packet going to the same egress router as this flow arriving within a period called maximum interpacket separation (MIS) since the last packet of this flow arrived. As soon as a flow ends and the last packet of the flow is sent, a lightpath release request is sent to tear down the lightpath for that flow. Clearly, the parameter MIS has an impact on the performance of OFS. On the one hand, a smaller MIS value results in shorter flows which in turn result in more frequent lightpath set-ups and releases and an increased signaling overhead. On the other hand, a larger MIS value results in longer idle gaps between packets in a flow and a decreased lightpath utilization.

Recall from earlier that OFS is able to optically bypass and thereby offload electronic routers and to provide guaranteed QoS. OFS can be viewed as a fast optical circuit switching technique where lightpaths are dynamically set up and released on demand. As a circuit switching technique, OFS suffers from the following two major drawbacks: (1) the lightpath set-up generally implies a *two-way reservation* between source and destination, resulting in a set-up delay of one round-trip time, and more importantly (2) OFS does not support statistical multiplexing (i.e., an established lightpath is dedicated to a single flow and cannot be shared among different flows). Unlike OFS, the so-called *optical burst switching* (OBS) technique avoids the two shortcomings of OFS, at the expense of guaranteed QoS. As opposed to OFS, OBS relies on one-way reservation and allows for the statistical sharing of the bandwidth of a wavelength channel among bursts belonging to different flows. In OBS, each ingress router assembles incoming IP packets going to the same egress router into a burst according to some burst assembly schemes. For each burst, a control packet is first sent out on a control wavelength channel to the egress router and the burst will follow on a separate data wavelength channel after a prespecified offset time. The control packet goes through OEO conversion at every intermediate node and attempts to reserve a data wavelength channel for just enough time to accomodate the following burst on the outgoing link. If the reservation succeeds, the optical switching fabric at each intermediate node is configured to switch the following burst. Otherwise, if the reservation fails due to the lack of wavelength resources, the burst will be dropped. If a burst arrives at the egress router, it will be disassembled into the individual IP packets. In OBS, the burst assembly scheme plays a major role. There exist several possibilities to design an appropriate burst assembly scheme. In the aforementioned simulative peformance evaluation of OFS in Xin and Qiao (2001), OFS was compared to OBS assuming the following burst assembly scheme. Packets going to the same egress router that arrived during a fixed period of time, called the burst assembly time (BAT), are assembled into a single burst. Packets arriving after the next assembly cylce begins will be assembled into a different burst, even though they may go to the same egress router as the packets in the previous burst. The value of the parameter BAT must be set carefully. A smaller BAT value will result in shorter bursts and thus more control packets. Conversely, a larger BAT value will result in a longer end-to-end delay due to the increased assembly delay. Note that this burst assembly scheme guarantees a bounded assembly delay, but not necessarily a guaranteed burst delivery due to possible collisions at intermediate nodes. We note that the comparison was done under the assumption that OFS requires a two-way reservation. OFS with a one-way reservation, similar to the TG reservation in Section 8.2.1, was not considered.

In Xin and Qiao (2001), OFS and OBS were compared using the previously described MIS-based flow detection and BAT-based burst assembly scheme, respectively, for a 10-node mesh WDM network with up to 100 wavelength channels per link and wavelength conversion at each node. It was assumed that one control wavelength channel is used for sending control packets in OBS or set-up/release/ACK/NAK packets in OFS. Furthermore, it was assumed that IP routing and multiprotocol label switching (MPLS) signaling protocols are used to control the WDM network. The obtained results indicate that OBS outperforms OFS in terms of percentage of dropped packets and mean end-to-end delay for a wide range of used wavelength channels and traffic loads. Specifically, under light and heavy traffic loads OBS achieves a significantly smaller percentage of dropped packets than OFS, especially for a small number of used wavelength channels. This is because in OBS the bandwidth of a single wavelength channel can be provisioned at a much smaller time scale (burst) than in OFS (flow), resulting in a significant statistical multiplexing gain. With respect to the mean end-to-end delay, the results confirm that using a small MIS value results in many short-duration flows, each of which encounters a large lightpath set-up delay of no less than the round-trip propagation delay, leading to an increased mean end-to-end delay. OFS can achieve a small mean end-to-end delay if most flows have a long duration such that the lightpath set-up delay is amortized among many packets. As for OBS, the obtained results show that the parameter BAT does not have a significant impact on the mean end-to-end delay. This is due to the fact that the BAT is several orders of magnitude smaller than the one-way propagation delay, which dominates the mean end-to-end delay in OBS. Note that the mean end-to-end delay in OFS can be lower than that in OBS by using a sufficiently large MIS value. However, the percentage of dropped packets will suffer unless there are sufficient wavelength channels such that each flow can be assigned a dedicated wavelength channel.

Given these results, we will explain and discuss OBS in greater detail in the subsequent chapter. Apart from the statistical multiplexing gain and smaller one-way reservation delay, we will highlight the unique features of OBS, which combine the respective merits of optical circuit switching (OCS) and OPS, and outline various approaches to provide service differentiation and QoS in OBS networks.

9 Optical burst switching

Optical burst switching (OBS) is one of the recently proposed optical switching techniques which probably received the greatest deal of attention (Chen et al., 2004). OBS may be viewed as a switching technique that combines the merits of optical circuit switching (OCS) and optical packet switching (OPS) while avoiding their respective shortcomings. The switching granularity at the burst rather than wavelength level allows for statistical multiplexing in OBS, which is not possible in OCS, while requiring a lower control overhead than OPS. More precisely, in OCS, the entire bandwidth of each lightpath is dedicated to one pair of source and destination nodes and unused bandwidth cannot be reclaimed by other nodes ready to send data. Thus, OCS does not allow for statistical multiplexing. On the other hand, in OCS networks no OEO conversion is needed at intermediate nodes. As a result, OCS networks provide all-optical circuits that are transparent in terms of bit rate, modulation scheme, and protocol. OCS is well suited for large data transmissions whose long connection holding time on the order of a few minutes, hours, days, weeks, or even months justify the involved two-way reservation overhead for setting up or releasing a lightpath, which may take a few hundred milliseconds. Since many applications require only subwavelength bandwidth and/or involve bursts that last only a few seconds or less, the coarse wavelength switching granularity of OCS becomes increasingly inefficient and impractical. Unlike OCS, OPS is able to provide a significant statistical multiplexing gain due to the fact that bandwidth is not dedicated to a single connection but may be shared by multiple data flows. As discussed in greater detail in Chapter 10, OPS is widely considered an optical switching technique that will not be economically viable in the near term. With current technology, OPS is difficult to implement because of its need for optical packet header processing. Furthermore, the lack of optical random access memory (RAM) prevents the implementation of efficient contention resolution in OPS networks. Alternative optical contention resolution schemes (e.g., the use of bulky fiber delay lines [FDLs]), are rather impractical solutions to build high-performance OPS networks. In OBS, only control packets carried on one or more control wavelength channels undergo OEO conversion at each intermediate node, whereas data bursts are transmitted on a separate set of data wavelength channels that are all-optically switched at intermediate nodes. Given that data is switched all-optically at the burst level, OBS combines the transparency of OCS with the statistical multiplexing gain of OPS, as described in greater detail in the following.

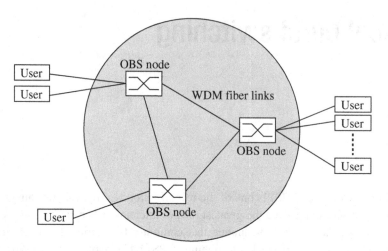

Figure 9.1 OBS network architecture. After Battestilli and Perros (2003). © 2003 IEEE.

9.1   OBS framework

An OBS network consists of OBS nodes interconnected by wavelength division multi-plexing (WDM) fiber links, as shown in Fig. 9.1. At the edge of the OBS network, one or more users are attached to an OBS node. Typically, a user at the edge of the OBS network consists of an electronic IP router equipped with an OBS interface, while an OBS node inside the OBS network comprises an optical switching fabric, a switch control unit, and routing and signaling processors. Following the approach in Battestilli and Perros (2003, 2004), the OBS network and its operation is best explained by discussing first the functions executed by users at the edge of the OBS network, followed by a description of the functions carried out by OBS nodes inside the OBS network.

9.1.1   OBS network edge

The four functions executed by edge OBS users can be categorized into (1) burst assembly, (2) signaling, (3) routing and wavelength assignment, and (4) the computation of the offset time for the control packet which is sent prior to the data burst. These four functions executed at the OBS network edge are explained next.

Burst assembly
OBS users collect traffic originating from upper layers (e.g., IP), sort it based on destination addresses, and aggregate it into variable-size data bursts by using appropriate burst assembly algorithms. The choice of the burst assembly algorithm can have a significant impact on the performance of the OBS network. Several burst assembly algorithms have been proposed and investigated up to date, as discussed at length in the next section. Generally, burst assembly algorithms have to take the following parameters into account: timer, minimum burst size, and maximum burst size. The timer is used by the OBS user to determine when to assemble a new burst. The minimum and maximum

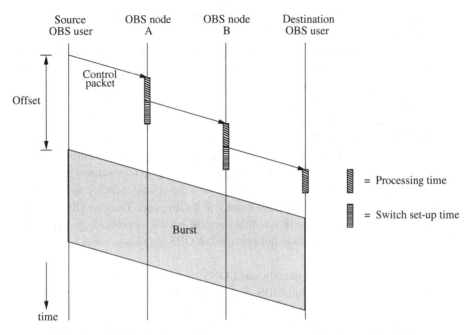

Figure 9.2 Distributed OBS signaling with one-way reservation. After Battestilli and Perros (2003). © 2003 IEEE.

burst size parameters determine the length of the assembled burst. These parameters must be carefully set since long bursts hold network resources for long time periods and may therefore cause higher burst losses, while short bursts cause an increased number of control packets since each burst is preceded by a separate control packet. Padding may be used by the burst assembly algorithm if an OBS user does not have enough traffic to assemble a minimum-size burst when the timer expires. It is interesting to note that burst assembly helps reduce the degree of self-similarity of higher-layer traffic. Reducing self-similarity is a desirable property since it makes the traffic less bursty and thus leads to a decreased queueing delay and smaller packet loss in OBS networks.

Signaling
Signaling is used to set up a connection in the OBS network for an assembled burst between a given pair of source and destination edge OBS users. In OBS networks, there are two types of signaling: (1) *distributed signaling with one-way reservation* or (2) *centralized signaling with end-to-end reservation*.

Most of the proposed OBS network architectures use the distributed signaling with one-way reservation, whose time-space diagram is shown in Fig. 9.2. In the one-way reservation scheme, a source OBS user sends a control packet on a separate out-of-band control channel to the ingress OBS node, say, OBS node A, prior to transmitting the corresponding burst. The out-of-band control channel may be a dedicated signaling wavelength channel or a separate control network. The control packet contains information about the burst (e.g., size). The control packet is OEO converted and processed in the electronic domain at each intermediate OBS node along the path to the destination OBS user. After a prespecified delay, called *offset*, the corresponding data burst is sent on one of the available data wavelength channels without waiting for an acknowledgment (ACK) that the connection between source and destination OBS users has been established successfully. Compared to the conventional two-way reservation scheme with acknowledgment, the resultant one-way reservation significantly decreases the connection set-up time, which effectively equals the one-way end-to-end propagation delay plus the time required to electronically process the control packet and configure the optical switch fabric at all intermediate OBS nodes. In the one-way reservation scheme, however, control packets may not be successful in setting up connections due to congestion on the data wavelength channels. As a result, the optical switch cannot be configured as needed and the corresponding burst will be dropped. Thus, in OBS networks with one-way reservation there is a certain nonzero burst loss probability. Typically, the retransmission of lost bursts is not provided in OBS networks, but is left to higher-layer protocols.

In the second less frequently used OBS signaling approach, centralized signaling with end-to-end reservation, OBS users send their connection set-up requests to ingress OBS nodes, which in turn inform a central request server about the received connection set-up requests. The central server has global knowledge about the current status of all OBS nodes and data wavelength channels throughout the OBS network. Based on this global knowledge, the central server processes all received connection set-up requests and sends ACKs to the requesting edge OBS users. Upon the receipt of their ACKs, OBS users transmit their bursts.

Routing and wavelength assignment

Routing in OBS networks can be done either on a hop-by-hop basis using fast routing table lookup algorithms at intermediate OBS nodes, similarly to IP networks, or by deploying generalized multiprotocol label switching (GMPLS) routing protocols to compute explicit or constraint-based routes at edge OBS users, as described in great detail in Chapter 5. Along the selected path, each link must be assigned a wavelength on which bursts are carried. In OBS networks, wavelength assignment with and without wavelength conversion at intermediate OBS nodes is possible, whereby wavelength conversion may be fixed, limited-range, full-range, or sparse. Wavelength conversion was discussed at length in Section 1.5.2.

Offset

As depicted above in Fig. 9.2, one of the salient features of OBS is the fact that an edge OBS user sends a control packet on a separate control channel prior to sending the corresponding data burst on one of the available data wavelength channels. More precisely, after sending the control packet an OBS user waits for a fixed or variable offset time until it starts transmitting the corresponding burst. The offset is used to let the control packet be processed, reserve the required resources, and configure the optical switching fabric at each intermediate OBS node along the selected path before the corresponding burst arrives. If the control packet is successful in reserving the needed resources and configuring the switch, the arriving burst can *cut through* each intermediate OBS node without requiring any buffering or processing. Clearly, the estimation and setting of the offset time is crucial. Ideally, the offset estimation should be based on the number of traversed OBS nodes and the processing and switch set-up times at each of them. In practice, however, the number of intermediate OBS nodes may not be known to the source OBS user or may change over time. Furthermore, the current level of congestion in the OBS network needs also to be taken into account in the offset estimation in order to achieve an acceptable burst loss probability. If the offset is estimated incorrectly, the burst may arrive at an OBS node before the optical switch has been configured properly. Clearly, the offset estimation in OBS networks is of utmost importance to achieve high resource utilization and low burst loss.

9.1.2 OBS network core

OBS nodes located in the core of OBS networks perform the following two functions: (1) scheduling of resources and (2) contention resolution if there are not enough resources to successfully schedule simultaneously arriving bursts. Both scheduling and contention resolution are discussed in the following.

Scheduling

Each control packet contains information about the corresponding burst (e.g., offset and size). Based on this information, OBS nodes schedule the resources inside the local optical switching fabrics such that bursts can cut through them. The resource scheduling schemes proposed for OBS networks can be classified based on the start and end times a burst occupies the resources inside the optical switching fabric of an OBS node as follows:

- **Explicit Set-up:** In explicit set-up, a wavelength channel is reserved and the optical switching fabric is configured immediately after receiving and processing the control packet.
- **Estimated Set-up:** In estimated set-up, the reservation of a wavelength channel is delayed and by using information provided by the control packet the OBS node estimates the arrival time of the corresponding burst. A wavelength channel is reserved and the optical switching fabric is configured at the OBS node right before the estimated arrival time of the burst.
- **Explicit Release:** In explicit release, the source OBS user sends an additional trailing control packet to indicate the end of the burst. After receiving the trailing control packet, the OBS node releases the reserved wavelength.
- **Estimated Release:** In estimated release, an OBS node estimates the end of the burst by using the offset and size information carried in the preceding control packet. Based on this information, an OBS node is able to estimate the time when to release the reserved wavelength after the corresponding burst has passed through the switching fabric.

There exist four possibilities to combine these four OBS resource scheduling schemes: explicit set-up/explicit release, explicit set-up/estimated release, estimated set-up/explicit release, and estimated set-up/estimated release. Each combination provides a different performance–complexity trade-off. Clearly, the estimated set-up/release schemes provide an improved resource utilization over their explicit counterparts. However, the explicit set-up/release schemes are simpler to implement, but they occupy the wavelength channels for longer time periods and therefore may lead to an increased burst loss probability.

Note that the choice of the resource scheduling scheme in OBS networks also depends on the burst assembly algorithm used by the OBS users at the OBS network edge. To see this, consider the following example. If an edge OBS user first assembles a burst and then sends a control packet with information about the offset and size of the burst, OBS nodes are able to apply estimated set-up and estimated release schemes. However, if the control packet is sent before the corresponding burst is completely assembled, OBS nodes have to deploy explicit release schemes (i.e., the edge OBS user has to send another trailing control packet which indicates the end of the burst transmission).

Contention resolution

Contention resolution is one of the main design objectives in OBS networks. Contention occurs if a burst arrives at an OBS node and all local resources are occupied or if two or more simultaneously arriving bursts contend for the same resource. Several techniques can be employed by OBS nodes to resolve contention. Contention resolution techniques may be applied in the time, wavelength, or space domains or any combination thereof. Among others, contention among simultaneously arriving bursts may be resolved by using FDLs where bursts are temporarily stored such that at a given output port of an OBS node contending bursts can be sent out sequentially without collision. Another way to resolve contention is deflection routing, where one burst is forwarded through the intended output port of an OBS node while the remaining contending bursts are sent through other output ports along different paths. Wavelength conversion is another approach to resolve contention where bursts leave the same output port of an OBS node on different wavelength channels at the same time.

9.1.3   OBS MAC layer

OBS introduces a number of specific network design issues that must be addressed properly in order to implement the aforementioned functions. To implement these functions, a medium access control (MAC) layer is required between the IP layer and the optical layer (Verma et al., 2000). Figure 9.3 depicts the functional blocks needed at the OBS MAC layer as well as at the optical layer for implementing OBS networks. The functional blocks correspond to the aforementioned functions that need to be executed by edge OBS users and core OBS nodes at the MAC layer and optical layer, respectively. At the OBS MAC layer, source and destination OBS users located at the edge of the OBS network perform the functions of burst assembly/disassembly, offset

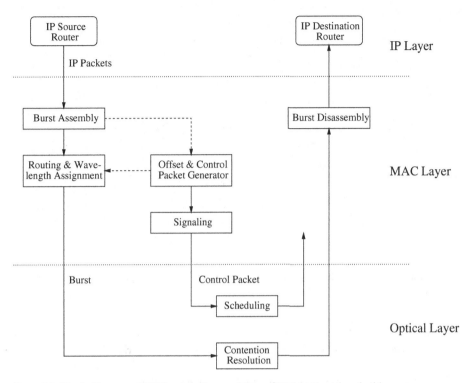

Figure 9.3 Block diagram of OBS networks consisting of IP, MAC, and optical layers. After Verma et al. (2000). © 2000 IEEE.

computation, control packet generation, routing and wavelength assignment (RWA), and signaling. At the optical layer, intermediate core OBS nodes are responsible for scheduling and contention resolution of in-transit bursts. The OBS MAC layer together with the underlying optical layer offers services that guarantee certain burst blocking probabilities to the higher-layer client IP routers. In the following, we describe the functional blocks of the OBS MAC and optical layers of Fig. 9.3 in greater detail and provide an in-depth discussion of state-of-the-art algorithms and techniques proposed for OBS networks.

9.2 Burst assembly algorithms

Various burst assembly algorithms have been investigated up to date. Most of the reported burst assembly algorithms use either burst assembly time or burst length or both as the criteria to aggregate bursts. Typically, the used parameters are a time threshold T and a burst length threshold B. The rationale behind these parameters is as follows. The time threshold limits the delay of packets buffered in the assembly queue within a maximum value T when traffic is light. The burst length threshold is used to launch the transmission of a burst as soon as the burst reaches or exceeds B. Both thresholds T and

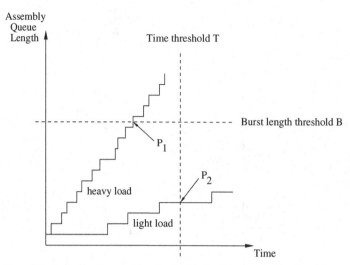

Figure 9.4 Burst length and time thresholds for burst assembly algorithms. After Yu et al. (2004). © 2004 IEEE.

B can be either fixed or adjusted dynamically. Based on these thresholds, burst assembly algorithms can be classified into the following four categories (Yu et al., 2004):

- **Time-Based Assembly Algorithms:** The first category contains time-based burst assembly algorithms where a fixed time threshold T is used as the criterion to send out a burst (i.e., the burst is transmitted after T time units).
- **Burst Length–Based Assembly Algorithms:** The second category contains burst length–based burst assembly algorithms where a burst length threshold B is used as the criterion to send out a burst (i.e., the burst is transmitted as soon as the aggregated burst reaches or exceeds B).

Each of the two aforementioned types of burst assembly algorithms suffer from some shortcomings at low and high traffic loads due to the fact that they consider only a single criterion (either time or burst length). To see this, consider Fig. 9.4 which illustrates the impact of burst length threshold B and time threshold T on the transmission of a burst under heavy and light loads. Under light load, a burst length–based assembly algorithm does not provide any constraints on the queueing delay of packets that wait for being aggregated into a burst of size B. As a result, burst length–based assembly algorithms are unable to guarantee any bound on the maximum or average queueing delay. As shown in Fig. 9.4, a time-based assembly algorithm could solve this problem since it will send out the burst after time T at point P_2 in the figure, no matter how many packets the burst contains. On the other hand, time-based assembly algorithms lead to longer average queueing delays than burst length assembly algorithms under heavy traffic. Unlike time-based assembly algorithms, under heavy traffic loads a burst length algorithm sends out the burst at point P_1 in Fig. 9.4 as soon as the burst length threshold B is crossed well before the

Table 9.1. Forward resource reservation (FRR) parameters

Parameter	Definition
T_b	Time when a new burst starts to be assembled
T_h	Time when corresponding control packet is sent
T_d	Time when burst is sent
τ_a	Duration of burst assembly
τ_o	Offset between control packet and burst

time threshold T is reached. Clearly, it is desirable to have a burst assembly algorithm that performs well under all traffic load conditions (i.e., a burst is sent out at P_1 under heavy load and at P_2 under light load), which gives rise to the third category of burst assembly algorithms.

- **Mixed Time/Burst Length–Based Assembly Algorithms:** The third category of burst assembly algorithms uses both a time threshold T and a burst length threshold B as criteria to send out a burst. Depending on the traffic loads and the values of parameters T and B, generally either threshold T or threshold B is crossed first and the burst is transmitted. In general, threshold T will be crossed before threshold B at light traffic loads and vice versa at heavy traffic loads.

- **Dynamic Assembly Algorithms:** The fourth category of burst assembly algorithms makes use of dynamic thresholds, where either the time threshold T or the burst length threshold B or both are set dynamically according to network traffic conditions. Dynamic burst assembly algorithms are adaptive and therefore achieve an improved performance at the expense of an increased computational complexity compared to the first three categories of burst assembly algorithms which use fixed (static) thresholds.

It is important to note that burst assembly algorithms executed by OBS users at the edge of OBS networks help smooth the input IP packet process and reduce the degree of self-similarity of IP traffic (Yu et al., 2004; Ge et al., 2000). This traffic smoothing effect of burst assembly algorithms simplifies traffic engineering (TE) and capacity planning of OBS networks.

An advanced burst assembly algorithm, called *forward resource reservation* (FRR), was studied in Liu et al. (2003). FRR deploys the following two performance-enhancing techniques: (1) *prediction* of the packet traffic arriving at edge OBS users, and (2) *pretransmission* of control packets in order to reduce the burst assembly delay incurred at the edge of OBS networks. FRR makes use of several parameters which are listed and explained in Table 9.1. FRR comprises three steps and works as follows:

1. **Prediction:** As soon as the previous burst is assembled, a new burst starts to be assembled at time T_b by a given OBS user. Based on a linear prediction method, the OBS user predicts the length of the new burst.

2. **Pretransmission:** Instead of waiting for the new burst to be assembled completely, the OBS user constructs a control packet upon completion of the prediction of the

previous step. Among other information, the control packet contains the predicted length of the corresponding burst. The control packet is sent out at time T_h, which is given by $T_h = \max\{T_b, T_b + \tau_a - \tau_o\}$.

3. **Examination:** Upon completion of the burst assembly, the actual burst length is compared with the length carried in the pretransmitted control packet to ensure that the OBS network resources reserved by the control packet are enough for the burst. The examination can have two possible outcomes:

   - If the actual burst length is less than or equal to the predicted length, the control packet is assumed to have reserved enough resources at all intermediate OBS nodes along the path between source and destination OBS users. In this case, the burst is transmitted at time T_d, which is given by $T_d = T_h + \tau_o$.

   - If the actual burst length exceeds the predicted length, the control packet is assumed to be unsuccessful in reserving enough resources at intermediate OBS nodes. In this case, the control packet for this burst is retransmitted at time $T_b + \tau_a$, carrying the actual (rather than predicted) burst length. Subsequently, the burst is sent after the offset τ_o.

The basic idea of FRR is to overlap burst assembly and signaling in time such that a control packet can be sent prior to completion of the burst assembly process. In doing so, part of the burst assembly delay can be effectively masked to higher layers, resulting in a decreased latency in OBS networks.

A mixed time/burst length–based assembly algorithm was used in the so-called *burst cluster* transmission technique that enables service differentiation in terms of burst blocking probability (Tachibana and Kasahara, 2006). The burst cluster transmission consists of a mixed time/burst length–based assembly algorithm to form burst clusters and a burst cluster transmission scheduling algorithm performed by edge OBS users. Specifically, each OBS user classifies arriving IP packets according to their egress OBS nodes and then sorts them in M separate queues depending on their service classes. The burst assembly algorithm generates M bursts from the queues, whereby each burst contains IP packets of the same service class. Based on their service classes, the M bursts are then sorted in an increasing order. The sorted M bursts are put together to form a burst cluster. Subsequently, the burst cluster is transmitted and is routed through the OBS network. If a burst cluster arrives at an intermediate OBS node whose resources are currently fully occupied, the first low-priority bursts at the head of the burst cluster are dropped until sufficient resources become available to optically switch the remaining high-priority bursts of the burst cluster. In doing so, low-priority bursts which contain low-priority IP packets may be dropped while high-priority bursts and thereby high-priority IP packets are switched and forwarded toward the egress OBS node. As a result, the burst cluster transmission technique enables service differentiation in OBS networks in terms of burst blocking probability in that low-priority bursts are subject to a higher burst loss probability than high-priority bursts. In Section 9.5, we take a closer look at service differentiation and elaborate on the operation and limitations of several available techniques to achieve service differentiation in OBS networks.

9.3   Signaling

Signaling in OBS networks can be done in two ways: (*i*) *just-in-time* (JIT) signaling, or (*ii*) *just-enough-time* (JET) scheduling. Both JIT and JET signaling approaches adopt the tell-and-go principle, as explained in the following.

In JIT signaling, an OBS node configures its optical switches for the incoming burst immediately after receiving and processing the corresponding control packet (Baldine et al., 2002). Thus, resources at the OBS node are made available before the actual arrival time of the burst. In JIT signaling, intermediate OBS nodes do not take the offset time information carried in control packets into account. JIT signaling is easy to implement. At the downside, however, JIT signaling does not use resources efficiently since wavelengths are reserved at OBS nodes prior to the burst arrival time. The overhead of configuring the switching elements for an unnecessarily long period of time leads to an increased burst loss probability.

To improve the bandwidth efficiency in OBS networks, JET signaling makes use of the offset time information carried in each control packet (Chen et al., 2004). JET signaling achieves a higher wavelength utilization by enabling OBS nodes to make the so-called *delayed reservation* for incoming bursts. With delayed reservation, the optical switches at a given OBS node are configured right before the expected arrival time of the burst. The major difference between JET signaling and JIT signaling is the time when optical switches are configured at OBS nodes. In JET signaling, an OBS node configures its optical switches right before the burst arrives, while in JIT signaling an OBS node configures its optical switches immediately after receiving and processing the preceding control packet. Another important feature of JET signaling is that the burst length information carried in the preceding control packet is used to enable close-ended reservation (i.e., only for the burst duration without explicit release) instead of open-ended reservation, which would require explicit release of the configured resources. The close-ended reservation helps OBS nodes make intelligent decisions about whether it is possible to schedule another newly arriving burst. As a result, JET signaling is able to outperform JIT signaling in terms of bandwidth utilization and burst loss probability, at the expense of increased computational complexity.

9.4   Scheduling

In OBS networks, bursts generally may have different offset times and may not arrive in the same order as their corresponding control packets. Therefore, each wavelength is likely to be fragmented with the so-called void (i.e., idle) intervals. An efficient scheduling algorithm for OBS networks should be able to utilize these voids for scheduling newly arriving bursts and reserving bandwidth on any suitable wavelength channel for them. Ideally, a scheduling algorithm should be able to process control packets fast and also to make efficient use of suitable void intervals.

To date, a number of burst scheduling algorithms have been proposed which can be roughly categorized into two categories: (1) *non-void-filling* and (2) *void-filling* scheduling algorithms (Xu et al., 2004). In general, a non-void-filling scheduling algorithm is fast but not bandwidth efficient, whereas a void-filling scheduling provides a better bandwidth utilization but has a larger computational complexity. For a survey and discussion of previously reported scheduling algorithms for OBS networks we refer the interested reader to Xu et al. (2004). For illustration, we briefly elaborate on two example burst scheduling algorithms which have received considerable attention. Let us first consider a non-void-filling scheduling algorithm called *Horizon* (Turner, 1999). In Horizon, a scheduler keeps track of the so-called horizon of each wavelength channel, which is the time after which no reservation has been made on that wavelength channel. The scheduler assigns each arriving burst to the wavelength channel with the latest horizon as long as it is still earlier than the arrival time of the burst. In doing so, the void interval between the current horizon and the starting time of the new reservation period is minimized. Horizon is relatively simple and has a short running time. However, Horizon suffers from low bandwidth utilization and high burst loss probability due to the fact that it does not take void intervals into account. Next, let us discuss a representative void-filling burst scheduling algorithm called *latest available unused channel with void filling* (LAUC-VF) (Xiong et al., 2000). LAUC-VF keeps track of all void intervals and assigns an arriving burst a large enough void interval whose starting time is the latest but still earlier than the burst arrival time. LAUC-VF provides a better bandwidth utilization and burst loss probability than Horizon. At the downside, however, the execution time of LAUC-VF is much longer than that of Horizon.

A set of novel burst scheduling algorithms that aim at providing low burst loss probability and low computation complexity at the same time were proposed and investigated in Xu et al. (2004). Unlike most existing burst scheduling algorithms (e.g., Horizon and LAUC-VF), which primarily aim at optimizing either the running time or burst loss probability, but not both, the proposed burst scheduling algorithms take both performance metrics into account. It was shown that most of the proposed burst scheduling algorithms for OBS networks have a short scheduling time for each incoming burst while maintaining a low burst loss rate.

9.5   Service differentiation

Several approaches exist to achieve service differentiation in OBS networks, which can be applied at the network edge and/or core. At the OBS network edge, OBS users can deploy an offset-time-based quality-of-service (QoS) scheme that uses an *extra offset* to isolate service classes from each other and thereby achieve service differentiation (Yoo et al., 2001). For illustration, we explain the extra-offset-based QoS scheme for only two service classes and extend it later to service differentiation of multiple classes. For now, let us consider service classes 0 and 1, whereby class 1 is assumed to have priority over class 0. To give class 1 a higher priority for resource reservation at core OBS

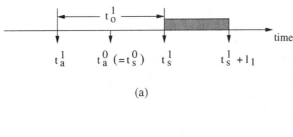

(a)

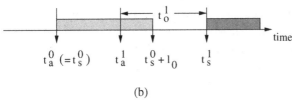

(b)

Figure 9.5 Service class isolation in extra-offset-based QoS scheme. After Yoo et al. (2001). © 2001 IEEE.

nodes, an extra offset, denoted t_o^1, is given to class 1 traffic, but not to class 0 traffic (i.e., $t_o^0 = 0$). Furthermore, for simplicity we assume that the normal offset is negligible compared to the extra offset and also assume that a link has only one wavelength for data and a separate wavelength for control. In addition, we assume *delayed reservation* of resources at intermediate core OBS nodes. With delayed reservation, the resources at an intermediate OBS node are reserved at the burst arrival time instead of at the time when the processing of the corresponding control packet finishes (which is known as *immediate reservation*). In the following explanation of the extra-offset-based QoS scheme, let t_a^i and t_s^i be the arrival times of the control packet and the corresponding data burst, respectively, for a class i request $req(i)$, where $i = 0, 1$. Furthermore, let l_i denote the burst length requested by $req(i)$. Figure 9.5 shows the isolation of service classes 0 and 1 by using an extra offset such that a class 1 request obtains a higher priority for resource reservation than a class 0 request. Figure 9.5 depicts two scenarios, (a) and (b), where a class 0 burst and a class 1 burst contend for resource reservation at a given OBS node. In scenario (a), the control packet of $req(1)$ arrives first and reserves the wavelength by means of delayed reservation. Next, the control packet of $req(0)$ arrives. Note that in this scenario $req(1)$ is successful in reserving the wavelength for the class 1 burst, but $req(0)$ will be blocked and the corresponding class 0 burst will be dropped if (1) $t_a^0 < t_s^1$ and $t_a^0 + l_0 > t_s^1$, or (2) $t_s^1 < t_a^0 < t_s^1 + l_1$. In scenario (b), the control packet of $req(0)$ arrives first, followed by the control packet of $req(1)$. If $t_a^1 < t_a^0 + l_0$ and there were no extra offset t_o^1, the control packet of $req(1)$ would fail in reserving the wavelength for the following burst. However, such a failure can be avoided by using a sufficiently large extra offset such that $t_s^1 = t_a^1 + t_o^1 > t_a^0 + l_0$. Since t_a^1 may occur right after t_a^0, t_o^1 needs to be larger than the maximum burst length of class 0 such that control packets of class 1 are not blocked by control packets of class 0. Note that, with a sufficiently large extra offset, the burst blocking probability of class 1 bursts is only a function of the offered load belonging to class 1, independent of the offered class 0 load, whereas the burst blocking probability of class 0 bursts is affected by the offered load of both classes 0 and 1. As a result, class 1 traffic can be isolated from class 0 traffic by setting the extra offset large enough.

We have seen that, in the case of two service classes, the extra offset assigned to class 1 needs to be at least as large as the maximum burst length of service class 0 in order to achieve complete class isolation. Note that by setting the extra offset to some value smaller than the maximum burst length of class 0 partial class isolation can be obtained, which can be used to achieve any arbitrary degree of isolation and thus satisfy variable service differentiation requirements between the two service classes 0 and 1. This approach can be easily extended to multiple service classes as follows. Let us consider two adjacent service classes i and $i - 1$. Furthermore, let t_{diff} denote the difference between the extra offsets assigned to the two classes (i.e., $t_{\text{diff}} = t_o^i - t_o^{i-1}$). To achieve a certain isolation degree between classes i and $i - 1$, t_{diff} must be set properly. For instance, if t_{diff} is larger than the maximum burst length of class $i - 1$, service class i is fully isolated from class $i - 1$. Setting t_{diff} to a smaller value leads to a partial isolation of classes i and $i - 1$.

The extra-offset-based QoS scheme is able to achieve service differentiation in that a service class with higher priority experiences lower burst loss probability than lower-priority service classes. At the downside, the extra-offset-based QoS scheme adds to the assembly delay whose impact depends on the number of service classes, the burst lengths of each service class, the desired degree of class isolation, and the end-to-end propagation delay of the OBS network. For a small number of service classes and carefully engineered burst lengths, the negative impact of the extra offset on the end-to-end latency is expected to become negligible, especially for large end-to-end propagation delays (Yoo et al., 2001). For a detailed analysis of the burst loss probability in multiclass OBS networks the interested reader is referred to Vu and Zukerman (2002) and Barakat and Sargent (2004).

It is important to note that the extra-offset-based QoS scheme does not mandate the use of FDLs at intermediate OBS users. But its performance can be improved by taking advantage of FDLs to resolve contention among multiple bursts such that a burst can be put on the FDL and delayed for a sufficiently long period of time in order to resolve the contention (Yoo et al., 2000). Note that the use of FDLs is one of several possibilities to resolve contention in the core of OBS networks, as discussed in greater detail in the following section.

Besides the increased burst assembly delay, the extra-offset-based QoS scheme suffers from unfairness against long bursts of low priority. This is due to the fact that it is difficult to find a long gap on any wavelength at an OBS node to serve a long burst of low priority in an almost full schedule table. As a consequence, long bursts of low priority are more likely dropped than short bursts belonging to the same traffic class. To avoid both shortcomings of the extra-offset-based QoS scheme, wavelength *preemption* techniques can be deployed instead in order to provide service differentiation in OBS networks (Phuritatkul et al., 2006). With wavelength preemption, an OBS node monitors the locally scheduled bandwidth allocation for each traffic class. An incoming high-priority burst which is unable to be scheduled on an available wavelength on the outgoing link of the OBS node is not immediately dropped but is rescheduled in order to preempt one or more low-priority bursts that were already scheduled.

Preemption techniques can be used in conjunction with wavelength *usage profiles* in order to efficiently provide service differentiation in OBS networks (Liao and Loi, 2004). Specifically, each traffic class is associated with a predefined usage limit, defined as a fraction of the wavelength resources the class is allowed to use at intermediate OBS nodes. Clearly, classes of higher priority are allowed to use more wavelength resources than classes of lower priority. Furthermore, OBS nodes maintain a wavelength usage profile for each class per output link and monitor the current wavelength usage for each class. Each OBS node keeps monitoring its usage profiles. Upon receiving a class i request, a given OBS node first attempts to find an appropriate wavelength on the intended output port. If the attempt succeeds, the corresponding burst is scheduled and the usage profile of class i is updated by the OBS node. Otherwise, the OBS node examines whether class i is in profile. Class i is considered in profile if its current usage does not exceed a predefined usage limit. If class i is in profile, a previously scheduled burst of an out-of-profile class is preempted, starting from the class with the lowest priority in an ascending order to the highest priority. Once a wavelength is preempted, the OBS node updates the usage profiles of both classes. If no out-of-profile scheduled bursts can be found to preempt a wavelength for the arriving class i burst, the class i request is rejected and the corresponding burst will be dropped. The obtained results showed that preemption together with usage profiles is able to outperform the extra-offset-based QoS scheme in terms of burst loss probability and wavelength utilization. It is important to note that the discussed preemption-based scheme provides only *relative* QoS. In relative QoS schemes, the performance of each traffic class is not specified in absolute terms. Instead, the QoS of each class is defined relatively with respect to other classes. For instance, in OBS networks a high-priority class experiences lower burst loss probability than a low-priority class. However, the actual burst loss probability of the high-priority class depends on the traffic loads of the low-priority class. Consequently, no upper bound on the burst loss probability can be guaranteed for the high-priority class in relative QoS schemes.

As opposed to relative QoS schemes, *absolute* QoS schemes for OBS networks are able to guarantee an upper bound for the burst loss probability of high-priority traffic. This guarantee is crucial to support applications with delay and bandwidth constraints. In Zhang et al. (2004), an absolute QoS scheme for OBS networks was studied, which ensures that the burst loss probability of the guaranteed traffic does not exceed a certain value. To provide burst loss guarantees in an OBS network, OBS nodes deploy an *early dropping* mechanism, which achieves service differentiation with absolute QoS by probabilistically dropping low-priority bursts in order to guarantee a prespecified burst loss probability of high-priority traffic. The proposed early dropping mechanism is similar to the concept of random early detection (RED) used by routers to avoid congestion in packet-switched networks. In RED, a router detects congestion by monitoring the average queue size. Note, however, that RED cannot be directly applied to OBS networks due to their inherently bufferless nature. Instead, the proposed early dropping mechanism monitors the burst loss probability rather than the average queue size by means of online measurements. More specifically, an OBS node computes an early dropping probability p_i^{ED} for each traffic class i based on the online measured burst loss probability and the maximum acceptable burst loss probability of the next higher traffic class, $i - 1$. In addition, an early dropping flag e_i is associated with each class i, where e_i is determined by generating a random number between 0 and 1. If the generated random number is less than p_i^{ED}, then e_i is set to 1. Otherwise, e_i is set to 0. Thus, e_i is equal to 1 with probability p_i^{ED}, and equal to 0 with the complementary probability $1 - p_i^{ED}$. To decide whether or not to intentionally drop an arriving class i burst, an OBS node considers not only the early dropping flag e_i of class i but also the early dropping flags e_j of all higher-priority classes, where $j = 1, \ldots, i - 1$. An OBS node intentionally drops an incoming class i burst if $e_1 \vee e_2 \vee \cdots \vee e_i = 1$.

9.6 Contention resolution

Contention in OBS networks occurs when two or more bursts arriving at a given OBS node request the same resources at the same time. To resolve contention at OBS nodes, several techniques were proposed and investigated (e.g., the aforementioned use of FDLs). Clearly, the major goal of these techniques is to resolve contention at intermediate OBS nodes such that bursts can be forwarded as efficiently as possible toward their destination. As we will see shortly, contention resolution techniques can also be deployed to enable service differentiation by taking the different service classes of contending bursts into account and giving bursts belonging to a higher service class priority over lower-class bursts in the process of contention resolution. In this section, we provide an overview of the various contention resolution techniques in OBS networks and discuss their ability of achieving service differentiation.

9.6.1 Fiber delay lines

To resolve contention in OBS networks, either fixed FDLs or switched delay lines (SDLs) can be used at OBS nodes. We have encountered SDLs already in Section 7.4 when discussing contention resolution in photonic slot routing (PSR) networks. SDLs make use of 2×2 space switches and are able to provide variable-delay buffering. Note that deploying SDLs adds another dimension to the wavelength reservation at OBS nodes. In Lu and Mark (2004), the assumed two-dimension reservation scheme consisted of the following two phases: (1) wavelength reservation and (2) SDL buffer reservation. During the wavelength reservation phase, the scheduler at an OBS node tries to reserve a wavelength for the arriving data burst. If no wavelength is available, the SDL buffer reservation phase begins where the scheduler of the OBS node tries to reserve an SDL buffer with appropriate delay value in order to store an arriving burst temporarily and thereby resolve contention at a given output port. Note that this implies that the scheduler knows both the arrival time and departure time of the incoming burst. Knowing both the arrival and departure times, the scheduler is able to compute the delay value required to store the arriving burst in an SDL buffer. If no SDL with an appropriate delay is available at the burst arrival time, the contending burst is dropped. For the performance evaluation of OBS networks with SDL buffers, it is important to understand the difference between electronic buffer and optical SDL buffer. An electronic buffer can accept an arriving burst if sufficient space is available. Importantly, after finding sufficient space an electronic buffer can store the burst for any arbitrary time period. In contrast, an SDL (and FDL) is able to store a burst only for a fixed maximum time period and thus provides a deterministic delay to an incoming burst. Apart from the deterministic delay, an SDL (and FDL) exhibits the so-called *balking* property (Lu and Mark, 2004). The balking property of an SDL refers to the fact that an incoming burst must be dropped if the maximum delay provided by the SDL is not sufficient to store the incoming burst and avoid contention with a burst that is currently being transmitted on a given output port. Taking both the deterministic delay and balking of SDLs into account, it was shown in Lu and Mark (2004) that the use of SDLs at OBS nodes decreases the burst loss probability of OBS networks significantly. The burst loss performance of OBS networks improves as the length B of SDLs increases, but the performance gain diminishes quickly after a certain threshold. For small B, contending bursts are unlikely to find an available SDL capable of providing a sufficiently long delay. Hence, burst losses during the wavelength reservation phase play the major role in the burst loss performance. Conversely, for large B most of the contending bursts can be temporarily stored in SDLs. As a result, the burst loss probability is decreased for large B, at the expense of increased queueing delays at intermediate OBS nodes.

9.6.2   Burst segmentation

We have seen in the previous section that the balking property of fiber delay lines leads to dropping a burst in its entirety if there are not sufficient wavelength and buffer resources available at intermediate OBS nodes. Instead of dropping bursts completely, it appears reasonable to drop only those parts of a burst which overlap with other contending bursts. In doing so, parts of the burst and all IP packets carried in those parts can be successfully forwarded, resulting in an improved *packet* loss probability of OBS networks. It is possible to drop only the overlapping parts of bursts by using so-called *burst segmentation* techniques at OBS core nodes, as explained in the following.

Burst segmentation as a means to resolve contention in OBS networks was first considered in the context of the so-called *optical composite burst switching* (OCBS) paradigm (Detti et al., 2002). OCBS is derived from the original OBS paradigm and the basic operation is similar in that in both paradigms several IP packets are aggregated into a burst. Unlike OBS, however, OCBS handles contention resolution by means of wavelength conversion and the burst-dropping technique. In traditional OBS networks, an entire burst is discarded when all wavelengths at a given output port are occupied at the burst arrival time. In contrast, a core node deploying the OCBS burst-dropping technique discards only the initial part of an arriving burst until a wavelength becomes free at the output port on which the remainder of the burst can be forwarded. More precisely, in OCBS a burst is forwarded by a core node if any wavelength channel is available at the burst arrival time by using wavelength conversion, if needed. Otherwise, if no wavelength channel is currently available the OCBS burst-dropping technique is deployed. The entire burst is lost if no wavelength channel becomes available before the burst departure time.

OCBS and the underlying burst-dropping technique increase the throughput of OBS networks in terms of IP packets since the remaining part of the forwarded burst contains IP packets that are successfully delivered to their destination. It was shown in Neuts et al. (2002) and Rosberg et al. (2003) that due to burst segmentation OCBS networks exhibit a decreased packet loss probability and for a given level of packet loss probability are able to support significantly more traffic than traditional OBS networks.

In burst segmentation, a burst is generally divided into basic transport units called *segments*. Each segment may contain a single IP packet or multiple IP packets. The boundaries of each segment represent the possible partitioning points of a burst when parts of the burst must be dropped at intermediate OBS nodes. In the event of contention, an OBS node must know which of the contending burst segments will be dropped. In Vokkarane and Jue (2005), two possible burst segment dropping policies were considered: (1) *tail dropping* and (2) *head dropping*. Figure 9.6 illustrates both burst segment dropping policies, where the burst that arrives at an OBS node first is referred to as the original burst and the burst that arrives later is referred to as the contending burst. In the first approach, tail dropping, the tail of the original burst is dropped (i.e., the last one or more segments of the previously arrived burst are dropped when another contending burst arrives). Note that if switching time at an OBS node is not negligible, one or more additional segments of the original burst may be lost when the output port is switched from one burst to another. In the second approach, head dropping, the head of the contending burst which arrives after the original burst is dropped. The dropped head consists of one or more burst segments, depending on the length of the contention region and switching time.

Burst segmentation can be used to resolve contention and reduce packet loss in OBS networks, while *prioritized* burst segmentation can also be used to provide service differentiation and QoS support by allowing high-priority bursts to preempt low-priority bursts. Prioritized burst segmentation allows for service differentiation without requiring any extra offset time. Prioritized burst segmentation was examined in Vokkarane and Jue (2003) under the assumption of tail dropping. Compared to head dropping, the advantage of tail dropping is that there is a better chance of in-order packet delivery at the destination OBS user, provided that dropped packets are retransmitted by the source OBS user at a later time. To provide QoS support, a new approach of assembling packets of different classes into the same burst was proposed and termed *composite burst assembly*. This new approach is motivated by the observation that, with tail dropping, packets toward the tail of a burst are more likely to be dropped than packets at the head of a burst. In the composite burst assembly approach, packets are placed in a single burst in descending order according to their traffic classes. Packets belonging to classes which have low loss and low delay requirements are placed toward the head of a burst while

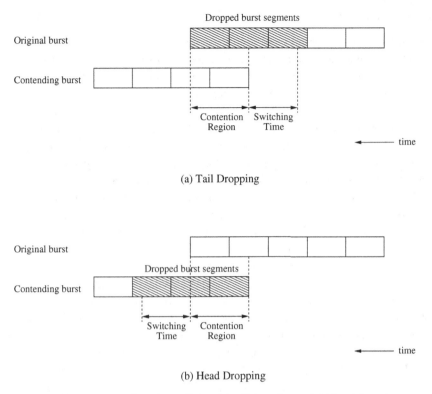

Figure 9.6 Burst segment dropping policies: (a) tail dropping and (b) head dropping.
After Vokkarane and Jue (2005). © 2005 IEEE.

packets belonging to classes which have a higher loss and delay tolerance are placed toward the tail of a burst. Based on the traffic classes of the assembled packets, bursts are assigned different priorities. The priority of each burst is included in the corresponding control packet. OBS nodes use the priorities to differentiate bursts with respect to tail dropping by letting high-priority bursts preempt low-priority bursts. Note that, like any other burst assembly algorithm, the composite burst assembly technique is implemented at OBS users located at the OBS network edge. To minimize the burst/packet loss, prioritized burst segmentation was done in conjunction with *deflection routing*. Rather than dropping the tail segment of a burst, either the entire burst or only the tail can be deflected. Deflection routing increases the probability that a burst's packets will reach the destination. To do so, at each OBS node one or more alternate deflection ports and routes must be specified for each destination. One of those alternate routes may be chosen if the OBS node port of the primary route is currently used. It was shown that with prioritized burst segmentation high-priority bursts have significantly lower packet losses and delay than low-priority bursts. Moreover, schemes which incorporate deflection routing tend to perform better than schemes with limited deflection or no deflection at all. Given the benefits of deflection routing to resolve contention in OBS networks, we discuss deflection routing in greater detail in the subsequent section.

9.6.3   Deflection routing

As we have seen in the previous section, deflection routing can be deployed together with burst segmentation to reduce the burst/packet loss performance of OBS networks. Similarly, deflection routing can also be used in conjunction with other contention resolution techniques (e.g., FDLs). In deflection routing, the use of FDLs is optional but not mandatory. As a matter of fact, deflection routing can work with limited optical buffering or no optical buffering at all. In the latter case, deflection routing is referred to as *hot-potato routing*. Hot-potato routing is used in networks where nodes have no buffer to store packets when multiple packets contend for a given output port. In hot-potato routing, a contending packet is forwarded immediately through another port without being stored, as opposed to store-and-forward routing which is widely used in routers that are equipped with plenty of electronic memory. The selection of output ports is usually made using local information (e.g., output port with the lowest delay). Alternatively, output ports may be selected based on preassigned port priorities.

Typically, deflection routing is triggered by an OBS node using the local status information of its own resources (e.g., wavelength availability or link congestion). Using local status information for selecting a deflection route does not take the global status of network resources into account and may lead to suboptimal network performance. To improve contention resolution by means of deflection routing, the so-called *contention-based limited deflection routing* (CLDR) protocol was proposed in Lee et al. (2005). In CLDR, all OBS nodes periodically exchange local status information about traffic load, burst contention rate, burst blocking probability, etc. For this periodic information exchange, the GMPLS OSPF/IS-IS routing protocols with appropriate TE extensions can be deployed, which were described at length in Section 5.2.3. The proposed CLDR algorithm runs at all OBS nodes and sequentially performs the following two steps:

1. **Deflection Routing versus Retransmission Decision:** First, the CLDR algorithm running at a given OBS node dynamically determines whether an arriving burst is deflection routed or dropped. In the latter case, this implies that the dropped burst needs to be retransmitted by the corresponding source OBS user. In CLDR, the use of deflection routing is limited, whereby the decision between deflection routing and retransmission of a burst is checked against a threshold function. More specifically, the threshold function used in CLDR includes the following variables: (1) traversed hop-count of the arriving burst and (2) burst blocking probability of all available paths from the congested OBS node to the destination OBS user. Both variables may be weighted, as needed. The traversed hop-count denotes the distance of the congested OBS node from the source OBS user in terms of hops. Note that this implies that the control packet preceding the burst needs to carry additional information about the number of hops traversed by the burst between source OBS user and congested OBS node. The burst blocking probability of all links is distributed throughout the OBS network by means of the aforementioned GMPLS routing protocols with appropriate TE extensions. The rationale behind the threshold function is as follows. Deflection routing should be performed at low traffic loads when the burst blocking probability

is small. Conversely, burst retransmission is more suitable at heavy traffic when the burst blocking probability is high. At medium traffic loads, either deflection routing or retransmission of a contending burst might be given higher priority, depending on the chosen weights of the threshold function.

2. **Alternate Path Selection:** In the case of deflection routing (i.e., if the decision was made to perform deflection routing rather than retransmission of the contending burst), the burst is sent along an alternate path that minimizes a cost function which is a weighted sum of the end-to-end burst blocking probability and the distance of the route. In general, the selected alternate path is not necessarily the shortest path. (If the contending burst was dropped, this step does not apply.)

The CLDR routing scheme is able to avoid injudicious deflection routing in OBS networks by using a threshold-based dynamic decision algorithm to decide whether to perform deflection routing or not. It is important to note that in CLDR (and any other deflection routing scheme) for bursts to arrive successfully at their destination over the alternate routes computed by the CLDR algorithm an additional offset time of the control packet or buffering delay of the burst must be provided. To see this, note that when deflection routing occurs due to contention at an OBS node, the required offset time on the alternate route is generally longer than that on the primary path. One solution to this problem is to let the corresponding edge OBS user add a sufficiently large extra offset time to the control packet, as discussed in Section 9.5 within the context of service differentiation in OBS networks. Alternatively, the control packet reserves FDL buffer to delay the subsequent burst at the OBS node for a sufficiently long time period.

9.6.4   Wavelength conversion

Wavelength conversion was introduced in Section 1.5.2 and discussed within the context of contention resolution in OBS networks in Section 9.6.2. Wavelength conversion is a powerful technique to effectively resolve contention in OBS networks. It can be used alone or in combination with other contention resolution techniques such as FDLs and burst segmentation, as discussed earlier. In practice, however, technological feasibility and economic issues do not allow full-range wavelength conversion at all OBS nodes. Instead, the deployment of limited-range and sparse wavelength conversion in OBS networks appears to be more realistic. Consequently, wavelength contention may occur more frequently in realistic OBS networks which can be resolved by using one or more of the aforementioned contention resolution techniques. Another interesting solution to mitigate contention in OBS networks without requiring any of the aforementioned contention resolution techniques is the careful wavelength assignment to bursts by OBS users at the network periphery (Teng and Rouskas, 2005). Several adaptive and nonadaptive wavelength assignment heuristics were examined for OBS networks in the absence of wavelength conversion and any other contention resolution technique discussed earlier. It was shown that intelligent choices in assigning wavelengths to bursts at edge OBS users can have a significant impact on the burst loss probability of OBS networks. Among the proposed heuristics, the best performing ones have the potential to improve the burst loss probability by as much as two orders of magnitude compared with the two rather simple random and first-fit wavelength assignment heuristics.

9.7   Multicasting

To support multicasting in OBS networks, OBS nodes should be able to split an incoming optical signal to multiple output ports by using an optical splitter. Optical splitters may be used by either all OBS nodes or only a subset of OBS nodes, resulting in sparse splitting. Clearly, the number and location of splitting capable OBS nodes have an impact on the construction of multicast trees in OBS networks.

Several multicasting approaches for OBS networks were proposed and investigated in Jeong et al. (2002). The first two proposed multicasting approaches are straightforward and are called separate multicast (S-MCAST) and multiple unicast (M-UCAST), respectively. In the first approach, S-MCAST, multicast traffic of a given multicast session is assembled at its source OBS user into multicast bursts. The OBS user maintains a separate burst assembly queue for each multicast session. After the burst assembly time is over, the multicast burst is sent out and delivered along the multicast tree which connects all destination OBS users belonging to the multicast group. In S-MCAST, each multicast session needs to construct its own source-specific multicast tree along which the assembled multicast bursts carrying multicast traffic for the corresponding multicast group are sent. Furthermore, multicast traffic is sent independent from unicast traffic in S-MCAST. In the second approach, M-UCAST, multicast traffic is treated as unicast traffic. Specifically, a source OBS user makes multiple copies of an arriving multicast packet belonging to a multicast group, one copy for each multicast group member, assembles them along with unicast packets destined to the same OBS user into unicast bursts. The assembled unicast bursts are sent out and delivered to the corresponding destination OBS users by means of unicasting. Note that due to the fact that multicast packets and unicast packets can be assembled together into a burst destined to the same receiving OBS user, the control overhead in M-UCAST is reduced since no separate control packets are needed for multicast bursts.

Besides these two approaches, another more bandwidth-efficient OBS multicasting approach, called tree-shared multicast (TS-MCAST), was examined. In TS-MCAST, the multicast sessions originating from the same source OBS user are decomposed into disjoint subsets according to a given tree sharing strategy, as described shortly. Each disjoint subset is called a multicast sharing class (MSC). TS-MCAST deploys a single multicast burst assembly queue for each MSC at a source OBS user. Each assembled multicast burst for an MSC is sent along a shared tree, which can be one of the multicast trees in the MSC or a newly constructed multicast tree. Tree sharing occurs when multiple multicast sessions use a single shared tree for the delivery of their multicast packets. Note that in TS-MCAST, the average multicast burst length will be longer than without tree sharing, where each multicast session assembles its packets independently from other multicast sessions originating from the same source OBS user. As a result, the control overhead is reduced since only one control packet needs to be sent for multiple multicast sessions. Unicast traffic is treated separately from multicast traffic in TS-MCAST.

To form an MSC in TS-MCAST, an OBS user adopts a tree-sharing strategy. Three tree-sharing strategies were proposed to decide which subset of multicast sessions belongs to the same MSC at a given source OBS user. The first strategy, called perfect overlap, groups all multicast sessions with the same multicast group members (i.e., the same destination OBS users). In the second strategy, called super overlap, the multicast membership constraint is relaxed in that a number of multicast sessions, whose multicast group members do not need to be exactly the same, can be grouped into one MSC. The third strategy, called arbitrary overlap, is a more generalized tree-sharing strategy in which multiple multicast sessions are allowed to be grouped into one MSC if they have sufficient overlap in terms of multicast destination OBS users, core OBS nodes, or links on their multicast trees. To decide whether the considered multicast sessions overlap sufficiently or not, a way of calculating the degree of overlap must be defined. For instance, the degree of overlap of multiple, say, s multicast sessions can be defined as follows. For each distinct destination OBS user e a fractional number $d_e \leq 1$ is calculated by dividing the number of multicast sessions having destination OBS user e as their member by the total number s of multicast sessions under consideration. The same process is repeated for all other distinct multicast destination OBS users. Then the degree of overlap for the s multicast sessions is calculated by taking the sum of all fractional numbers d_e over all distinct destination OBS users and dividing the sum by the number of all distinct destination OBS users belonging to the s multicast sessions. Similarly, the degree of overlap in terms of core OBS nodes and links can be calculated. To decompose a set of multicast sessions using the arbitrary overlap tree-sharing strategy, a source OBS user selects the subset of multicast sessions that has the highest degree of overlap in terms of destination OBS users, core OBS nodes, and/or links.

The performance of TS-MCAST was investigated in Jeong et al. (2003). The obtained results show that TS-MCAST outperforms S-MCAST and M-UCAST in terms of bandwidth efficiency and control overhead (i.e., number of control packets) for a given amount of multicast traffic under the same unicast traffic load with static multicast sessions and multicast group membership.

9.8   Protection

The protection schemes proposed for OBS networks can be roughly classified into two categories. The first scheme, $1 + 1$ path protection switching, is based on the concept of a session created between a pair of source and destination OBS users located at the network periphery (Griffith and Lee, 2003). The $1 + 1$ path protection switching scheme does not require any special functionality at intermediate OBS nodes. The second scheme, segment protection switching, is performed by intermediate OBS nodes in the core of an OBS network by enabling them with the aforementioned burst segmentation and deflection routing techniques (Griffith et al., 2005). Both OBS protection schemes are explained in greater detail in the following.

The 1 + 1 path protection switching is an extension of the GMPLS 1 + 1 protection scheme to OBS networks. Recall from Section 5.2.5 that in the GMPLS 1 + 1 protection scheme two preprovisioned disjoint paths are used in parallel to transmit data simultaneously. Both paths should be disjoint with respect to links and nodes, as well as shared risk link group (SRLG). The receiving node uses a selector to choose the best signal (e.g., based on the measured optical signal-to-noise ratio [OSNR]). If the selected path fails or the OSNR degrades, the destination node switches over to the other path instantaneously. The proposed 1 + 1 path protection switching scheme is applicable only in the context of a *session*. A session denotes a persistent route connection between a pair of source and destination OBS users, where all bursts associated with a given session follow the same route from source OBS user to destination OBS user. To achieve this, output ports need to be reserved at all traversed OBS nodes in order to pin down the path. After computing and pinning down two disjoint paths, the source OBS user duplicates each burst, assigns both bursts with a separate control packet, and sends the control packets followed by the bursts along both paths. Note that the intermediate OBS nodes on each path do not need any modification or special functionality in order to support 1+1 protection. They merely schedule and forward the incoming bursts, as done in conventional OBS networks. (The pinning down of explicit routes in OBS networks will be described in greater detail shortly in Section 9.9.1.)

We have seen in Section 9.6 that the two techniques, deflection routing and burst segmentation, can be used to resolve contention in OBS networks. Apart from contention resolution, both techniques can also be deployed for protection in OBS networks. Deflection routing can be used for failure recovery by sending bursts on alternate paths to their intended destinations. As soon as a failure is detected, the OBS node immediately upstream from the failure sends bursts that are destined for the affected output port to other output ports. If burst segmentation is used by the OBS node to resolve contention, deflection routing can be used in combination with burst segmentation to reduce packet loss due to OBS network failures.

9.9   OBS derivatives

There have been several suggestions how to extend the functionality of OBS. Among the most important OBS derivatives are the so-called labeled OBS (LOBS), wavelength-routed OBS (WR-OBS), and dual-header OBS (DOBS), which are described in this section.

9.9.1   Labeled OBS

We have seen in Section 9.6.3 that deflection routing can be used at OBS nodes to reroute a control packet and its corresponding data burst on an alternate outgoing link in the event of contention (or network failure), preventing both control packet and data burst from being dropped. If the alternate path has so many more hops than the original path it might happen that the deflection routed data burst overtakes its corresponding control packet due to the fact that the offset time set by the source OBS user is no longer sufficient to cover the total processing delays encountered by the control packet at intermediate OBS nodes that are equipped with no or not sufficient FDLs to delay the data burst. To avoid such a situation, OBS network resources (e.g., alternate paths for contention resolution and failure recovery) must be carefully engineered, which can be done by using GMPLS (see Chapter 5).

By augmenting each OBS node with a GMPLS-based controller, OBS is extended to *labeled OBS* (LOBS) (Qiao, 2000). A LOBS node (i.e., an OBS node equipped with a GMPLS-based controller) is similar to a label switched router (LSR) in conventional GMPLS networks. In LOBS, each control packet can be sent as an IP packet containing a label as part of its control information on the OBS control channel along a preestablished LOBS path, similar to a GMPLS label switched path (LSP). The corresponding burst can be sent along the same LOBS path used by the control packet. By capitalizing on the explicit and constraint-based routing capabilities of GMPLS, routes can be pinned down and traffic can be engineered in LOBS networks.

An established LOBS path makes use of the same or multiple different wavelengths along the involved links, depending on whether or not all-optical wavelength conversion is applied at intermediate OBS nodes, similar to MPλS where a wavelength represents the label. However, unlike circuit-switched MPλS where each LSP is a wavelength path and different λ LSPs cannot be optically aggregated into one common λ LSP due to the lack of wavelength-merging techniques, LOBS allows bursts belonging to two or more LSPs (i.e., LOBS paths) to be aggregated without having to undergo OEO conversion at LOBS nodes. This is because in LOBS the label information is carried by control packets which are processed in the electrical domain at all intermediate LOBS nodes. Note that in LOBS not only the label information in the control packet but also the information about the wavelength on which the corresponding burst will arrive and the offset time are electronically processed at each core LOBS node. Whereas at edge LOBS users the burst assembly process comprises not only the aggregation of multiple IP packets into one burst but also GMPLS-related functions such as label stacking and LSP aggregation. Thus, a source edge LOBS user is able to merge multiple (electronic) LSPs over a LOBS path, which are subsequently separated at the destination edge LOBS user.

9.9.2   Wavelength-routed OBS

WR-OBS combines OBS with centralized dynamic wavelength allocation, where bursts are routed across a fast reconfigurable optical network (Dueser and Bayvel, 2002a,b). WR-OBS may be viewed as a fast circuit switching technique. Figure 9.7 depicts the WR-OBS network architecture which differs from the OBS network architecture of Fig. 9.1 in that a central control node is deployed in WR-OBS networks. Unlike OBS, WR-OBS uses a *two-way* reservation mechanism with ACK between each OBS user

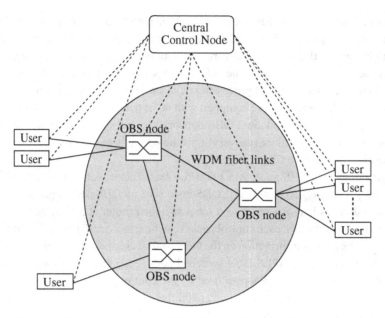

Figure 9.7 Wavelength-routed OBS (WR-OBS) network architecture.

and the central control node for the dynamic set-up of end-to-end lightpaths across the OBS network core. More precisely, after assembling a burst an edge OBS user requests an end-to-end wavelength channel from the central control node for transmission of the assembled burst to its intended destination OBS user. Upon reception of the request, the central node executes a given routing and wavelength assignment (RWA) algorithm to assign a free wavelength channel across the OBS core network and sends an ACK to the source OBS user to inform it about the assigned wavelength channel. After receiving the ACK, the source OBS user starts transmitting the burst on the assigned wavelength channel. Following transmission, the wavelength channel is released and becomes available for other connections.

WR-OBS provides the benefits of circuit-switched networks (i.e., once the lightpath is established the end-to-end delay is essentially deterministic and equal to the end-to-end propagation delay), as opposed to conventional OBS where the end-to-end delay may be nondeterministic due to further delays arising from contention at intermediate OBS nodes. At the downside, WR-OBS suffers from a larger control overhead than conventional OBS since one control packet per request and ACK, respectively, between each edge OBS user and the central node, plus several control packets from the central control node to core OBS nodes for switch configuration, need to be sent. Furthermore, WR-OBS may suffer from a small wavelength utilization. This is because in WR-OBS the central control node reserves a wavelength channel after executing the RWA algorithm. However, it takes a finite propagation delay $t_{\mathrm{prop,ack}}$ for the ACK to arrive at the corresponding edge OBS user and another finite end-to-end propagation delay $t_{\mathrm{prop,network}}$ for the first transmitted bit of the burst to arrive at the destination OBS user. Therefore, the reserved wavelength is idle and not used for data burst transmission for the time period $t_{idle} = t_{prop,ack} + t_{prop,network}$, which is a key parameter in the design of WR-OBS networks. Clearly, for increasing t_{idle} the utilization of a wavelength channel decreases. As a consequence, WR-OBS is more suitable for metropolitan area and regional networks with small to medium network diameters, which translate into a small to medium t_{idle} and an acceptable wavelength utilization.

To improve the wavelength utilization of large WR-OBS networks, a *prebooking* mechanism that excludes the ACK propagation delay $t_{prop,ack}$ from the idle time period t_{idle} was studied in Kong and Phillips (2006). Prebooking makes use of estimated traffic information to proactively reserve wavelength resources. Specifically, a lightpath pre-booking request is initialized by a source OBS user at the beginning (instead of end) of the burst aggregation. The burst aggregation starts when the source OBS user receives the first bit of the new burst and continues until the burst transmission begins. Based on the observed traffic characteristics, the request includes the earliest burst transmission time, the maximum tolerable burst assembly delay, and the predicted burst length information in the request message that is sent to the central control node for dynamic lightpath set-up. After receiving the request message and executing its RWA algorithm, the central control node sends a prebooking ACK back to the source OBS user. The prebooking ACK comprises information about the assigned wavelength channel and reservation time window. The prebooking has to be completed before the burst transmission starts. The burst transmission takes place on the assigned wavelength channel at the allocated reservation time. Note that prebooking is done during the burst assembly process. In doing so, prebooking enables the central control node to run its RWA algorithm while the source OBS user is assembling the burst. The established lightpath is held only for the burst transmission time plus end-to-end propagation delay $t_{prop,network}$, resulting in a reduced idle period time $t_{idle} = t_{prop,network}$. If the reservation window is long enough, the entire burst can be sent. Otherwise, if insufficient wavelength resources are reserved due to erroneous prediction, the burst can be sent only in part and the remainder of the burst is dropped. It was shown that prebooking is able to improve not only the wavelength utilization but also the burst loss probability of WR-OBS networks significantly, while avoiding high traffic losses due to erroneous prediction.

9.9.3   Dual-header OBS

In conventional OBS networks, control packets are OEO converted and processed at each intermediate OBS node while the corresponding bursts cut through all-optically. Thus, each control packet incurs processing delay at intermediate OBS nodes and the size of the offset time shrinks as the corresponding burst travels through the OBS network. As a result, offsets are of *variable* size in conventional OBS networks and vary as a function of their path lengths. This offset variability complicates the task of burst scheduling at intermediate OBS nodes and may lead to a decreased network performance. These complexity and performance issues in conventional OBS networks could be avoided if one could precisely control the size of offsets at core OBS nodes. Toward this end, a novel signaling approach called *dual-header optical burst switching* (DOBS) was introduced and investigated in Barakat and Sargent (2006). DOBS enables core OBS nodes to precisely control the offsets of incoming bursts without requiring any FDLs and thereby provides core OBS nodes with much more control over the order and manner in which bursts are scheduled, as explained in greater detail in the following.

In conventional OBS networks using either JIT or JET signaling, bursts are immediately scheduled as soon as their control packets arrive at an OBS node, albeit resource reservation might be delayed in JET (delayed reservation). By separating the control information of each burst into two control packets and decoupling the resource request and scheduling operations at OBS nodes, DOBS removes the requirement that bursts be scheduled immediately after their reservation requests are received, giving rise to *delayed scheduling* which leads to decreased scheduling complexity and improved network performance, as we will see shortly.

Apart from reduced scheduling complexity and improved network performance, DOBS also helps mitigate unfairness in JET signaling-based OBS networks. To see this, note that there are two sources of unfairness in JET-based OBS networks. First, in JET-based OBS networks bursts with longer offsets experience less blocking than bursts with smaller offsets. Since offsets shrink in JET-based OBS networks, as bursts travel through the network the burst loss probability is a function of the path length. Bursts that are close to their destination OBS users face a higher loss probability than bursts that are far from their destinations, resulting in unfairness. This unfairness is undesirable because in general bursts that are close to their destinations have already consumed a large amount of network resources when they are dropped. Second, in JET-based OBS networks that deploy void-filling scheduling algorithms, burst loss probability is generally an increasing function of burst length because shorter bursts are more likely to successfully fit into the voids between previous burst reservations, leading again to unfairness. This unfairness is also disadvantageous in terms of throughput since long bursts contain more payload than short bursts. It is important to note that both types of unfairness do not exist in JIT signaling-based OBS networks which deploy immediate reservation, where the void between a control packet and the corresponding burst is not available for other burst reservation. Therefore, JIT-based OBS networks do not suffer from the aforementioned path-length and burst-length unfairness associated with variable-size offsets and void-filling scheduling algorithms, at the expense of a decreased wavelength utilization due to immediate reservation. DOBS is able to avoid both types of unfairness without sacrificing bandwidth, as explained in the following.

DOBS signaling is characterized by the use of two different types of control packets for each burst: (1) service request packet (SRP) and (2) resources allocated packet (RAP). The SRP contains information about the service requirements of the burst (e.g., routing and class-of-service (CoS) information) as well as the offset and length of the corresponding burst, similar to a control packet in conventional OBS. A single persistent SRP precedes the burst and communicates the burst's service requirements, offset, and length to each OBS node along the path. The RAP contains additional physical information about the burst (e.g., assigned wavelength channel), which is used when configuring the optical switching fabric at an OBS node. The SRP is processed immediately after arriving at a given OBS node i. The processing determines the required resources of the corresponding burst. The burst's resource request is communicated to the burst scheduler module at OBS node i and the SRP is then immediately forwarded to the downstream OBS node $i + 1$ without waiting for the burst scheduling operation to be executed at OBS node i. That is, the SRP has already been processed and forwarded while the corresponding burst waits to be scheduled at OBS node i. This salient feature of DOBS is referred to as *delayed scheduling*. The burst scheduling algorithm is executed at OBS node i sometime later at time t_{BS}^i after sending the SRP but prior to the arrival time of the corresponding burst. After the burst scheduling is complete, OBS node i transmits a RAP downstream to inform OBS node $i + 1$ about the wavelength channel assigned to the burst. After receiving in turn the RAP from upstream OBS node $i - 1$ and using the result of its burst scheduling algorithm executed at time t_{BS}^i, OBS node i configures its optical switch fabric just before the burst from upstream OBS node $i - 1$ arrives at OBS node i at time t_b^i. Note that similar to conventional OBS the offset between the SRP and the corresponding burst shrinks due to the processing delay incurred at each intermediate OBS node. However, the so-called *scheduling offset* Ω_{BS}^i, which denotes the time period between the burst scheduling time t_{BS}^i and the burst arrival time t_b^i at a given OBS node i (i.e., $\Omega_{BS}^i = t_b^i - t_{BS}^i$), can be chosen arbitrarily at OBS node i and all other OBS nodes within a certain range. Since the order in which incoming bursts are serviced at an OBS node is a function of their scheduling offset sizes, DOBS provides OBS nodes with complete control over the order in which arriving bursts are processed. The ability to independently select the scheduling offset at each OBS node can be exploited in several ways. For instance, an OBS node may select a larger scheduling offset than the other OBS nodes in order to accommodate the slow switching time of its optical switching fabric.

The benefits of a possible DOBS variant, called *constant-scheduling-offset* (CSO) DOBS, were examined in Barakat and Sargent (2006). In CSO DOBS, each link in the OBS network is associated with a CSO value. By setting the scheduling offset of a given link to a fixed value, CSO DOBS ensures that all arriving bursts have the same offset and can be scheduled in a first-come-first-served (FCFS) manner. It was shown that CSO DOBS has the same low scheduling complexity as JIT signaling, while avoiding wasted bandwidth due to the immediate reservation in JIT signaling. Furthermore, CSO DOBS achieves an improved throughput-delay performance compared to both JIT and JET signaling approaches. Finally, CSO DOBS outperforms JET signaling in terms of fairness since useful voids are not created during burst scheduling in CSO DOBS, so void-filling is unnecessary and bursts of all lengths will experience the same blocking probability on a given link (i.e., there is no burst-length unfairness in CSO DOBS). Further, since in CSO DOBS the scheduling offset size is generally not a function of the path length, path-length unfairness is also avoided.

9.10 Implementation

As we have seen earlier, the operation of OBS networks involves a number of different functions. Most of the previous research efforts on OBS networks focused on their theoretical investigation by means of analysis and/or simulation. To examine possible implementation issues and demonstrate a proof of concept, part of the pre-viously described functions of OBS networks were experimentally verified, whereby particular attention was paid to the implementation of JIT signaling, wavelength assign-ment, deflection routing, and LOBS.

9.10.1   JIT signaling

Under the *multiwavelength optical networking* (MONET) program sponsored by the U.S. government's Defense Advanced Research Project Agency (DARPA), a software proto-type of the JIT signaling protocol was successfully demonstrated (Wei and McFarland, 2000). In the MONET testbed, an ATM-based control channel, referred to as the data communication network (DCN), was implemented on the ITU-T standard signaling wavelength channel at 1510 nm, which is also referred to as the optical supervisory channel (OSC). The ATM-based control channel provided bidirectional communica-tions and carried the JIT signaling messages between neighboring OBS users and OBS nodes. The following signaling scenarios were considered in the MONET testbed:

- **Connection Establishment:** JIT signaling was used to successfully set up an end-to-end optical burst switched connection.
- **Connection Tear-Down:** After transmitting a data burst, a sequence of messages was exchanged between source and destination OBS users and the OBS network to release resources reserved at intermediate OBS nodes.
- **Connection Failure due to Request Rejection:** As a JIT signaling message travels downstream to reserve resources at intermediate OBS nodes, a collision may occur along the intended path due to unavailable wavelengths, ports, etc. In this event, a call-clearing signaling message is created by the respective OBS node and sent back to the source OBS user.
- **Connection Failure due to Message Loss:** A connection attempt may also fail if the JIT signaling message is dropped or lost. This might be caused by a failed control channel or crashed JIT signaling agent at an intermediate OBS node. The connec-tion failure can be detected by the source OBS user based on the timeout of timers associated with the connection set-up attempt.

An OBS network testbed which deploys JIT signaling on a control wavelength channel at 1310 nm was reported in Xinwan et al. (2005). The experimental set-up consists of one core OBS node and three edge OBS users (routers), which use a time-based burst assembly algorithm. The implemented JIT signaling protocol enables not only the reservation and release of wavelength resources at the core OBS node but also the retransmission of failed control packets. More precisely, an edge OBS user explicitly releases wavelength resources reserved at the core OBS node by sending a so-called burst end packet (BEP) after the data burst. The BEP travels on the same path as the corresponding burst toward the destination OBS user. Upon its reception, the destination OBS user returns a so-called BEP-ACK acknowledgment to the source OBS user on the reverse path in order to release the reserved wavelength resources at the core OBS node, a process called *reverse deletion*. In case either BEP or BEP-ACK is lost in the wavelength reservation or release process, respectively, a timer at the source OBS user expires which indicates the control packet loss. Subsequently, the source OBS user resets its timer and retransmits the BEP.

JIT was also the signaling protocol of choice in the *JumpStart* architecture and protocols implemented in several OBS networking testbeds on the U.S. east coast (Baldine et al., 2005). JumpStart supports the ultrafast lightpath provisioning by implementing JIT signaling in hardware (i.e., control packets are processed in hardware at intermediate OBS nodes). In JumpStart, the dynamic lightpath set-up might be initiated by the operator, user, application, or protocol. JumpStart supports a variety of emerging applications, as discussed in greater detail in Section 9.11.

9.10.2   Wavelength assignment and deflection routing

An OBS network testbed with an interesting edge OBS user node architecture was reported in Sun et al. (2005). The edge OBS users deploy a mixed time/burst length–based assembly algorithm which classifies packets by destination OBS user and CoS. Apart from burst assembly and service differentiation, the investigated contention resolution scheme is of particular interest in that it combines deflection routing at the core OBS node with a smart wavelength assignment algorithm called *priority-based wavelength assignment* (PWA) at the edge OBS user. Recall from Section 9.6.4 that the contention and burst loss probability of OBS networks can be significantly reduced by means of intelligent choices in assigning wavelengths to bursts at edge OBS users without requiring any wavelength converters at core OBS nodes. In PWA, each edge OBS user maintains a dynamically updated wavelength priority database where every wavelength channel of the OBS network is prioritized for each destination OBS user by learning from the wavelength utilization history. More specifically, when a source OBS user learns from the destination OBS user about the successful burst delivery on a given wavelength channel, it increases the priority of that wavelength. Otherwise, it decreases the priority of the wavelength. The source OBS user assigns the wavelength with the highest priority to an assembled burst in order to reduce burst loss probability and mitigate contention. If the transmitted burst experiences congestion at an intermediate OBS node, it is deflection routed. The experimental results showed that both PWA and deflection routing reduce the burst loss probability, whereby deflection routing was more effective than PWA.

9.10.3   Labeled OBS

A new concept of optically labeling bursts suitable for LOBS networks was experimentally demonstrated in Vlachos et al. (2003). By using different modulation schemes, the optical label and corresponding burst were modulated orthogonally to each other on the same wavelength. It was shown that the optical label can be easily separated from the burst at intermediate OBS nodes. Furthermore, optical label swapping at OBS nodes was successfully performed by first erasing the incoming label and then inserting a new outgoing label by using wavelength converters and electro-absorption modulators (EAMs), respectively.

9.11   Application

OBS can be used to support various applications. For instance, in the aforementioned JumpStart project OBS was used to realize dynamic lightpath set-up for sending uncompressed high-definition television (HDTV) signals, Grid applications, file transfers without requiring transport-layer sequence numbering and reassembly, and low-latency, zero-jittery supercomputing applications (e.g., interactive visualization of high-volume imagery) (Baldine et al., 2005). OBS can also be used to realize consumer-oriented Grids (Leenheer et al., 2006). A consumer-oriented Grid is a Grid platform that is accessible by a large number of highly dynamic users. A typical consumer-oriented Grid application might be a multimedia editing application in which integrated audio and video manipulation programs allow users to manipulate video clips, add effects, restore films, etc. Advances in recording, visualization, and effects technology will demand more computational and storage capacity which may not be locally available at each user. A Grid user could create an optical burst containing the multimedia material to be processed/stored and hand over this burst to the network, which is responsible for delivering the burst to a remote site with sufficient resources.

10 Optical packet switching

Optical fiber provides huge amounts of bandwidth which can be tapped into by means of dense wavelength division multiplexing (DWDM), where each fiber may carry tens or even hundreds of wavelength channels, each operating at electronic peak rate (e.g., 40 Gb/s). Given this huge number of high-speed wavelength channels, one may think that network capacity will not be an issue in future optical networks and it seems reasonable to deploy dynamic optical circuit switching (OCS) to meet future service requirements in support of existing and emerging applications. Typically, these optical circuits may be lightpaths that are dynamically set up and torn down by using a generalized multiprotocol label switching (GMPLS) based control plane to realize reconfigurable optical transport networks, leading to multiprotocol lambda switching (MPλS), as discussed at length in Chapter 5. While OCS may be considered a viable solution that can be realized using mature optics and photonics technologies, economics will ultimately demand that network resources are used more efficiently by decreasing the switching granularity from optical wavelengths to optical packets, giving rise to *optical packet switching* (OPS) (O'Mahony et al., 2001). Especially given the fact that networks increasingly become IP data-centric, OPS naturally appears to be a promising candidate to support bursty data traffic more efficiently than OCS by capitalizing on the statistical multiplexing gain. Furthermore, the connectionless service offered by OPS helps reduce the network latency in that OPS avoids the two-way reservation overhead of OCS. Note that in Chapter 9 we have seen that the same holds for optical burst switching (OBS) as well. In fact, both OBS and OPS face a number of similar challenges (e.g., contention resolution in the optical domain at intermediate nodes). Unlike OBS, however, OPS does not require edge routers to perform any burst assembly and disassembly algorithms at the network periphery. Moreover, in OPS the header is generally encoded together with the payload on the same wavelength channel and does not need to be sent on a separate control wavelength channel, as is done in OBS. As a result, OPS not only saves on the number of required wavelength channels and associated transceivers but also, and more importantly, does not require any offset time and thus completely avoids the issue of setting the offset time properly, which might pose significant problems if the number of intermediate nodes is unknown to the source node or when, in the event of a network failure, intermediate nodes need to route bursts along secondary backup paths whose length and/or hop count differ from the primary working paths.

Most of the concepts used in OPS are borrowed from its electronic packet switching counterparts such as asynchronous transfer mode (ATM) switches and IP routers. OPS may be viewed as an attempt to mimic electronic packet switching in the optical domain while taking the shortcomings and limitations of current optics and photonics technology into account. The challenge faced in OPS involves developing an elegant solution to the mismatch between the transmission capacity offered by the wavelength division multiplexing (WDM) optical layer and the processing power of electronic switches and routers (Hunter and Andonovic, 2000). This mismatch, which is also widely referred to as the *electro-optical bottleneck,* can be in principle alleviated by increasing the degree of parallelism at the electronic layer. In practice, however, a higher level of parallelism not only increases the complexity of electronic switches and routers but also results in a significantly increased power consumption, cost, and footprint (Jourdan et al., 2001). By using low-cost optics and photonics technology and performing part of the switching and/or routing in the optical domain, electronic switches and routers can be offloaded, resulting in reduced complexity, footprint, and power consumption, improved performance, as well as significant cost savings.

The research and experimental demonstration of the feasibility of OPS networks already started in the mid 1990s. Among the most notable OPS research projects are the European RACE ATMOS (ATM optical switching) and ACTS KEOPS (keys to optical packet switching) projects. The RACE ATMOS project studied the possible application of photonic technologies to ATM switching systems to enhance node throughput, speed, and flexibility (Renaud et al., 1997). The ACTS KEOPS project was based on the outcome of ATMOS and focused on the development and assessment of OPS network node architectures using optical packets of fixed duration (Gambini et al., 1998). At present, the implementation of OPS where both the packet header and payload are processed and routed entirely in the optical domain is challenging due to the lack of optical random access memory (RAM) and the difficulty to execute complex computations and logical operations using only optics and photonics without requiring any OEO conversion at OPS routers. An interesting approach that allows practical OPS networks to be realized in the near term is the so-called *optical label switching* (OLS) technique (Chang et al., 2006). OLS may be viewed as a particular implementation of OPS. In OLS, only the packet header, referred to as *label,* is processed electronically for routing purposes while the payload remains in the optical domain. The label can be differentiated from the payload in a number of ways (e.g., diversities in time, wavelength, and modulation format). In general, the label is encoded at a lower bit rate compared to the payload in order to simplify OE conversion during the label extraction and rewriting process. The basic functions performed by each OLS router involve the following steps: (1) extraction of the label from the optical packet, (2) electronic processing of the label to obtain routing information, (3) optical routing of the payload and resolving contention if necessary, and (4) rewriting of the label and recombining it with the optical payload. Several key enabling technologies for OLS networks have been developed over the last few years, including optical label generation, optical label swapping, optical buffering, clock recovery, and wavelength conversion. In the following, we elaborate on how these technologies can be exploited to build future OPS networks.

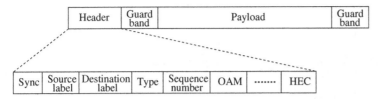

Figure 10.1 Generic optical packet format. After El-Bawab and Shin (2002). © 2002 IEEE.

10.1 Optical packet switches

10.1.1 Generic packet format

Figure 10.1 depicts a generic optical packet format consisting of a packet header and payload with additional guard bands before and after the payload (El-Bawab and Shin, 2002). The guard bands are used to cope with timing uncertainties of optical packets arriving at OPS nodes. The payload is the user data and should contribute to most of the packet size. The packet header should provide a reasonable trade-off between overhead size and the number of supported control functions that allow OPS nodes to route the packet across the OPS network. Among others, the header may comprise the following fields:

- **Sync:** Delineation and synchronization bits.
- **Source Label:** Source node address.
- **Destination Label:** Destination node address.
- **Type:** Type and priority of packet and carried payload.
- **Sequence Number:** Packet sequence number to reorder packets arriving out of order and guarantee in-order packet delivery.
- **OAM:** Operation, administration, and maintenance functions.
- **HEC:** Header error correction.

Several methods exist for encoding packet headers. For example, the packet header can be encoded at a lower bit rate than the payload in order to simplify electronic processing of the header. Alternatively, the header can be subcarrier multiplexed (SCM) with the payload. Typically, in practical OPS networks a small portion of the optical power of an arriving packet is tapped off at each intermediate OPS node. The packet header is then OE converted and processed electronically. As mentioned earlier, this kind of implementation of OPS is also known as OLS. All-optical packet header processing, which avoids OEO conversion of the header, is still in its infancy and currently allows only for simple operations such as label matching by using *optical correlators*. Optical correlators are used to recognize addresses and work as follows. When an arriving packet's destination address matches the signature of an optical correlator, autocorrelation pulses are generated that are above a given threshold. Otherwise, cross-correlation pulses below the threshold are generated. By deploying simple threshold detection at the

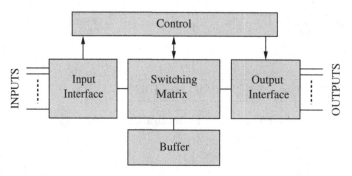

Figure 10.2 Generic OPS node architecture. After El-Bawab and Shin (2002). © 2002 IEEE.

output of the optical correlator the packet's destination address can be recognized and the packet can be forwarded to the appropriate OPS output port.

10.1.2   Generic switch architecture

An OPS node will not be restricted to a simple switching matrix. Figure 10.2 depicts a generic functional block diagram of an OPS node. In general, an OPS node has multiple input and output ports and consists of an input interface, switching matrix, buffer, output interface, and an electronic control unit (Gambini et al., 1998; El-Bawab and Shin, 2002). Each of these blocks performs different functions:

- **Input Interface:** The input interface performs packet delineation to identify the beginning and end of each arriving packet's header and payload. In case of synchronous switches, synchronization is done for performing phase alignment of packets arriving on different wavelength channels and input ports. Note that synchronization is needed only in synchronous switches but not in asynchronous switches, as explained in greater detail shortly. Next, the packet header is extracted, OE converted, decoded, and forwarded to the control unit where it is processed electronically. The control unit processes the routing information in each packet header and configures the switch accordingly. In addition, it updates the header information and forwards the header to the output interface. If necessary, the external wavelength of an arriving optical packet may need to be converted to an internal wavelength for use in the switching matrix.
- **Switching Matrix:** The switching matrix performs the switching operation of the payload in the optical domain according to the commands received from the control unit. It also resolves contention by using optical buffers in conjunction with other contention resolution techniques such as wavelength conversion and deflection routing.
- **Output Interface:** The output interface performs 3R (reamplification, reshaping, re-timing) regenerative functions to compensate for degradation of the extinction ratio (ER) and signal-to-noise ratio (SNR), power variation between packets, and jitter accumulation due to different paths and insertion losses incurred in the switching matrix. Furthermore, the output interface attaches the updated header to the corresponding

optical packet and performs packet delineation. In synchronous switches, the output interface also carries out packet resynchronization. Finally, internal wavelengths are converted back to external wavelengths, if necessary.

10.1.3   Synchronous versus asynchronous switches

OPS networks can be categorized into slotted and unslotted networks (Yao et al., 2001). In slotted OPS networks, packets are of fixed size and are placed in time slots. The duration of each time slot equals the size of a single packet consisting of header, payload, and additional guard bands. The guard bands are used to compensate for timing uncertainties of optical packets arriving at OPS nodes. In slotted OPS networks, OPS nodes operate in a synchronous fashion where the slot boundaries of incoming optical packets are aligned before entering the switching matrix. Synchronization can be realized by using switched delay lines (SDLs) to create different delays with limited resolution.

In unslotted OPS networks, time is not divided into slots and packets can be of variable size. Furthermore, packets do not need to be aligned at the input interface since OPS nodes operate asynchronously in unslotted OPS networks. Asynchronous OPS nodes switch each optical packet on the fly without requiring any alignment of slot boundaries.

Slotted OPS networks using synchronous switches exhibit fewer packet contentions than unslotted OPS networks due to the fact that in slotted OPS networks packets are of fixed size and are switched together with their slot boundaries aligned. At the downside, synchronous switches require packet alignment and synchronization stages, as opposed to asynchronous switches. Furthermore, in unslotted OPS networks using asynchronous switches, packet segmentation and reassembly are not necessary at the ingress and egress OPS network nodes. Therefore, asynchronous switches are more suitable to carry variable-size IP packets, whereas synchronous switches appear to be a viable choice for supporting natively fixed-size packets (e.g., ATM cells).

Initial research on OPS networks focused on slotted networks using synchronous OPS nodes for switching fixed-size packets (e.g., the ACTS KEOPS project) (Gambini et al., 1998). Recently, the design of OPS nodes capable of switching variable-size optical packets has been attracting considerable attention. An OPS node using an asynchronous input interface and a synchronous switching matrix was proposed and investigated in Pattavina (2005). More specifically, the OPS network operation is assumed to be asynchronous; that is, optical packets can be received by OPS nodes at any instant without requiring packet alignment. However, the internal operation of each OPS node is synchronous (i.e., slotted), meaning that the optical packet switching must start at the beginning of a time slot, whereby the duration of each time slot is set to the amount of time needed to transmit a 40-byte optical packet. The choice of the slot duration is motivated by the fact that the minimum-size packet in IP networks is a TCP acknowledgment whose length equals 40 bytes. Optically transporting variable-size packets in such an OPS network is made possible by allowing an optical packet to cover several consecutive slots. For example, an IP packet of 1500 bytes is switched by such OPS nodes using 38 consecutive slots. Another good example of recent research efforts on the design of high-speed asynchronous optical packet switches is the *optical packet switched network*

(OPSnet) project (Klonidis et al., 2005; Vanderbauwhede and Harle, 2005). In OPSnet, an asynchronous OPS node capable of switching variable-size optical packets at 40 Gb/s with technology scalable to beyond 100 Gb/s was presented and demonstrated. In the considered OPS node, the optical header is extracted, OE converted, and electronically processed to determine the output port. Space switching of the optical packet inside the switching matrix is achieved based on wavelength conversion and selection through an arrayed waveguide grating (AWG). Routing of the optical packet to the designated output port is achieved by converting the incoming optical packet to the appropriate internal wavelength, thereby establishing an optical path to the corresponding output port according to the wavelength routing properties of the AWG. Thus, the wavelength conversion determines the path from a given AWG input port to the desired AWG output port. A second stage of wavelength converters is deployed at the output ports of the switching matrix in order to reconvert optical packets to the external wavelengths compatible to the OPS network. Finally, the updated header is inserted in the optical packet.

10.2   Contention resolution

Similar to OBS networks, discussed in Chapter 9, a contention may occur in OPS networks whenever two or more packets try to leave an OPS node through the same output port on the same wavelength at the same time. Contention can be resolved by exploiting the time, wavelength, and space dimensions or any combination thereof. More precisely, contention in OPS nodes can be resolved by using buffering (time dimension), wavelength conversion (wavelength dimension), and/or deflection routing (space dimension). Note that the latter one may be also viewed as a special case of buffering where the OPS network is used as a buffer to store deflection-routed OPS packets. Deflection routing simplifies the OPS node architecture in that no buffers are needed at any OPS node. At the downside, however, deflection-routed packets generally consume more network resources, incur higher delays, and may require packet reordering mechanisms at the destination nodes in order to achieve in-order packet delivery. In the following, we elaborate on the various contention resolution techniques used in OPS networks.

10.2.1   Buffering

According to the position of the optical buffer, OPS nodes can be classified into four major configurations: (1) output buffering, (2) shared buffering, (3) recirculation buffering, and (4) input buffering (Hunter et al., 1998). The advantages and disadvantages of all four buffering schemes are described in the following.

- **Output Buffering:** An output-buffered OPS node consists of a space switch with a buffer on each output port, as shown in Fig. 10.3(a). At any given time, zero or more packets may arrive destined for a particular output port, whereby all of them are placed

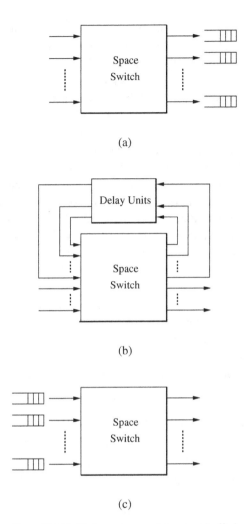

(a)

(b)

(c)

Figure 10.3 Buffering schemes: (a) output buffering, (b) recirculation buffering, and (c) input buffering.

in the corresponding output buffer. If a packet arrives at a full output buffer, the packet is discarded, resulting in packet loss. Typically, acceptable probabilities for a given packet being lost at a single OPS node are in the range of 10^{-10} to 10^{-11}, depending on the application. A packet needs to be output buffered and experiences queueing delay due to contention when more than one packet is destined for the same output port at the same time. Many OPS node architectures are based on output buffering, although OPS nodes usually emulate an output-buffered space switch by means of *virtual output queueing* (VOQ). VOQ is usually deployed in input-buffered OPS nodes, as discussed shortly.

- **Shared Buffering:** Shared buffering may be viewed as a form of output buffering, where all output buffers share the same memory space. In doing so, the buffering capacity is not restricted to the number of packets in each individual buffer but to the total number of packets in all buffers together. Shared buffering is commonly used in electronic switches using RAM. In the optical domain, shared buffering may be realized by using fiber delay lines (FDLs) that are shared among all output ports. Shared buffered OPS nodes are able to achieve a significantly reduced packet loss performance with much smaller switch sizes and fewer FDLs than their output-buffered counterparts (T. Zhang et al., 2006).

- **Recirculation Buffering:** In recirculation buffering, a number of recirculating optical loops from some of the switch output ports are fed back into the switch input ports, as shown in Fig. 10.3(b). Each optical loop has a certain delay (e.g., one packet). In recirculating buffered OPS nodes contention is resolved by placing all but one packet into the recirculating loops whenever more than one packet simultaneously arrive at the switch input ports destined for the same switch output port. Recirculating packets are forwarded onto the intended switch output port as soon as contention clears. Recirculation buffering helps resolve contention at the expense of optical signal degradation since looped-back optical packets need to pass the delay units and space switch more than once.

- **Input Buffering:** Figure 10.3(c) depicts the final buffering scheme, input buffering. It consists of a space switch with a buffer attached to each separate input port. The fundamental drawback of input buffering is *head-of-line* (HOL) *blocking*. HOL blocking occurs when the packet at the head of an input queue cannot be forwarded to its intended output port due to current contention. The packet has to be stored in the input queue until there is no more contention at the intended output port. As a consequence, the HOL packet blocks other packets within the same input buffer whose intended output ports are free of contention. Those packets need to wait until the HOL packet is forwarded. As result, input-buffered OPS nodes suffer from a decreased throughput and increased delay and packet loss. Input buffering is usually never proposed for optical networks, primarily because of its poor performance. In principle, it is possible to enhance input buffering with a so-called *look-ahead capability* which allows for selecting packets other than those at the head of each input buffer to be forwarded to their corresponding output ports. However, the optical implementation of look-ahead-capable input buffers for OPS networks appears to be too complex. Alternatively, VOQ could be deployed in input-buffered OPS nodes, where each input buffer is replaced with multiple VOQ buffers, one for each switch output port. Arriving optical packets are placed in the corresponding VOQ buffers according to their intended switch output port. Consequently, each VOQ buffer contains only packets destined for the same switch output port, thus avoiding HOL blocking.

Typically, optical buffers are implemented by using an array of FDLs of different lengths or SDLs. Using fibers of variable length to store variable-size optical packets is somewhat tricky since each optical delay line is of fixed length. Once a packet has entered the optical delay line, it can be retrieved only a fixed time period later equal to the propagation delay of the delay line. This constraint poses some limitations on the realization of optical buffers and resource efficiency of optical variable-size packet switching networks.

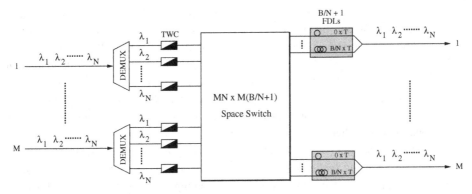

Figure 10.4 OPS node architecture with input tunable wavelength converters (TWCs) and output fiber delay lines (FDLs). After Danielsen et al. (1998). © 1998 IEEE.

All the aforementioned optical buffering schemes can be implemented in either *single-stage* or *multiple-stage* OPS nodes in a *feed-forward* or *feedback* configuration. In a feed-forward configuration, an FDL feeds forward to the next stage of the OPS node. Thus, in a feed-forward configuration, optical packets travel from one end of the OPS node to the other, involving a constant number of FDL traversals. Whereas in a feedback configuration, an FDL sends optical packets back to the input of the same stage, implying that the number of FDL traversals generally differs between optical packets (Hunter et al., 1998).

10.2.2   Wavelength conversion

Besides buffering, wavelength conversion is another approach to resolve contention in OPS network nodes by converting optical packets destined for the same output port to different wavelengths. Wavelength conversion can be applied either in conjunction with buffering or without. Figure 10.4 depicts an OPS node architecture that deploys tunable wavelength converters (TWCs) at each input port of the space switch and optical output buffers realized by means of FDLs of different lengths, proposed and investigated in Danielsen et al. (1998). The OPS node consists of three main blocks: (1) the first block uses a wavelength demultiplexer (DEMUX) at each of the M input fibers to separate optical packets arriving on wavelengths $\lambda_1, \lambda_2, \ldots, \lambda_N$, and feeds them into separate TWCs, one for each wavelength; the TWCs address free space available in the optical output buffers by converting incoming packets to the appropriate wavelengths; (2) the second block consists of a nonblocking space switch with $M \cdot N$ input ports and $M \cdot (B/N + 1)$ output ports, where B is assumed to be an integer multiple of N and $B/N + 1$ denotes the number of FDLs used in each optical output buffer; and (3) the third block comprises M FDL arrays, one for each of the M output ports. Each FDL array consists of $B/N + 1$ delay lines of different lengths. Specifically, each array is built of B/N FDLs whose lengths differ by the transmission time of a packet, T, as well as one additional fiber with an infinitesimal small length. It was shown in Danielsen et al. (1998) that by using TWCs the packet loss performance of OPS nodes is improved, especially for an increasing number of wavelengths. Furthermore, by exploiting the wavelength dimension by means of TWCs, the required number of FDLs in OPS nodes can be reduced. In fact, the flexibility gained by using TWCs allows for realizing OPS nodes without any FDLs, resulting in bufferless OPS node and network architectures.

Most of today's available optical wavelength conversion technologies, for example, four-wave mixing (FWM), cross-gain modulation (XGM), or cross-phase modulation (XPM) in semiconductor optical amplifiers (SOAs), suffer from physical limitations. Due to the fact that the performance of optical wavelength converters strongly depends on the combination of input and output wavelengths, more realistic OPS networks can be built by using *limited-range wavelength converters* (LRWCs) (Eramo et al., 2005). LRWCs allow for the conversion of any given input wavelength only to a limited range of output wavelengths. An LRWC translates the wavelength of an arriving optical packet to a limited set of adjacent output wavelengths. Similar to buffers, LRWCs can be shared in order to reduce the number of required LRWCs, at the expense of increased control algorithm complexity. In Eramo et al. (2005), the effectiveness of sharing LRWCs was examined for two different bufferless OPS node architectures. In the first proposed OPS node architecture, called shared per node, all LRWCs are placed together in a converter bank at the output of the space switch. The converter bank can be accessed by any arriving optical packet by configuring the space switch accordingly. Only packets requiring wavelength conversion are directed to the converter bank. After wavelength conversion, the optical packets are switched to the intended output port by using another space switch of smaller degree. In the second considered OPS node architecture, referred to as shared per output fiber, each output fiber is equipped with a dedicated converter bank which can only be accessed by optical packets directed to the corresponding output fiber. The shared-per-output-fiber approach provides smaller savings of used LRWCs than the shared-per-node approach, but requires less complex control algorithms. It was shown in Eramo et al. (2005) that it is sufficient to deploy a reduced number of LRWCs with a small wavelength conversion range at the OPS node in order to achieve the packet loss performance of an OPS node that is fully equipped with full-range wavelength converters.

The use of wavelength converters gives rise to the question of which of the available output wavelengths a given input wavelength should be converted to. Heuristics might be used to match input and output wavelength sets of wavelength converters (e.g., first-fit or random wavelength assignment algorithms). The output wavelength assignment problem for an OPS node using a reduced number of shared LRWCs was studied in greater detail in Gordon and Chen (2006). The proposed optimal matching algorithm is based on the well-known Hungarian algorithm and it was shown to yield minimum packet loss probabilities.

10.2.3   Unified contention resolution

We have seen that optical buffering and wavelength conversion are effective approaches to resolve contention in OPS networks. Both techniques can be applied jointly or separately and have their respective merits and shortcomings. Unlike electrical RAM, optical buffers realized by FDLs offer only fixed and finite amounts of delay. To realize optical buffers of large capacity, OPS nodes needed to deploy a large number of FDLs. Using FDLs not only deteriorates the optical signal quality but also increases delay and may cause packet reordering. On the other hand, wavelength conversion does not introduce any significant delay increase and avoids packet reordering. However, optical buffering and wavelength conversion have in common that they require additional hardware at the OPS node. Optical buffering and wavelength conversion require the OPS node to be equipped with FDLs and TWCs, respectively.

Another approach to resolve contention in OPS networks without requiring hardware upgrades at any OPS node is deflection routing, which can be easily done in software. However, with deflection routing, optical packets may arrive out of order. Furthermore, deflection routing is less resource efficient than buffering and wavelength conversion and its effectiveness strongly depends on the given network topology and traffic condition. Furthermore, to prevent deflected packets from looping endlessly throughout the OPS network certain mechanisms must be deployed; for example, a maximum-hop-count field may be used in the packet header which is decremented at each intermediate OPS node such that the packet is discarded when the value of the field equals zero after the packet has traveled the maximum number of hops.

In Yao et al. (2003a), a unified contention resolution across wavelength, time, and space domains for irregular mesh network topologies was examined. The proposed unified contention resolution explored various combinations of all three dimensions in terms of throughput, packet loss, and average hop distance. The obtained results indicate that wavelength conversion is the preferred technique to resolve contention in OPS networks, especially for an increasing number of wavelengths and under heavy traffic loads. Importantly, the study showed that deflection routing can be a good approach to resolve contention in OPS networks with high-connectivity topologies, but it is less effective in low-connectivity topologies. In general, deploying only deflection routing does not lead to significant performance improvements and it was shown to be the least effective approach to resolve contention in OPS networks. However, deflection routing combined with wavelength conversion and possibly also optical buffering is a powerful approach to build high-performance OPS networks. Among all the considered combinational schemes, wavelength conversion, combined with carefully designed optical buffering and deflection routing at selected OPS nodes, appears to be the best approach to resolve contention in OPS networks.

10.3   Service differentiation

Similar to contention resolution, service differentiation in OPS networks can be achieved by exploiting the time, wavelength, and/or space dimensions. Service differentiation in OPS networks is closely related to the techniques chosen to resolve contention and may be achieved by using one or more of the aforementioned dimensions.

Several quality-of-service (QoS) differentiation schemes for asynchronous OPS networks that exclusively utilize the wavelength dimension for contention resolution without resorting to buffering and deflection routing were studied in Øverby et al. (2006). The proposed schemes focused on per-class classification of the traffic. The following three QoS differentiation schemes were examined:

- **Access Restriction:** With access restriction, a subset of the available resources (e.g., wavelengths, wavelength converters, or FDLs) is exclusively reserved for high-priority traffic. For example, let us consider two traffic classes, high-priority traffic class 0 and low-priority traffic class 1. Furthermore, let N denote the number of wavelengths on each output fiber, whereby $n < N$ wavelengths are exclusively reserved for traffic class 0. As long as less than $N - n$ wavelengths on a given output fiber are occupied both new class 0 and class 1 packet arrivals destined for that output fiber are accepted. Otherwise, only class 0 packets are accepted, while class 1 packets are discarded, resulting in a lower packet loss probability for traffic class 0 than for traffic class 1.
- **Preemption:** With preemption, all free resources are available to all traffic classes. However, when all resources are occupied, a high-priority packet is allowed to preempt a resource currently occupied by a low-priority packet, which is then discarded. Consequently, on average fewer resources are available to low-priority packets than to high-priority packets, resulting in a lower packet loss probability for high-priority packets.
- **Packet Dropping:** With packet dropping, an OPS node drops low-priority packets with a certain probability before attempting to utilize any resources. This results in an increased packet loss probability for low-priority traffic, but also a decreased packet loss of high-priority traffic since the total traffic load is reduced.

The obtained results showed that preemption yields the best performance in terms of packet loss probability, followed by access restriction and packet dropping schemes, at the expense of an increased implementation complexity.

We note that a slightly modified version of the aforementioned access restriction scheme was applied to achieve QoS differentiation in bufferless synchronous OPS networks using full-range wavelength converters (Øverby, 2005).

Several methods to implement different levels of QoS in asynchronous OPS networks that deploy not only wavelength converters but also FDLs for contention resolution were examined in Callegati et al. (2002). The proposed schemes combined the wavelength and time dimensions to differentiate the packet loss probability between different traffic classes. It was shown that the wavelength domain is more effective than the time domain to realize QoS differentiation in OPS networks. We made a similar observation earlier when discussing contention resolution in OPS networks.

10.4   Self-routing

Current optical logic devices are able to execute only simple logic functions such as AND, OR, and XOR. Due to their limited capability and the fact that most of them are bulky and difficult to integrate, pure OPS networks, where the packet header is processed in the optical domain rather than being OEO converted, may be built in the near future by deploying simple single-bit optical processing schemes that avoid complex optical logic circuits. One such approach that simplifies routing in OPS networks and requires optical single-bit header processing is the so-called *self-routing* address scheme (Yuan et al., 2003). In the self-routing address scheme, each output port of all OPS nodes is associated with a bit in the optical packet header. Let N denote the number of nodes and K denote the number of bidirectional links in the considered OPS network. OPS nodes are labeled from 1 to N. The output ports of OPS node i are labeled from 1 to $d(i)$, where $d(i)$ denotes the number of output ports of node i. Thus, the total number of output ports in the OPS network is given by $\sum_{i=1}^{N} d(i) = 2K = N \cdot D$, where $D = 2K/N$ denotes the average output degree of an OPS node. Each output port of all OPS nodes has a one-to-one correspondence with a bit in the address carried in the packet header. The address contains $2K$ bits, which are grouped into N address subfields. Each subfield corresponds to a different OPS node. Let us assume OPS node i sends a packet to OPS node j. In this case, all bits in the address subfields of OPS nodes i and j are set to 0. In the address subfield of intermediate OPS node k, $k \neq i, j$, the lth bit is set to 1 to indicate that the optical packet sent by OPS node i exits OPS node k through output port l on its way to OPS node j. The other bits in the address subfield of OPS node k are set to 0. Each OPS node optically processes the header of an arriving packet bit by bit and forwards the packet to the output port with its corresponding address bit set to 1. More precisely, when an intermediate OPS node receives a packet, it only processes its own address subfield and forwards the packet to the specified local output port. OPS node j recognizes that a packet is destined for it since its address subfield consists of all zeros, indicating that the packet does not need to be forwarded.

Note that the self-routing address scheme allows multiple addresses for the same OPS node, whereby each address encodes a different path. The multiple paths may be used for traffic engineering and fast rerouting in case of network failures. The self-routing address scheme is simple and can be implemented in arbitrary network topologies using any routing protocol. However, the length of the address limits the self-routing approach to networks with a small to medium number of OPS nodes.

10.5 Example OPS node architectures

Various OPS node architectures have been proposed and investigated over the past years. In this section, we will highlight some example OPS node architectures. Based on the switching fabric, they can be classified into the following three categories: *space switch*, *broadcast-and-select*, and *wavelength-routing* OPS node architectures (Xu et al., 2001).

10.5.1 Space switch architecture

Figure 10.5 depicts an example synchronous space switch OPS node architecture with N incoming and N outgoing fiber links, each carrying W wavelength channels. The

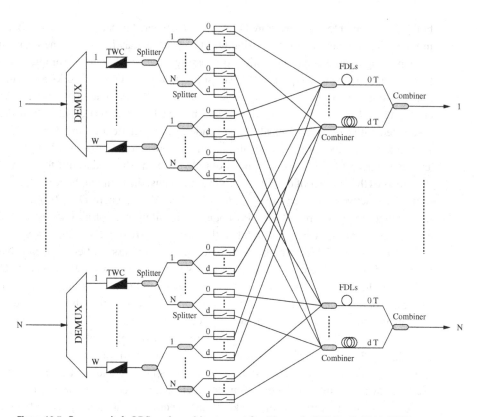

Figure 10.5 Space switch OPS node architecture. After Xu et al. (2001). © 2001 IEEE.

operation of the switch is slotted. The length of each slot is equal to the transmission time T of a fixed-size packet. For each incoming fiber link a wavelength DEMUX is used to separate the incoming optical signal into W different wavelengths. Each wavelength is fed to a different TWC which converts the wavelength of the arriving optical packet to a wavelength that is free at the intended optical output buffer. The optical packet is then switched to the intended optical output buffer through the space switch fabric which consists of optical splitters and optical gates. Specifically, the output signal of a TWC is fed into a splitter which distributes the signal to N different output fibers, one per OPS node output port. The signal on each of these output fibers goes through another optical splitter which distributes it to $d + 1$ different output fibers. Each of these output fibers is connected through an optical gate to one of the FDLs of the destination output port. An optical packet is forwarded to a specific FDL by keeping the corresponding optical gate open while closing the remaining gates. The information to which wavelength the TWC should convert the wavelength of an arriving optical packet and to which FDL the packet should be switched to is provided by the control unit of the OPS node. Each output buffer consists of $d + 1$ FDLs, numbered from 0 to d. FDL i delays an optical packet for a fixed period of time equal to i slots, each of duration T, where $i = 0, 1, \ldots, d$. Note that FDL 0 introduces zero delay, and an optical packet arriving at this FDL is immediately transmitted out of the OPS node output port. Also, note that each FDL can delay optical

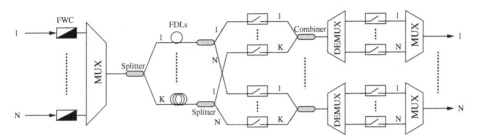

Figure 10.6 Broadcast-and-select OPS node architecture. After Xu et al. (2001). © 2001 IEEE.

packets on any of the W wavelengths by using an optical combiner placed in front of it. The combiner is used to collect optical signals on all W wavelengths. Thus, each FDL can accept up to W optical packets simultaneously arriving in the same slot.

10.5.2   Broadcast-and-select architecture

Figure 10.6 shows an example synchronous broadcast-and-select OPS node architecture. Each of the N input and N output fibers carries only one wavelength. The wavelength at an output port is not fixed and may vary from packet to packet. Therefore, an output interface (not shown in the figure) is required to translate it to the desired transmission wavelength. On each input fiber a fixed wavelength converter (FWC) is used. At the beginning of a slot, each FWC converts the wavelength of an arriving optical packet to a fixed wavelength. After wavelength conversion, a wavelength multiplexer (MUX) collects the optical signals and feeds them into a common fiber, followed by an optical splitter. The splitter distributes the combined optical signal to K different FDLs. Each FDL has a different delay equal to an integer multiple of slots. At the beginning of the next slot, a maximum of $N \cdot K$ optical packets exit from the K FDLs and up to N of them are directed to their destination output ports without any collisions. This is achieved through a combination of optical splitters, optical gates, wavelength DEMUXs, and wavelength MUXs. Specifically, the output signal of each FDL goes through a splitter which distributes it to N outputs. Each of these output signals consists of up to N optical packets, one on each wavelength. The signal from output i of each splitter is directed to OPS node output port i. Since there are K optical splitters, one is selected from the K output signals by using optical gates. The selected output signal is then directed to output port i. The selected output signal is fed into a wavelength DEMUX which separates it into N wavelengths, of which only one is transmitted out by opening the corresponding gate while closing the remaining gates. The operation of the optical gates is managed by the control unit of the OPS node.

The aforementioned broadcast-and-select OPS node architecture can be extended to M input and M output fibers, whereby each input and output fiber carries $W > 1$ wavelengths. This is done by demultiplexing the optical WDM signal arriving on each input fiber into W wavelengths and then using the switch as if it had $M \cdot W$ input wavelengths instead of N. At the output ports of the extended broadcast-and-select OPS

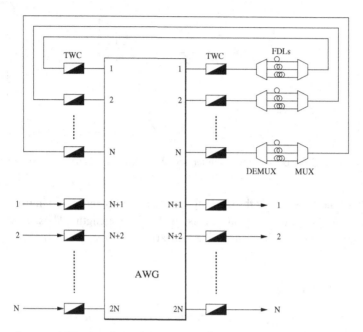

Figure 10.7 Wavelength-routing OPS node architecture. After Xu et al. (2001). © 2001 IEEE.

node, each set of W wavelengths is combined onto the same output fiber by using an additional combiner.

10.5.3   Wavelength-routing architecture

Figure 10.7 depicts an example wavelength-routing OPS node architecture with N single-wavelength input ports and N single-wavelength output ports. The switch consists of a $2N \times 2N$ AWG, N FDL arrays, and $4N$ TWCs. In general, the switching takes place in two phases. First, optical packets are routed to the FDLs to resolve contention, if applicable. Second, optical packets are routed to the intended output ports. These two phases are implemented by using a $2N \times 2N$ AWG together with N FDL arrays. The $2N$ TWCs on the left-hand side of the AWG are used to select the intended output port of the AWG. The upper N TWCs on the right-hand side of the AWG are used to select the FDL with the correct delay for recirculating optical packets. The lower N TWCs on the right-hand side of the AWG are required to convert the wavelengths of outgoing optical packets to the corresponding transmission wavelengths since there are more internal wavelengths than external wavelengths. It is worthwhile to note that the considered switch is able to provide service differentiation by prioritizing optical packets. After leaving the FDL, an optical packet may be delayed again because of preemption by a higher-priority optical packet.

   The preceding wavelength-routing OPS node architecture can be extended to WDM input/output ports by using additional wavelength DEMUXs and combiners to realize multiple wavelength planes. Each wavelength plane consists of a switching fabric with single-wavelength inputs and outputs. The WDM extended wavelength-routing OPS node architecture has in total N input and N output fibers, each carrying W wavelengths, and comprises W wavelength planes.

10.6   Implementation

Some of the aforementioned OPS node architectures have been successfully implemented in a number of proof-of-concept demonstrators (e.g., broadcast-and-select architecture). In this section, we highlight the implementation of a few other OPS network architectures and elaborate on their feasibility. We also provide a brief discussion on the viability of OPS networks in general and their expected application in the near, mid, and long term.

A synchronous slotted unidirectional OPS ring network using a single data wavelength channel (i.e., no WDM) and a separate control wavelength for out-of-band signaling was experimentally demonstrated in Takada and Park (2002). Due to its unidirectional ring topology and the fact that no WDM is used for data transmission, wavelength conversion and deflection routing cannot be applied to resolve contention in the proposed OPS ring network, leaving buffers as the only viable option. Instead of using optical buffers, the presented OPS network deploys electronic buffers at each source node to store locally generated data packets. The OPS ring network works as follows. When a source node detects an empty slot on the ring, the source node can use the empty slot to send a data packet on the data wavelength channel, accompanied by a control packet simultaneously transmitted on the separate control wavelength. The control packet contains the destination address of the optical data packet. While transmitted optical data packets travel all the way to the intended destination without undergoing OEO conversion at any intermediate ring node, control packets are OEO converted at each traversed ring node in order to extract the destination address of the corresponding data packet and decide whether to drop the optical packet or forward it together with its accompanying control packet. If the destination address of an incoming control packet matches that of the receiving ring node, the corresponding optical packet is dropped by setting a 2×2 optical cross-bar switch to the cross state. Otherwise, the 2×2 switch is set to the bar state to pass the optical packet through the ring node. It was experimentally demonstrated that the OPS ring network can operate error-free at a line rate of 40 Gb/s using readily available optical and electronic technologies. Furthermore, by letting one ring node be the master node, bandwidth guarantees and fairness can be achieved in the OPS ring network. This is done by letting the master node create reserved slots for individual ring nodes by writing the addresses of the individual nodes in the corresponding control packets. The reserved slots are allowed to be used only by the addressed ring nodes. In doing so, guaranteed bandwidth can be assigned to individual ring nodes and fair access can be controlled network-wide by the master node. Clearly, by increasing the number of slots reserved for individual nodes fewer empty slots are available to the remaining ring nodes for random access. To alleviate this problem, a reserved slot may be used by intermediate ring nodes other than the addressed ring node if both source and destination intermediate ring nodes are located between the master node and the addressed ring node. In this case, the slot becomes again empty at the destination intermediate node before arriving at the addressed ring node. Thus, reserved slots may be spatially reused, resulting in an increased network throughput and a decreased network access delay.

A synchronous slotted bufferless OPS ring network that makes use of WDM was studied within the European IST project DAVID (data and voice integration over DWDM) (Dittmann et al., 2003). In DAVID, multiple unidirectional slotted WDM rings are interconnected via an optical hub that is used to optically switch packets between rings. The hub comprises synchronization stages, a space switch, and regeneration stages, if necessary. Similarly to the aforementioned single-channel OPS ring, each WDM OPS ring in DAVID deploys a separate control wavelength which carries the status information of all data WDM channels in the same slot. The control information includes the state (empty or occupied) of each wavelength in the slot and the destination ring of the entire WDM slot. The destination ring of each WDM slot is set by the hub based on either explicit reservations or traffic measurements. Unlike the data wavelength channels, the control wavelength undergoes OEO conversion at all ring nodes. After OE converting and processing the control information, each ring node is able to determine the destination ring of the entire slot and which wavelengths therein are empty. Based on this information, a ring node might use an empty slot to transmit a data packet to the destination ring of the current slot. The hub acts as a nonblocking switch which is reconfigured in every time slot to forward slots between different WDM rings. Note that the operation of the interconnected WDM OPS rings resembles that of photonic slot routing (PSR) networks, where all packets in a given WDM slot are forwarded to the same destination node. PSR was described at length in Chapter 7.

The viability and cost-effectiveness of WDM OPS ring networks using currently available electronic and optical technologies were assessed and compared to alternative ring technologies such as SONET/SDH ring and Resilient Packet Ring (RPR) in Develder et al. (2004) (RPR will be explained in great detail in Chapter 11). The findings can be summarized as follows. For current and near-term capacity requirements of a few tens of Gb/s, WDM-enhanced RPR networks with OEO conversion of all wavelengths at each ring node appear to be the most advantageous solution due to their capability of spatial bandwidth reuse. In the medium to long term, however, as capacity requirements increase from a few tens to a few hundreds of Gb/s, WDM OPS ring networks are expected to become competitive due to their optical transparency.

In conclusion, OPS appears a natural candidate to carry the ever-increasingly vast amounts of data traffic directly in the optical domain without requiring complex adaptation layers to electronic clients and costly OEO conversions which largely contribute to the costs of today's communications networks. A number of different OPS node architectures have been proposed and investigated up to date, ranging from node architectures with OEO conversion of all wavelengths, OLS nodes where only control packets undergo

OEO conversion while data packets remain in the optical domain, to all-optical node architectures that provide optical packet processing capabilities. Recent studies indicate that the introduction of all-optical nodes is unlikely to take place in the near term. Apart from the lack of optical RAM, the cost of optical components must be reduced significantly to make OPS networks competitive with current state-of-the-art OLS networks, which provide a better cost-performance trade-off for the time being (Caenegem et al., 2006).

Part III

Optical metropolitan area networks

Overview

In this part, we explore a wide range of different optical metropolitan area network (MAN) architectures and protocols. MANs are found at the metro level of the network hierarchy between wide area networks (WANs) and access networks. Typically, MANs have a ring topology and are deployed in interconnected ring architectures that are composed of metro core and metro edge rings, as depicted in Fig. III.1. Each metro core ring interconnects several metro edge rings with the long-haul backbone networks. Apart from inter-metro-edge-ring traffic, metro core rings also carry traffic from and to the long-haul backbone networks. Metro edge rings in turn carry traffic between metro core rings and access networks, for example, hybrid fiber coax (HFC), fiber-to-the-home (FTTH), fiber-to-the-building (FTTB) networks, and passive optical networks (PONs). Ring networks offer simplicity in terms of operation, administration, and maintenance (OAM). Moreover, ring networks provide fast protection switching in the event of a single link or node failure.

Optical metro ring networks can be either single-channel or multichannel wavelength division multiplexing (WDM) systems. Optical ring networks were initially single-channel systems, where each fiber link carries a single wavelength channel (e.g., IEEE 802.5 Token Ring and ANSI Fiber Distributed Data Interface (FDDI)). Optical single-channel ring networks belong to the first generation of opaque optical networks where OEO conversion takes place at each node. Opaque ring networks have come a long way. Among others, the so-called Cambridge ring is a unidirectional ring network whose channel access is based on the empty-slot principle (Hopper and Williamson, 1983). The Cambridge ring deploys *source stripping*, where the source node takes the transmitted packet from the ring. Another example of opaque single-channel optical ring networks is the so-called buffer insertion ring (Huber et al., 1983). The buffer insertion ring is a unidirectional ring network where each node temporarily stores the incoming ring traffic in an electrical insertion buffer in order to allow the local node to transmit a packet on the ring. As opposed to the Cambridge ring, packets are removed from the buffer insertion ring by the receiving node (rather than transmitting node). This so-called *destination stripping* enables nodes downstream from the destination node to *spatially reuse* bandwidth, resulting in an increased network capacity.

Unlike the two aforementioned ring networks, the so-called MetaRing is a dual-fiber *bidirectional* full-duplex ring operating either in buffer insertion mode for variable-size packets or slotted mode for fixed-size packets (cells), where the slot size equals the transmission time of a fixed-size packet (cell) (Cidon and Ofek, 1993). In MetaRing,

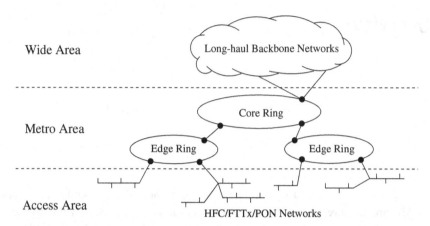

Figure III.1 Metro area networks: metro core rings interconnect metro edge rings and connect them to long-haul backbone networks.

nodes deploy destination stripping and shortest path routing by choosing the appropriate directional ring.

Chapter 11

The interest in single-channel optical rings is still high, as witnessed by the recently approved standard IEEE 802.17 Resilient Packet Ring (RPR), which represents the state of the art of optical single-channel ring networks for high-performance packet-switched optical single-channel MANs. Various aspects of the aforementioned ring networks can be found in RPR. In Chapter 11, we will describe RPR in detail, including recent research activities on performance enhancements of RPR networks.

Chapter 12

In Chapter 12, we will provide an overview of WDM ring networks. Multichannel upgraded optical ring networks have been receiving a great deal of attention, where each fiber link carries multiple wavelength channels by using WDM. WDM ring networks leverage on the existing fiber ring infrastructure and thus do not require additional fiber. Furthermore, WDM rings allow the design of all-optical (OOO) node architectures in which a part of the optical WDM comb signal, namely all wavelengths except the locally dropped wavelength(s), remains in the optical domain and does not need to be converted into the electrical domain, electronically stored and processed, and converted back to an optical signal. The resultant OOO node structures provide transparency against protocol, data rate, and modulation format. This transparency facilitates the support of a wide variety of both legacy and future traffic types, services, and applications.

Chapter 13

Both single-channel and multichannel WDM optical ring networks suffer from a number of shortcomings. Their limited recovery against only a single failure might be insufficient for MANs, which have to be extremely survivable. Survivability of optical ring networks becomes crucial in particular for storage networking protocols, which are one of the important applications without built-in adequate survivability that rely almost entirely on the failure recovery techniques of the optical layer. In addition, ring networks are poorly suited to support unpredictable traffic, which stems from events that are hard to predict by current traffic forecasting techniques (e.g., breaking news, flash crowd events, and denial-of-service attacks). These shortcomings call for topological modifications of ring networks. In Chapter 13, we will introduce an alternative approach to multichannel upgrade optical single-channel ring networks, where a subset of ring nodes is interconnected by a single-hop star WDM subnetwork. The resultant hybrid ring–star network, called RINGOSTAR, is able to mitigate most shortcomings of conventional optical single-channel and WDM ring networks, as discussed in greater detail in Chapter 13.

11　Resilient packet ring

The IEEE standard 802.17 Resilient Packet Ring (RPR) aims at combining SONET/SDH's carrier-class functionalities of high availability, reliability, and profitable TDM service (voice) support with Ethernet's high bandwidth utilization, low equipment cost, and simplicity (Davik et al., 2004; Yuan et al., 2004; Spadaro et al., 2004). RPR is a ring-based architecture consisting of two counterdirectional optical fiber rings with up to 255 nodes. Similar to SONET/SDH, RPR is able to provide fast recovery from a single link or node failure within 50 ms, and carry legacy TDM traffic with a high level of quality of service (QoS). Similar to Ethernet, RPR provides advantages of low equipment cost and simplicity and exhibits an improved bandwidth utilization due to statistical multiplexing. The bandwidth utilization is further increased by means of spatial reuse. In RPR, packets are removed from the ring by the corresponding destination node (destination stripping). The destination stripping enables nodes in different ring segments to transmit simultaneously, resulting in spatial reuse of bandwidth and an increased bandwidth utilization. Furthermore, RPR provides fairness, as opposed to today's Ethernet, and allows the full ring bandwidth to be utilized under normal (failure-free) operation conditions, as opposed to today's SONET/SDH rings where 50% of the available bandwidth is reserved for protection. Current RPR networks are single-channel systems (i.e., each fiber ring carries a single wavelength channel) and are expected to be primarily deployed in metro edge and metro core areas.

In the following sections, we explain RPR in greater detail, paying particular attention to its architecture, access control, fairness control, and protection. We thereby not only highlight the salient features of RPR as specified in IEEE 802.17 but also review recently proposed performance enhancements of RPR's fairness control and protection.

11.1　Architecture

RPR is a bidirectional packet-switched ring network consisting of two counterrotating fiber rings, called ringlet 0 and ringlet 1. RPR is a MAC layer protocol which is specified for several physical layers such as Ethernet and SONET/SDH. RPR accommodates up to $N = 255$ ring nodes. Figure 11.1 illustrates the destination stripping and spatial reuse of an RPR network with $N = 8$, where node 1 sends data packets to node 3 and node 6 sends data packets to node 8 on the clockwise fiber ring simultaneously.

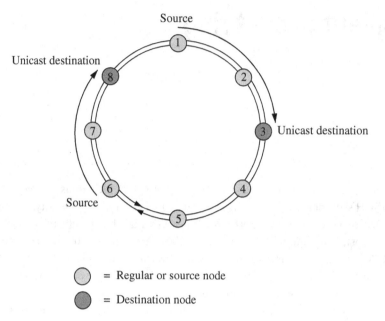

Figure 11.1 Bidirectional RPR network with destination stripping and spatial reuse.

For unicast traffic, the source node decides on which of the two ringlets it sends locally generated data packets to the corresponding destination node. Typically, a source node selects the ringlet that provides the shortest path to the destination node in terms of hops (i.e., the source node applies shortest path routing). In general, however, a source node is free to choose either ringlet unless the ringlet is specified by the medium access control (MAC) client. The transmitted data packet travels to the destination node on the selected ringlet. If an intermediate ring node does not recognize the destination MAC address in the packet header, the packet is forwarded to the next ring node. RPR supports both cut-through and store-and-forward methods. In the former method, an intermediate ring node starts to forward the packet before it is completely received. In the latter method, the packet is completely received by an intermediate ring node before it starts to forward the packet to the downstream ring node. To prevent packets whose destination MAC address is not recognized by any ring node from circulating on the ring forever, packets carry a supplementary 1-byte time-to-live (TTL) field which is added to each packet by the RPR MAC control entity and whose value is decremented by each intermediate ring node. When the destination node recognizes the destination MAC address it takes the packet from the ring (destination stripping), thereby enabling downstream nodes to spatially reuse the bandwidth. For example, in Fig. 11.1 destination node 3 takes packets destined for it from the ring and the available bandwidth is spatially reused by node 6 to send packets to node 8.

Multicasting is realized in RPR by means of broadcasting. More specifically, a multicast packet is transmitted to a set of nodes whose multicast group membership is identified by a group MAC address carried within the destination address field of the

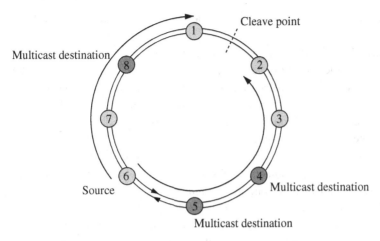

Figure 11.2 Multicasting in an RPR network.

multicast packet. A multicast packet is sent in one of two ways: (1) *unidirectional flooded* or (2) *bidirectional flooded*. A flooded packet traverses a sequence of nodes and is stripped from the ring based on the expiration of the TTL field. The TTL field expires when it reaches zero, which indicates that the packet has passed through the intended number of nodes. In unidirectional flooded multicast transmission, the multicast packet is sent to all other nodes on either the clockwise or counterclockwise ringlet. The multicast packet is taken from the ring by either the source node based on a matching source MAC address or an expired TTL field, whichever comes first, or the node just short of the source node based on an expired TTL field. In bidirectional flooded multicast transmission, the multicast packet is generally replicated and sent to all other nodes on both clockwise and counterclockwise ringlets. A bidirectional flooded multicast packet is stripped from the ring based on the expiration of the TTL field at an agreed upon span, called a *cleave point*. Both multicast packets are not allowed to overlap their delivery and therefore cannot be sent beyond the cleave point. The overlap is prevented by setting the TTL values in both multicast packets to the number of hops to the nodes just short of the cleave point. In general, the cleave point can be put on any span of the ring. Figure 11.2 depicts an example multicast scenario, where source node 6 sends a multicast packet to destination nodes 4, 5, and 8 by means of bidirectional flooding. The two counterpropagating multicast packets are taken from the bidirectional ring by nodes 1 and 2 short of the cleave point. Note that the cleave point can also be put immediately adjacent to the source node. In this case, a given multicast packet does not need to be replicated since it is sent only in one direction. Obviously, broadcasting does not allow for spatial reuse. To achieve multicasting with spatial reuse, the TTL values of multicast packets must be set such that they expire after the two final multicast destination nodes are reached. In doing so, the multicast packets are stripped from the ring by both final multicast destination nodes instead of being forwarded onward to the cleave point to all remaining nodes that do not belong to the respective multicast group. As a result, the multicast packets consume bandwidth only up to the final multicast destination node in either direction, enabling spatial reuse on the remaining part of the ring (Maier et al., 2006).

To set the TTL field of a given packet to the appropriate value a ring node may use the information provided by RPR's *topology discovery protocol* (Davik et al., 2004). The topology discovery protocol determines the connectivity and ordering of RPR nodes around the ring. The topology information is stored in the topology database of each node. At system initialization, all nodes broadcast topology discovery control packets on both ringlets with an initial TTL value equal to 255 (i.e., the maximum number of nodes). The topology discovery packet contains information about the status of the corresponding node and its attached links. Each topology discovery packet is received by all nodes, enabling them to compute a complete topology image of the RPR network. The topology image provides each node not only with the number and ordering of all ring nodes but also the status of each link. When a new node is inserted into the ring or a node detects a link/node failure, it will immediately transmit a topology discovery packet. If a node receives a topology discovery packet that is inconsistent with its current topology image, it will also immediately transmit a new topology discovery packet that contains its own status. Thus, the first node that notices a change starts a ripple effect, resulting in all other nodes sending their updated topology discovery packets and recomputing their topology images. Once the topology image is stable (i.e., it does not change during a specified time period), the node performs a consistency check. For example, the node will make sure that the topology information collected on each ringlet matches each other. After achieving stable and consistent conditions, each node continues to periodically send topology discovery packets to guarantee robust operation. Among others, the acquired topology database can be used by each source node to determine the number of hops and appropriate TTL value or to select the shortest path to any destination node.

Figure 11.3 depicts the architecture of an RPR node for one ringlet; the node architecture is replicated for the other ringlet (Yuan et al., 2004). Local ingress traffic, which can be of class A, B, or C, as described in Section 11.2, entering an RPR node is first throttled by rate controllers, which are token bucket traffic shapers. Each RPR node uses a separate rate controller for each destination node. That is, RPR nodes deploy virtual output queueing (VOQ) in order to avoid head-of-line blocking. Ingress traffic is throttled by the RPR node if it traverses a congested link along the selected path. Ingress traffic ready to be transmitted is placed in the so-called stage queue. The transit path of each RPR node consists of one or two transit queues which temporarily store packets transiting the RPR node. In the single-queue mode, the transit path consists of a single first-in/first-out (FIFO) queue called the primary transit queue (PTQ). In the dual-queue mode, the transit path consists of the PTQ and an additional FIFO queue referred to as the secondary transit queue (STQ). The PTQ stores class A in-transit traffic, while the STQ stores class B and class C in-transit traffic. Note that an RPR network may consist of both single-queue and dual-queue nodes; that is, RPR nodes are not required to have the same type of transit queues. The checker at the transit input port of each RPR node is used to perform various operations for incoming packets. Among others, the checker performs error detection, decides whether an arriving packet

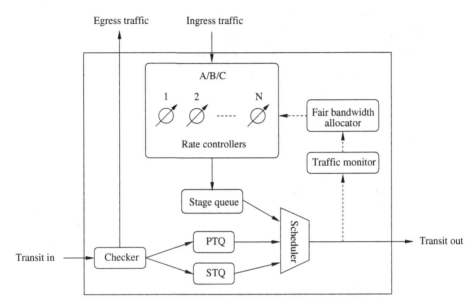

Figure 11.3 RPR node architecture. After Yuan et al. (2004). © 2004 IEEE.

is locally dropped or forwarded, decrements the TTL field, and places the packet in the correct transit queue. The scheduler arbitrates service among the stage queue, PTQ, and STQ, as discussed in detail in the subsequent section. Each RPR node deploys a traffic monitor (byte counter) which measures serviced in-transit traffic and serviced local ingress traffic. The traffic measurements are used by the fairness control algorithm to compute a feedback control signal for the local rate controllers as well as the rate controllers of upstream RPR nodes in order to alleviate congestion. Fairness control will be discussed at length in Section 11.3. Note that one of the major design objectives of RPR was to ensure hardware simplicity. This design goal was achieved by using one or two simple FIFO transit queues that are able to provide a lossless transit path; that is, a packet once injected on the ring is not dropped at intermediate nodes, as explained in the following section.

11.2 Access control

As briefly mentioned in the previous section, RPR supports three traffic classes, A, B, and C. The three traffic classes are assigned different priorities: class A is given high priority, class B is given medium priority, and class C is given low priority. The objectives of the class-based priority scheme are to achieve QoS support and service differentiation in RPR. Specifically, class A offers a low latency and low jitter service with guaranteed bandwidth, class B is used to provide service with predictable latency and jitter, and class C is used for best-effort traffic. These services may be used to support a wide range of applications in RPR networks. Class A service is designed to support real-time

Table 11.1. Traffic classes in RPR

Traffic class	Subclass	Bandwidth preallocation
A	A0	Reserved
	A1	Reclaimable
B	B-CIR	Reclaimable
	B-EIR	No preallocation
C	–	No preallocation

applications that require guaranteed bandwidth and low latency and low jitter (e.g., voice). Class B service is dedicated to near-real-time applications that are less delay sensitive but require some bandwidth guarantees (e.g., video on demand). And finally, class C service implements best-effort data transport (e.g., web browsing) (Spadaro et al., 2004).

Traffic classes A and B are further divided into two subclasses, as shown in Table 11.1. Class A traffic is divided into classes A0 and A1, and class B traffic is divided into classes B-CIR (committed information rate) and B-EIR (excess information rate). Traffic classes B-CIR and B-EIR differ in that B-CIR provides guaranteed service at the committed information rate and best-effort service for excess traffic beyond the committed information rate. Thus, traffic classes A0, A1, and B-CIR provide service guarantees, while traffic classes B-EIR and C provide best-effort transport. To fulfill the service guarantees, bandwidth is preallocated for traffic classes A0, A1, and B-CIR in RPR (Davik et al., 2004). Specifically, bandwidth preallocated for class A0 traffic is called *reserved*. The bandwidth reservation for class A0 traffic of a given RPR node is broadcast on the ring using RPR's topology discovery protocol. After receiving the topology discovery packets from all ring nodes, every RPR node calculates how much bandwidth needs to be reserved for class A0 traffic. Reserved bandwidth can be used only by the RPR node which has made the reservation. If the RPR node does not use its reserved bandwidth, it cannot be used by other nodes and is therefore wasted. The remaining bandwidth is called *unreserved* and can be used by all other traffic classes. Unreserved bandwidth is in part preallocated for class A1 and B-CIR traffic, which is referred to as *reclaimable* bandwidth. Reclaimable bandwidth not used by class A1 and class B-CIR traffic as well as the remaining bandwidth not preallocated may be used by class B-EIR and class C traffic. Table 11.1 summarizes the different RPR traffic classes and their preallocated bandwidth. Note that in RPR no bandwidth is preallocated for class B-EIR and class C traffic. The two traffic classes B-EIR and C are called *fairness eligible* (FE), because traffic belonging to these two classes is controlled by the RPR fairness control algorithm, as described in Section 11.3.

Access control in RPR networks works as follows. In single-queue mode, highest priority is given to local control traffic if the PTQ is not full. In the absence of local control traffic, priority is given to in-transit ring traffic over local ingress traffic. In dual-queue mode, if both PTQ and STQ are not full, highest priority is given to local control traffic (similar to single-queue mode). If there is no local control traffic, PTQ traffic is

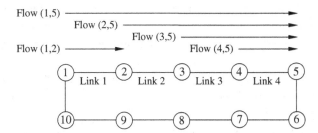

Figure 11.4 Fairness and spatial reuse illustrated by the parallel parking lot scenario. After Yuan et al. (2004). © 2004 IEEE.

always served first. If the PTQ is empty, local ingress traffic is served until the STQ reaches a certain queue threshold. If the STQ reaches that threshold, STQ in-transit ring traffic is given priority over local ingress traffic such that in-transit packets are not lost due to buffer overflow. Thus, under stable network conditions the transit path is lossless and each packet put on the ring, no matter which traffic class it belongs to, is not dropped at downstream ring nodes. In particular, note that a class A packet will mostly experience only the propagation delay and occasional queueing delay while waiting for local ingress traffic to be completely transmitted by intermediate ring nodes. In single-queue mode, where RPR nodes have only one transit queue (PTQ), in order for class A traffic to move quickly around the ring, the transit queues in all ring nodes should be almost empty. This can be achieved by letting all class A traffic be of subclass A0 such that there will be sufficient bandwidth reserved along the selected path (Davik et al., 2004).

11.3   Fairness control

We have seen that in stable operation RPR provides a lossless transit path, which ensures that a packet put on the ring is not dropped at intermediate ring nodes and is successfully received by the intended destination node. On the downside, however, an intermediate node ready to send data has to wait for the transit path to become empty before it can send its data. As a consequence, upstream nodes can easily starve downstream nodes by continuously sending data and thereby preventing downstream nodes from accessing the ring network, giving rise to fairness problems. To resolve them, RPR deploys a distributed fairness control protocol that dynamically throttles traffic in order to achieve fairness while maintaining spatial reuse. The following description of the RPR fairness control is explained in greater detail in Yuan et al. (2004).

To illustrate the design objectives of RPR's fairness control protocol, Fig. 11.4 depicts the so-called parallel parking lot scenario, where four infinite demand flows are assumed to share link 4 en route to destination node 5. To achieve fairness, each of these four flows should be equally allocated 1/4 of the bandwidth of link 4. Moreover, to fully exploit spatial reuse, flow (1,2) should use the entire excess bandwidth on link 1 which amounts to 3/4 of the bandwidth of link 1; the remaining 1/4 of the bandwidth of link 1 is assigned to flow (1,5). Note that in our example we assumed that each flow is assigned an equal share of bandwidth on link 4. In general, however, the RPR fairness control algorithm is able to provide not only equal fairness but also weighted fairness, where ring nodes are assigned different fractions of bandwidth according to their given weights.

To achieve fairness with maximum spatial reuse in RPR, a distributed fairness control algorithm that dynamically allocates bandwidth to ring nodes is deployed according to the *ring ingress aggregated with spatial reuse* (RIAS) reference model. In RIAS, the level of traffic granularity for fairness determination at a link is defined as an ingress aggregated (IA) flow, that is, the aggregate of all flows originating from a given ingress ring node. Furthermore, in RIAS bandwidth can be reclaimed by IA flows when it is unused to ensure maximum spatial reuse. The goal of the RPR fairness control protocol is to throttle flows at their ring ingress nodes to their network-wide fair rates and thereby alleviate congestion. The fairness control in RPR is realized by enabling a backlogged ring node to send fairness control packets based on its local measurements to upstream ring nodes.

The RPR fairness control algorithm operates in either of the following two operation modes: (1) *aggressive mode* (AM) or (2) *conservative mode (CM)*. Both modes operate within the same framework. A congested downstream node conveys its congestion state to upstream nodes such that they will throttle their traffic and ensure that there is sufficient spare capacity for the downstream node to send its traffic. Toward this end, a congested node transmits its local fair rate upstream and all upstream nodes sending on this link have to throttle to this rate. After a convergence period, congestion is alleviated once all nodes' rates are set to the fair rate. When congestion clears all ring nodes periodically increase their sending rates to ensure that they receive their maximum bandwidth share and to enable maximum spatial reuse.

RPR's fairness control is based on the two measured traffic rates *forward_rate* and *add_rate*. Both traffic rates are measured as byte counts at the output of the scheduler and thus denote serviced rates rather than offered rates (see also Fig. 11.3). More precisely, *forward_rate* denotes the serviced rate of all in-transit traffic while *add_rate* denotes the serviced rate of all local traffic. Both traffic rate measurements are taken over a prespecified time period called *aging_interval* and are low-pass filtered by means of exponential averaging using parameter $1/LPCOEFF$ for the current measurement and $1 - 1/LPCOEFF$ for the previous average.

At each *aging_interval*, every ring node checks its congestion status based on different conditions depending on whether aggressive mode (AM) or conservative mode (CM) is deployed, as explained shortly. When node n is congested, it calculates its *local_fair_rate*[n] which equals the fair rate of each ingress-based flow destined for node n (i.e., the minimum rate at which the corresponding source ring nodes are supposed to transmit via intermediate node n). Node n then sends a fairness control packet containing *local_fair_rate*[n] to upstream ring node $n - 1$ to inform it about the calculated fair rate. If upstream ring node $n - 1$ is also congested, it will forward the fairness control packet containing the minimum of the received *local_fair_rate*[n] and its own *local_fair_rate*[$n - 1$] to upstream node $n - 2$. If node $n - 1$ is not congested but its *forward_rate* is greater than the received *local_fair_rate*[n], it forwards the fairness control packet containing *local_fair_rate*[n] upstream since this situation implies that congestion is caused by node $n - 2$ or other ring nodes farther upstream. Otherwise, node $n - 1$ sends a null-value fairness control packet to indicate a lack of congestion. When an upstream node i receives the fairness control packet advertising *local_fair_rate*[n], it decreases the rate controller value of all flows traversing the congested node n to *allowed_rate_congested*, which equals the sum of serviced rates of all flows (i, j) originating from node i and traversing intermediate node n on their path toward destination node j. In doing so, source node i throttles its local traffic to the minimum rate it can send along the path to destination node j. As a consequence, local traffic rates of upstream node i will not exceed the advertised *local_fair_rate*[n] of the congested node n. Otherwise, if upstream node i receives a null-value fairness control packet, it increases *allowed_rate_congested* by a prespecified value such that it can reclaim additional bandwidth if one of the downstream flows reduces its rate.

As mentioned earlier, AM and CM differ in how an RPR node detects congestion and calculates its local fair rate. Both operation modes are explained in greater detail in the following:

- **Aggressive Mode (AM):** AM is the default operation mode of the RPR fairness control. By default, AM deploys dual-queue mode and works as follows. In AM, an RPR node n is considered congested whenever

$$STQ\_depth[n] > low\_threshold \qquad (11.1)$$

  or

$$forward\_rate[n] + add\_rate[n] > unreserved\_rate \qquad (11.2)$$

  hold, whereby *low_threshold* denotes a prespecified fraction of the STQ size and *unreserved_rate* denotes the link capacity minus the reserved bandwidth (see Table 11.1). A congested node n then calculates its *local_fair_rate*[n] as the serviced rate of its own local traffic *add_rate* and transmits a fairness control packet containing *add_rate* upstream. Upon receiving the fairness control packet, upstream nodes throttle their local traffic to *add_rate*. As a result, node n becomes uncongested so that flows will increase their *allowed_rate_congested*. This process leads to the targeted fair rate after a certain convergence period.

- **Conservative Mode (CM):** By default, CM deploys single-queue mode. In CM, each RPR node uses an access timer that measures the time between two consecutive transmissions of a given ring node. The access timer prevents a ring node from being starved. In CM, an RPR node n is said to be congested if the access timer expires or

$$forward\_rate[n] + add\_rate[n] > low\_threshold, \qquad (11.3)$$

where *low_threshold*, as opposed to AM, is a rate-based parameter set to a prespecified value below the link capacity. Apart from measuring *forward_rate*[n] and *add_rate*[n], node n also measures the number of active ring nodes that have at least one packet served at node n in the past *aging_interval*. If node n is congested in the current *aging_interval*, but was not congested in the previous one, it computes its *local_fair_rate*[*n*] as the *unreserved_rate* divided by the number of active ring nodes. If node *n* is continuously congested, the *local_fair_rate*[*n*] depends on the sum of *forward_rate*[*n*] and *add_rate*[*n*]. If this sum is smaller than *low_threshold*, *local_fair_rate*[*n*] ramps up. If this sum is above a prespecified *high_threshold*, which is larger than *low_threshold*, *local_fair_rate*[*n*] ramps down. In CM, the maximum utilization of the link bandwidth is given by *high_threshold*, which is typically smaller than the link capacity. Accordingly, this operation mode is called conservative.

It is important to note that the aforementioned CM and AM may suffer from significant performance limitations under unbalanced traffic scenarios where flows do not demand the same bandwidth. It was shown in Gambiroza et al. (2004) that under unbalanced and constant-rate traffic inputs the RPR fairness algorithm exhibits severe and permanent throughput oscillations that span nearly the entire range of the link capacity. These throughput oscillations not only decrease throughput but also prevent full spatial reuse and fairness as well as increase delay jitter and convergence. Based on this observation, most of the recent research efforts related to RPR focused on the enhancement of fairness control. Various alternative fairness algorithms whose fair rate calculation differs from that of AM and CM were shown to mitigate oscillations, increase spatial reuse, and improve convergence (Gambiroza et al., 2004; Wang et al., 2004; Shokrani et al., 2005; Davik et al., 2005; Robichaud and Huang, 2005). A fairness algorithm for RPR with very low computational complexity was studied in Alharbi and Ansari (2004b,a). Apart from the design of alternative fairness algorithms for performance enhancements of RPR networks, a generalized RIAS fairness model, referred to as *ring ingress aggregated max-min*, was considered in Khorsandi et al. (2005).

11.4   Protection

Given its bidirectional dual-fiber ring topology, RPR is able to recover from a single link or node failure. In the presence of two or more simultaneous failures, an RPR network would be divided into two or more disjoint subnetworks and thus become unable to preserve full network connectivity. Fault detection in RPR can be done in two ways. The underlying physical layer technology issues an alarm signal to the higher layer if connectivity is lost (e.g., loss-of-signal (LOS) alert in SONET/SDH). Alternatively, an RPR node assumes a link or node failure has occurred if it does not receive any keep-alive messages from its neighboring RPR node for a specified period of time, which is typically set to 3 ms. As soon as a ring node detects that one of its attached links or an adjacent node has failed, it broadcasts a topology discovery update packet on the ring to inform all remaining ring nodes about the failure. After failure detection and notification, RPR nodes perform one or two protection techniques to recover from the failure. RPR deploys the following two protection techniques: (1) *wrapping* and (2) *steering*, whereby in RPR wrapping is optional and steering is mandatory. To guarantee proper failure recovery network wide, all RPR nodes must adhere to the same

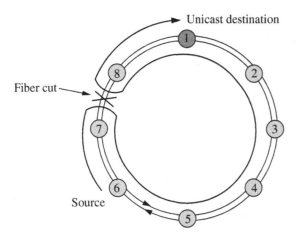

Unicast destination

Fiber cut

Source

Figure 11.5 Wrapping (optional in RPR).

protection technique (i.e., if wrapping is deployed in a given RPR network it must be supported by all nodes).

Let us first take a look at wrapping. Figure 11.5 illustrates wrapping for an RPR network with $N = 8$ nodes, where source node 6 is assumed to send unicast data packets to destination node 1 on the clockwise ringlet along the shortest path in terms of hops. Furthermore, we assume that a fiber cut occurs between nodes 7 and 8. After detecting the link failure, node 7 next to the fiber cut wraps the ring and immediately sends packets arriving on the clockwise ringlet back in the other direction on the counterclockwise ringlet. Wrapped packets travel along the counterclockwise ringlet until node 8 which is located on the other side of the fiber cut opposite to node 7. Node 8 wraps the packets arriving on the counterclockwise ringlet away from the fiber cut such that node 1 receives the packets destined for it on the clockwise ringlet. Note that wrapped packets arrive at the destination node on the ringlet they were originally sent by the source node. Wrapping is a relatively simple protection technique which is similar to automatic protection switching (APS) used in SONET/SDH-based self-healing rings (SHRs). Wrapping provides fast protection switching time and helps minimize packet loss in the event of a link or node failure since wrapping is performed locally by the two ring nodes next to the failure whose detection takes only a few milliseconds, as discussed earlier. Packets arriving during the time interval between failure event and its detection are lost, but all subsequent packets are wrapped away from the failure and thus are prevented from being dropped. On the downside, wrapping is not bandwidth efficient since wrapped traffic traverses all ring nodes and spans except the failed ones and thus consumes much network bandwidth along the backup path, which in general is significantly longer than the original shortest path. Note that wrapping is performed only on packets whose *wrap eligible* bit in the packet header is set. Packets not marked wrap eligible are discarded by the wrapping node.

Next, let us consider steering, which is mandatory in RPR networks. When an RPR node receives a topology discovery update packet sent by the wrapping node, it steers

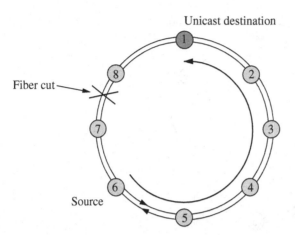

Figure 11.6 Steering (mandatory in RPR).

its local ingress traffic away from the fiber cut and starts to transmit packets on the counterclockwise ringlet in the failure-free direction to the destination node, as depicted in Fig. 11.6. Steering achieves a higher bandwidth efficiency than wrapping since steered packets in general need to traverse fewer ring nodes and links between source and destination.

If both wrapping and steering techniques are used simultaneously in an RPR network, it is recommended to first wrap and then steer. This two-step *wrap-then-steer* protection strategy tries to combine the reduced packet loss of wrapping with the improved bandwidth efficiency of steering, but it might introduce packet reordering. By default, RPR deploys the *strict* packet mode, where packets are guaranteed to arrive at the destination node in the same order as they were sent by the source node. Strict order packets are identified by setting the *strict order* bit in the packet header. Note that strict order packets cannot be marked wrap eligible. Consequently, strict order packets are never wrapped and are discarded when reaching the wrap point. To achieve in-order packet delivery in the event of a link or node failure, all nodes stop sending strict order packets after learning about the failure and discard all strict order in-transit packets until the topology stabilization timer expires. The value of the topology stabilization timer can be configured between 10 and 100 ms, and is set to 40 ms by default. The topology stabilization timer must be set properly such that each ring node is able to empty its transit queues. After the topology stabilization timer expires and each node acquires a stable and consistent updated topology image, ring nodes start to steer strict order packets onto the corresponding ringlet. It is worthwhile to mention that RPR optionally supports a *relaxed* packet mode, where packets are allowed to arrive out of order. Relaxed packets may be steered immediately after failure detection and notification without waiting for the updated topology image to become stable and consistent. Relaxed packets are not discarded by RPR nodes from their transit queues.

RPR aims at achieving a restoration time below 50 ms. The restoration time denotes the time period from when a ring node detects a failure until all remaining ring nodes have a stable and consistent updated topology image and begin steering. Recently, it was shown that the topology stabilization timer, which adds to the restoration time, in many situations may prevent real sub-50-ms restoration for strict order packets. An interesting approach for maintaining in-order delivery of packets and improving the service disruption time was recently proposed in Kvalbein and Gjessing (2004). In this approach, ring nodes do not have to wait for the updated topology to be stable. Unlike that specified in IEEE 802.17, strict order packets are allowed to be marked wrap eligible and are not discarded at transit queues during the topology stabilization period. The receiver is responsible for dropping strict order packets that arrive out of order. When a failure occurs, the receiver still accepts packets from the primary ringlet, even after the discovery of a topology change, as long as no strict order packets are received on the backup ringlet. Once the first strict order packet arrives on the backup ringlet, all subsequent packets on the primary ringlet are discarded. The obtained results show that this alternative protection approach is able to provide sub-50-ms restoration for all traffic.

Another drawback of the topology stabilization timer, set to 40 ms by default, is the fact that it leads to service disruption and packet loss for strict order traffic. Recall that during the topology stabilization period all strict order packets in transit are discarded and new strict order packets are not allowed to be transmitted on the ring. As a result, strict order traffic may suffer from packet loss and service disruption in the event of a link or node failure. Recently, three different improvements to the IEEE 802.17 standard that aim to reduce strict order packet loss and service disruption time were proposed in Kvalbein and Gjessing (2005). The so-called automatic method seeks to minimize the period when packets are discarded without compromising the in-order delivery guarantee. This method implies setting the topology stabilization timer lower than the minimum value allowed in IEEE 802.17 but otherwise does not demand any further changes to the standard. It was shown by means of simulation that this method is able to reduce packet loss by about 60% compared to the default setting of the topology stabilization timer. The other two proposed methods move some logic from the transit path to the destination node. The so-called receiver method does not require any state information to be maintained at the destination, while the so-called selective discard method requires the destination node to know which source nodes it receives packets from after a topology update, which requires an extra entry in the topology database of each RPR node. The presented simulation results indicate that the receiver and selective discard methods are superior to the automatic method and both reduce the packet loss by roughly 90% compared to IEEE 802.17. Given the similar performance of both methods, it appears reasonable to deploy the receiver method in practical RPR networks due to its smaller complexity.

12 WDM ring networks

Most of the wavelength division multiplexing (WDM) ring networks presented in this chapter are based on a unidirectional optical fiber ring, as shown in Fig. 12.1. At each node an optical add-drop multiplexer (OADM) drops a certain wavelength from the ring and allows data to be added at any arbitrary wavelength. A node transmits data on the added wavelength while it receives data on the dropped wavelength. Data on the dropped wavelength is removed from the ring in that the dropped wavelength is optoelectronically converted at an OADM. If the number of nodes, N, is equal to the number of wavelengths, W, as depicted in Fig. 12.1 for $N = W = 4$, each node has a dedicated *home channel* for reception. However, in general $N \geq W$ since the number of available wavelengths is limited (e.g., for cost reasons or finite transceiver tuning ranges). With $N \geq W$ the system becomes scalable since the number of nodes is independent of the number of available wavelengths.

Each node is equipped with one or more fixed-tuned and/or tunable transmitters and receivers. We adopt the FT^i–TT^j–FR^m–TR^n notation to describe the node architecture, where $i, j, m, n \geq 0$ (Mukherjee, 1992). That is, each node is equipped with i fixed-tuned transmitters, j tunable transmitters, m fixed-tuned receivers, and n tunable receivers. For example, a TT–FR node structure means that each node has one tunable transmitter and one fixed-tuned receiver.

When a node inserts a packet on a given wavelength while another packet is currently passing the ring on the same wavelength a *channel collision* occurs and both packets are disrupted. With tunable receivers so-called *receiver collisions*, which are also known as destination conflicts, can occur when a node's receiver is not tuned to the wavelength of an incoming packet. This can happen if the destination node does not know about the transmission or another packet is currently received on a different wavelength. Clearly, both channel and receiver collisions have a detrimental impact on the throughput-delay performance of the network. The degradation of the network performance due to channel or receiver collisions can be mitigated or completely avoided at the architecture and/or protocol level. For example, equipping each node with a receiver fixed-tuned to a home channel (either dedicated to a single node or shared by multiple nodes) prevents receiver collisions. Similarly, allocating each node a separate home channel for transmission avoids channel collision at the expense of scalability. In scalable systems, however, each wavelength channel is typically shared by multiple nodes giving rise to channel collisions. Clearly, medium access control (MAC) protocols are needed to govern the

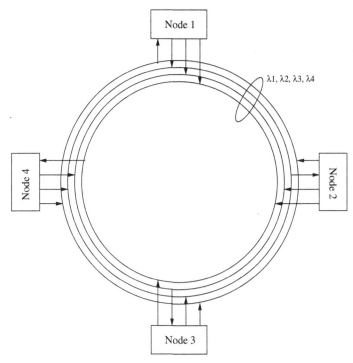

Figure 12.1 Unidirectional WDM ring network with $N = 4$ nodes and $W = 4$ wavelengths.

access to the wavelength channels and to mitigate or prevent channel (and receiver) collisions.

In this chapter, we provide a comprehensive survey on WDM ring network architectures and MAC protocols for WDM ring networks. WDM ring networks can be classified according to a number of different criteria, for example, unidirectional versus bidirectional rings or dedicated versus shared protection (Johansson et al., 1998). In this chapter, we categorize WDM ring networks with respect to the applied MAC protocol. As shown in Fig. 12.2, we categorize MAC protocols for WDM ring networks into protocols for slotted rings, multitoken rings, and meshed rings. Slotted ring protocols can be further subdivided into protocols without and with channel inspection and those making use of a separate control channel. MAC protocols with channel inspection use one of two different access strategies: either an *a priori* or an *a posteriori* access strategy. The different WDM ring types are discussed in greater detail in the remainder of this chapter.

12.1   Slotted rings without channel inspection

A simple way to avoid channel and receiver collisions is the deployment of time division multiple access (TDMA). Time is divided into slots equal to the packet transmission time. Typically, these time slots are of fixed size with multiple slots circulating at each wavelength on the ring, as illustrated in Fig. 12.3. The slots at different wavelengths

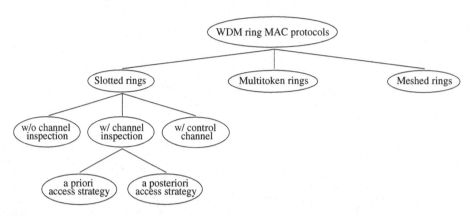

Figure 12.2 Classification of WDM ring network MAC protocols.

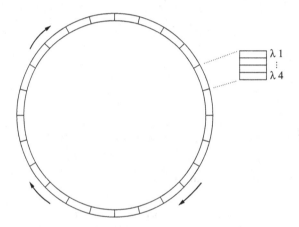

Figure 12.3 Slotted unidirectional WDM ring with $W = 4$ wavelengths.

are typically aligned. With TDMA, channel and receiver collisions are avoided by statically assigning each slot to a certain source–destination pair. Thus, a fixed amount of bandwidth is allocated to each pair of nodes, which is well suited for uniform regular traffic at medium to high loads but leads to wasted bandwidth and low channel utilization in the case of bursty and low traffic loads.

12.1.1 MAWSON

The metropolitan area wavelength switched optical network (MAWSON), presented in Summerfield (1997), Fransson et al. (1998), and Spencer and Summerfield (2000), is based on a FT^W–FR or alternatively TT–FR node architecture. N nodes are connected to the ring via passive OADMs that use fiber Bragg gratings (FBGs) for dropping a different wavelength for reception at each node. Since each node has a dedicated home channel ($W = N$), receiver collisions cannot occur. With the FT^W–FR node structure, broadcasting and multicasting can be achieved by turning on multiple lasers simultaneously, but only unicasting is considered in the evaluation of the MAC protocol.

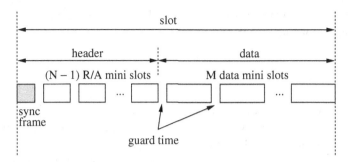

Figure 12.4 Slot structure of Request/Allocation Protocol (RAP) in MAWSON. After Fransson et al. (1998). © 1998 IEEE.

Time is divided into fixed-size slots, which are assumed to be aligned across all W wavelengths. Each slot is further subdivided into header and data fields, as shown in Fig. 12.4. Slots are assigned dynamically on demand. To this end, the header of each slot consists of $N − 1$ so-called Request/Allocation (R/A) minislots which are statically preassigned in a TDMA fashion to $N − 1$ source nodes. Each R/A minislot essentially comprises two fields, one for requests and one for allocations. More precisely, node i, ready to send variable-size data packets to node j, uses the request field of its assigned R/A minislot on j's home wavelength channel to make a request. Upon receipt of node i's request, node j allocates one or more data minislots to node i by using the allocation field of its assigned R/A minislot on i's home wavelength. After one end-to-end round-trip propagation delay node i transmits the data packet using the allocated data minislots but no more than M data minislots of the slot.

To save costs the node architecture and protocol are kept simple (e.g., no carrier sensing capabilities are required). Due to in-band signaling no additional control channel and control transceivers are needed. The protocol completely avoids both channel and receiver collisions, achieves good throughput performance, and provides fairness by allocating slots in a round-robin manner. However, the R/A procedure introduces some overhead and additional delay.

12.2   Slotted rings with channel inspection

In most slotted WDM rings channel collisions are avoided by enabling the nodes to check the status (used/unused) of each slot. Generally, this is done by tapping off some power from the fiber and delaying the slot while the status of each wavelength is inspected in the tapped-off signal. A packet can then be inserted in a slot at an unused wavelength. Packets waiting for transmission are stored in virtual output queues (VOQs). Typically, a node maintains separate VOQs either for each destination or for each wavelength. In the latter case, packets arriving at a node from the higher layer are put in the VOQ associated with the drop wavelength of the packet's destination. The MAC protocol has to select the appropriate VOQ to send a packet in a time slot according to a given access strategy. This can be done a priori without or a posteriori with taking into account the status of the slots. In the a priori access strategy each node selects a VOQ prior to inspecting the slot status, whereas in the a posteriori strategy each node first checks the status of a slot and then selects an appropriate (nonempty) VOQ.

12.2.1   RINGO

The ring optical (RINGO) network uses an FT^W–FR node architecture. Each node has channel inspection capability built with commercially available components. Nodes execute a multichannel empty-slot MAC protocol.

A MAC protocol with a posteriori queue selection has been implemented in the testbed described in Carena et al. (2002). The number of wavelengths is assumed to be equal to the number of nodes and each node has one VOQ with first-come-first-served queueing discipline for each wavelength. Only the VOQs where the correspondent wavelengths have been found to be empty (unused) are allowed to send data packets in the free time slot. In tie situations the longest among those queues is chosen and the oldest packet is sent.

The overhead of the RINGO empty-slot MAC protocol is very small. To identify the status of a given slot a single bit is sufficient. All wavelengths are used for data transmission and no separate control channel and control transceivers are required. It was demonstrated that all-optical packet switched WDM ring networks are feasible with currently available technology. However, owing to the slot size packets have to be of fixed size. However, note that variable-size packets can be transmitted in slotted rings without segmentation and reassembly by means of buffer insertion techniques exploiting optical fiber delay lines (FDLs) (Bianco et al., 2002).

12.2.2   Synchronous round robin (SRR)

Synchronous round robin (SRR) is another empty-slot MAC protocol for a unidirectional WDM ring network with fixed-size time slots and destination stripping (Marsan et al., 1996b,a, 2000). Each node is equipped with one tunable transmitter and one fixed-tuned receiver (TT–FR), where the transmitter is assumed to be tunable across all W wavelengths on a per-slot basis. If $N = W$ each node has its own home wavelength channel for reception. In the more general case $N > W$, each wavelength is shared by multiple destination nodes.

In SRR, each node has $N - 1$ separate first-in/first-out (FIFO) VOQs, one for each destination, as shown in Fig. 12.5. SRR uses an a priori access strategy. Specifically, each node cyclically scans the VOQs in a round-robin manner on a per-slot basis, looking for a packet to transmit. If such a deterministically selected VOQ is nonempty, the first (oldest) packet is transmitted, provided the current slot was sensed to be empty. If the selected VOQ is empty the transmission of the first packet from the longest queue of the remaining VOQs is sent, again provided the current slot is unused. In any case, if transmission in an occupied slot is not possible (because it would cause a channel collision), the next VOQ is selected for the transmission attempt in the subsequent slot according to the round-robin scanning of SRR. In doing so, under heavy uniform load

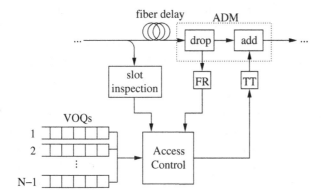

Figure 12.5 SRR node architecture with VOQs and channel inspection capability.

conditions when all VOQs are nonempty the SRR scheduling algorithm converges to round-robin TDMA.

For uniform traffic, SRR asymptotically achieves a bandwidth utilization of 100%. However, the presence of unbalanced traffic leads to wasted bandwidth due to the nonzero probability that the a priori access strategy selects a wavelength channel whose slot is occupied while leaving free slots unused. It was shown in Bianco et al. (1998) that a posteriori access strategies avoid this drawback, resulting in an improved throughput-delay performance at the expense of an increased complexity.

SRR achieves good performance requiring only local information on the backlog of the VOQs, which also avoid the well-known head-of-line (HOL) blocking problem. On one hand, owing to destination stripping, slots can be spatially reused several times as they propagate along the ring. On the other hand, slot reuse raises fairness control problems, particularly under nonuniform traffic. A node to which a large amount of slots is directed generates a large amount of free slots, and nodes immediately downstream are in a favorable position with respect to other nodes. We will address this fairness problem in Section 12.6.1. Note that in order to provide quality of service (QoS) SRR requires additional modifications, which are discussed in greater detail in Section 12.6.2.

12.2.3   HORNET

The hybrid optoelectronic ring network (HORNET) is a unidirectional WDM ring network using destination stripping, where nodes have a TT–FR structure (Shrikhande et al., 2000). Similar to SRR, to avoid HOL blocking each node uses VOQs, one for each wavelength, and both a priori and a posteriori access strategies can be used. Nodes sense the availability of each slot by monitoring *subcarrier multiplexed* (SCM) tones. The SCM-based carrier-sensing scheme is more cost-effective than demultiplexing, separately monitoring, and subsequently multiplexing all wavelengths of the WDM comb, as is done, for instance, in RINGO. Instead of the multiplexer, photodiode array, and demultiplexer, the HORNET channel inspection scheme needs only one single photodiode.

HORNET deploys a carrier sense multiple access with collision avoidance (CSMA/CA) MAC protocol, initially assuming fixed-size slots which are well suited for the transport of fixed-size packets (e.g., ATM cells) (Gemelos et al., 1999). The CSMA/CA MAC protocol can be extended to support IP packets of variable size. Two CSMA/CA MAC protocols both supporting variable-size packets are proposed and investigated in Shrikande et al. (2000). In the first protocol, multiple different slot sizes circulate along the ring. The slot sizes are chosen according to the predominant IP packet lengths as typically found in traffic measurements. For example, three slots sizes can be chosen such that 40-, 552-, and 1500-Byte-long IP packets are accommodated. A dedicated node controls the size and number of slots such that they match the packet size distribution. (A variant of this protocol for a TT–FR^W architecture has been proposed in Chen and Hwang [2002].) The second protocol is unslotted and operates similarly to CSMA/CD with collision detection and backoff. More precisely, when a wavelength is sensed idle, a given node begins to transmit a packet. When another packet arrives on the same wavelength before the transmission is complete, the packet transmission is aborted. In this case, the incomplete packet is marked by adding a jamming signal to the end of the packet. Aborted transmissions are resumed after some backoff time interval. (A more bandwidth-efficient modification of the latter unslotted CSMA/CA protocol was examined in Li et al. [2002]. In this *carrier sense multiple access with collision preemption* protocol, variable-size IP packets do not necessarily have to be transmitted in one single attempt. Instead, packets are allowed to be transmitted and received as fragments by simply interrupting the packet transmission. Thus, successfully transmitted parts of the original IP packet are not retransmitted resulting in a higher channel utilization.)

Besides demonstrating the cost-effective SCM-based channel inspection approach, the HORNET project also proved the feasibility of fast tunable transmitters. This allows for replacing arrays of multiple fixed-tuned transmitters with a single tunable transmitter. HORNET scales well since the number of nodes is independent of the number of wavelengths. Generally, each wavelength is allowed to be shared by multiple destination nodes with packet forwarding at intermediate nodes resulting in multihop networks. Note that intermediate nodes not only forward packets toward the destination but also provide signal regeneration in the electronic domain. On the other hand, the CSMA/CA random access protocol does not provide QoS and the destination stripping gives rise to fairness problems.

12.2.4   A posteriori buffer selection schemes

Several a posteriori buffer selection schemes are studied in Bengi and van As (2002) and Bengi (2002b). The access protocol is intended to be run on the HORNET architecture described earlier. Recall that in an empty-slot protocol each unused slot on any wavelength channel can be used for packet transmission by a source node. However, when more than one wavelength channel carries an empty slot in the current slot period, one packet (or equivalently, one VOQ) corresponding to one of the empty channels has to be chosen according to a certain selection rule. As a consequence, the a posteriori packet selection process has to be performed at a higher speed in the electronic domain, which increases the processing complexity compared to an a priori packet selection scheme. Five different VOQ selection strategies are described and examined in Bengi and van As (2002):

- **Random Selection:** The VOQ from which a packet is to be transmitted is selected randomly according to a uniform distribution.
- **Longest Queue Selection:** The longest VOQ is chosen upon buffer contention.
- **Round-Robin Selection:** The VOQ is chosen in a round-robin fashion.
- **Maximum Hop Selection:** The packet (VOQ) associated with the maximum hop distance between source and destination node is selected when buffer contention arises.
- **C-TDMA Selection:** In this so-called channel-oriented TDMA (C-TDMA) scheme each VOQ is allocated a certain slot within a TDMA frame of size W, where W denotes the number of wavelengths in the system. Similar to SRR, the slots on the ring have to be successively numbered with the slot number, requiring a global sychronization of all nodes in the network.

It was found that the random and round-robin buffer selection schemes provide a satisfactory compromise between performance and implementational complexity.

12.2.5   FT–TR rings

A unidirectional empty-slot WDM ring network that uses source stripping was proposed in Jelger and Elmirghani (2001). Each node is equipped with one fixed-tuned transmitter and one tunable receiver (FT–TR). Packets are buffered in a single FIFO transmit queue at each node. In the applied source-stripping scheme a sender must not reuse the slot it just marked empty. It was shown that this scheme introduces a very simple fairness mechanism where a node cannot starve the entire network (note that this holds only for source-stripping rings and does not apply for destination-stripping implementations). The performance of the network was compared for both source and destination stripping in Jelger and Elmirghani (2002a). By means of simulation it was shown that destination stripping clearly outperforms source stripping in terms of throughput, delay, and packet dropping probability.

Clearly, with a single tunable receiver at each node, receiver collisions can occur. Receiver collisions can be avoided in a number of ways. In one approach, arriving packets which find the destination's receiver busy recirculate on the ring until the receiver of the destination is free, that is, tuned to the corresponding wavelength (Jelger and Elmirghani, 2002a). Alternatively, receiver collisions can be completely avoided at the architecture level by replacing each node's tunable receiver with an array of W fixed-tuned receivers, each operating at a different wavelength (FT–FR^{W}) (Jelger and Elmirghani, 2002b). (Another proposal to resolve receiver contention is based on optical switched delay lines; see Wong et al. [1995]. A destination node puts all simultaneously arriving packets but one into optical delay lines such that packets can be received sequentially.)

12.3   Slotted rings with control channel

In some networks the status of the slots is transmitted on a separate control channel (CC) wavelength. To this end, each node is typically equipped with an additional transmitter and receiver, both fixed-tuned to the control wavelength. A separate CC wavelength enables nodes to exchange control information at high line rates and eases the implementation of enhanced access protocols with fairness control and QoS support, as we will see shortly.

12.3.1   Bidirectional HORNET – SAR-OD

An extended version of the original unidirectional TT–FR HORNET ring architecture in which SCM is replaced with a separate CC wavelength is investigated in White et al. (2002b). Transmission on the CC (and data wavelengths) is divided into fixed-size slots. The CC conveys the wavelength availability information such that nodes are able to "see" one slot into the future. Two counterdirectional fiber rings each carrying W data wavelengths and an additional CC wavelength operate in parallel. On each ring every node deploys one fast tunable transmitter and one fixed-tuned receiver for data, and one transceiver fixed-tuned to the CC wavelength. Thus, the CC-based HORNET is a $CC–FT^2–TT^2–FR^4$ system.

A modified MAC protocol able to efficiently support variable-size packets over the bidirectional ring network was examined. This so-called *segmentation and reassembly on demand* (SAR-OD) access protocol aims at reducing the number of segmentation and reassembly operations of variable-size packets. Specifically, the transmission of a packet from a given VOQ starts in an empty slot. If the packet is larger than a single slot the transmission continues until it is complete or the following slot is occupied (i.e., the packet is segmented only if required to avoid channel collisions). If a packet has to be segmented, it is marked incomplete and the transmission of the remaining packet segment(s) continues in the next empty slot(s) on the corresponding wavelength. By means of simulation it was shown that SAR-OD reduces the segmentation/reassembly overhead by approximately 15% compared to a less intelligent approach where all packets larger than one slot are segmented irrespective of the state of successive slots.

The CC-based bidirectional HORNET preserves the advantages of the original unidirectional HORNET (e.g., scalability and cost-effectiveness). Bidirectional dual-fiber rings provide an improved fault tolerance against node/fiber failures and survivability over unidirectional single-fiber rings (White et al., 2000, 2002a). Furthermore, the CC can also be used to achieve efficient fairness control, as described in greater detail in Section 12.6.1.

12.3.2   Segmentation/reassembly

An access protocol for a CC-based slotted ring WDM network that completely avoids segmentation and reassembly of variable-size packets was studied in Bengi (2002a,c).

The access protocol is an extended version of the original protocol (described in Section 12.2.4). The architecture differs from the CC-based HORNET in that a unidirectional ring is deployed and each node uses an additional transmitter fixed-tuned to the node's drop wavelength, resulting in a CC–FT^2–TT–FR^2 system. The additional transmitter is used to forward dropped packets that are destined to downstream nodes which share the same drop wavelength.

The extended MAC protocol relies on a frame-based slot reservation strategy including reservation of successive slots for data packets longer than the given slot size and immediate access for packets shorter than the slot length. Each node is equipped with two VOQs for each wavelength, one for short packets and one for long packets. The ring is subdivided into multiple *reservation frames* with the frame size equal to the largest possible packet length. In these frames multiple consecutive slots are reserved to transmit long packets without segmentation. A single reservation control packet containing all reservations circulates on the CC. Each node maintains a table in which the reservations of all nodes are stored. When the control packet passes, a node updates its table and is allowed to make a reservation. The additional fixed-tuned transmitter is used to forward packets concurrently with transmitting long packets within multiple contiguous slots. Besides the support of long packets via reservation, short packets fitting into one slot are accommodated by means of immediate access of empty and unreserved slots.

The proposed protocol provides immediate medium access for packets shorter than one slot and completely avoids the segmentation and reassembly of longer variable-size packets, resulting in a reduced complexity. The reservation protocol also enables QoS support, as discussed in greater detail in Section 12.6.2. On the other hand, the reservation protocol introduces some delay overhead and reserved slots on their way back to the source node cannot be spatially reused after destination stripping.

12.3.3   Wavelength stacking

The proposal in Smiljanic et al. (2001, 2002) uses the so-called *wavelength stacking* technique to transmit multiple packets in one slot of a CC-based slotted unidirectional WDM ring. Recall that wavelength stacking was already discussed in the context of photonic slot routing in Section 7.6. Each node is equipped with one fast-tunable transmitter and one photodiode. Time is divided into slots of duration T_p. A fast-tunable laser at a given node starts transmission W time slots before its scheduled time slot. As illustrated in Fig. 12.6, in each following time slot it transmits data on a different wavelength. The signal passes through the array of fiber gratings separated by delay lines so that the data packets transmitted at different wavelengths are aligned in time. The resulting packet is then transmitted to the network on all wavelengths in parallel by setting switch S to the cross state. On the receiver side, the reverse procedure is performed. A packet is received when switch S is in the cross state and then unstacked when passing through the same array of fiber gratings and delay lines.

Because wavelength stacking takes W time slots, a node should decide in advance when to access the medium. A separate wavelength is used as a CC for the reservations.

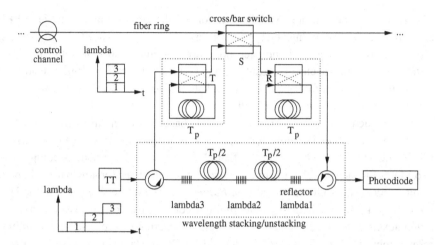

Figure 12.6 Node architecture for wavelength stacking. After Smiljanic et al. (2002).

Time slots are grouped into cycles of length W. Each node may transmit and receive at most one packet within each cycle. Switches T and R in Fig. 12.6 synchronize wavelength stacking and unstacking. Wavelength stacking is completed in the last time slot of the cycle and the packet is stored in the delay line by setting T in the cross state. A packet is stored as long as switch T is in the bar state. The packet is transmitted to the network by setting switches T and S in the cross state exactly $2W$ time slots after the reservation. Whenever a node recognizes its address on the control channel, it stores the packet in the delay line by setting switches S and R in the cross state $2W$ time slots after the address notification. The node starts unstacking the packet at the beginning of the next cycle by setting switch R in the cross state. Each node removes a packet that it receives as well as its reservation.

The wavelength stacking/unstacking allows a node to simultaneouly send and receive multiple packets in one slot on the ring despite the fact that the node has only one transceiver. However, the quality of the optical signal might suffer from passing the numerous delay lines and switches in a node.

12.3.4   Virtual circles with DWADMs

In the unidirectional slotted ring WDM network presented in Cho and Mukherjee (2001) each node is equipped with a *dynamic wavelength add-drop multiplexer (DWADM)*. As opposed to tunable transmitters and receivers which can operate independently, the input and output wavelengths of a DWADM must be the same; that is, if the wavelength to receive at a given node s is λ_i, the wavelength to transmit must be the same wavelength λ_i. Furthermore, if the node has to send to another node d, then node d has to use wavelength λ_i to receive and to send a packet. As a consequence, *virtual circles* are created, as depicted in Fig. 12.7, which can be changed dynamically according to varying traffic demands.

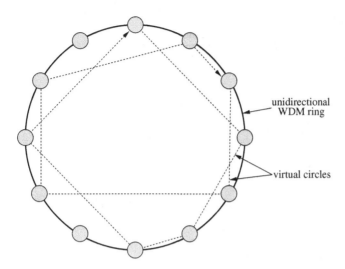

Figure 12.7 Virtual circles comprising nodes whose DWADMs are tuned to the same wavelength.

The ring network uses W data wavelength channels and a separate control wavelength channel. Nodes communicate over the control channel in a TDM fashion to exchange (1) transmission requests and (2) acknowledgments. The $W + 1$ wavelengths are divided into three cycles, which are repeated periodically. In the first cycle a control packet sent by a server node collects transmission requests from all nodes. These are processed by the server node and wavelength assignments/acknowledgments are sent back to the nodes in the second cycle. In the third cycle, each node that has been assigned a wavelength tunes the DWADM appropriately and starts the data transmission.

DWADMs are expected to be less expensive than tunable transceivers and can be used to realize cost-effective metro networks. However, due to their reduced flexibility, the wavelength utilization is expected to be smaller than in TT–TR systems, where transmitters and receivers can be tuned to any arbitrary wavelength independently.

12.4   Multitoken rings

Slotted WDM ring networks have a number of advantages such as easy synchronization of nodes even at high data rates, high channel utilization, low access delay, and simple access schemes. However, variable-size packets are difficult to handle and explicit fairness control is needed. In contrast, variable-size packets can be transported in a reasonably fair manner in token rings where the access is controlled by means of a special control packet – the token – which circulates around the ring. The token is passed downstream from node to node. Each node can hold the token for a certain amount of time during which the node is allowed to send (fixed-size or variable-size) packets. Due to the limited token holding time fairness is achieved. Furthermore, as opposed to slotted rings, nodes do not have to be synchronized.

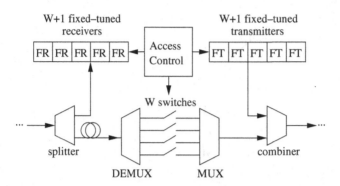

Figure 12.8 MTIT node architecture. After Cai et al. (2000). © 2000 IEEE.

12.4.1   MTIT

A token-based access scheme for a unidirectional WDM ring network with a CC–FT^{W+1}–FR^{W+1} node structure, the so-called *multitoken interarrival time* (MTIT) access protocol, was examined in Fumagalli et al. (1999) and Cai et al. (2000). For each data channel every node has one fixed-tuned transmitter, one fixed-tuned receiver, and one on-off optical switch, as shown in Fig. 12.8. The on-off switches are used to control the flow of optical signals through the ring and prevent recirculation of the same packet in the ring. Once transmitted by the source node, the packet makes one round trip in the ring and it is removed from the network by the same source node (i.e., MTIT deploys source stripping). A separate wavelength is used as a CC for the purpose of access control and ring management. The optical signal on the CC is separately handled by an additional fixed-tuned transceiver.

Channel access is regulated by a multitoken approach. Each channel is associated with one token that circulates among the nodes on the control channel and regulates the access to the corresponding data channel. The MTIT protocol controls the token holding time by means of a *target token interarrival time* with value *TTIT*. *TTIT* is agreed upon by all nodes connected to the ring at the configuration time of the system. The *token interarrival time TIAT* is defined as the time elapsed between two consecutive token arrivals at the node. Upon a token arrival, the node is allowed to hold that token for a period of time equal to *TTIT − TIAT*. When the token holding time is up, the node must release the token as soon as the current ongoing packet transmission is completed. A token can also be released earlier if no more packets are left in the node's transmission buffer. Note that concurrent transmissions on distinct channels are possible at the same node when two or more tokens are simultaneously held at the node.

Given the node structure, MTIT avoids receiver collisions and allows each node for simultaneously using multiple data wavelenth channels. However, the number of transceivers at each node is rather large. MTIT achieves low access delay due to the fact that a node has the opportunity to grab a token more frequently than in conventional token rings where a node has to wait one round-trip time for the next token. A unique feature of MTIT is its capability to self-adjust the relative positions of tokens along the ring circumference and maintain an even distribution of them. As a result, the variance of the token interarrival time is low, guaranteeing to every node a consistent channel access delay in support of high-priority traffic. On the other hand, the capacity of MTIT is expected to be smaller than that of destination-stripping ring networks since source stripping does not allow for spatial wavelength reuse.

12.5   Meshed rings

In unidirectional WDM ring networks with source stripping, packets are removed by the source node and each transmission requires a full circulation of the packet on the ring. The network capacity is limited by the aggregate capacity of all wavelengths. The network capacity of unidirectional ring networks can be increased with destination stripping where a transmission is bounded to a ring segment between the corresponding pair of source and destination nodes. Due to spatial reuse multiple simultaneous transmissions can take place at each wavelength. For uniform traffic the mean distance between source and destination is half the ring circumference. As a consequence, two simultaneous transmissions can take place at each wavelength on average, resulting in a network capacity that is 200% larger than that of unidirectional rings with source stripping. In bidirectional rings the network capacity can be further increased by means of shortest path routing, where a given packet is sent on that ring which provides the shortest distance to the corresponding destination in terms of hops. For uniform traffic the mean distance between source and destination is only a quarter of the ring circumference. Therefore, the aggregate capacity of bidirectional destination-stripping ring networks is increased by 400% on each directional ring compared to unidirectional source-stripping ring networks, which translates into a total capacity increase of 800% when both counterdirectional fiber rings are taken into account. The capacity of bidirectional ring WDM networks can be further increased in so-called *meshed rings*, which are discussed next.

12.5.1   SMARTNet

The *scalable multichannel adaptable ring terabit network* (SMARTNet) provides a significant increase in network capacity (Rubin and Hua, 1995a,b, 1997; Rubin and Ling, 1999). SMARTNet is based on a bidirectional slotted ring network with shortest path routing and destination stripping. Each node is connected to both rings and has a FT^W–FR^W structure, which allows a node to transmit and receive data on W different wavelengths simultaneously. All wavelengths are divided into fixed-size slots whose length is equal to the transmission time of a fixed-size packet (and a header for indicating the slot status). Medium access is governed by means of an empty-slot protocol.

In addition to the N nodes, K equally spaced wavelength routers, each with four pairs of input/output ports, are deployed in the bidirectional ring. Wavelength routers are used to provide short-cuts in that data packets do not have to pass through the ring nodes that are between two interconnected routers. Specifically, two input/output ports are used to

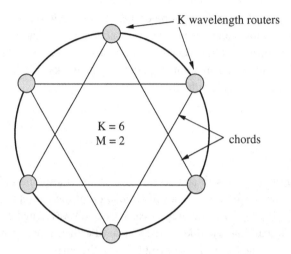

Figure 12.9 SMARTNet: Meshed ring with $K = 6$ wavelength routers, each connected to its $M = 2$nd neighboring routers.

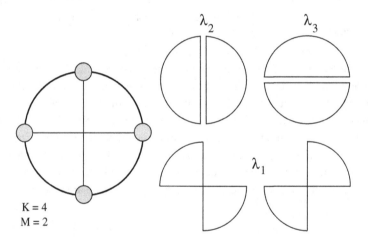

Figure 12.10 Wavelength paths in a meshed ring with $K = 4$ and $M = 2$, using $W = 3$ wavelengths.

insert a router into the bidirectional ring; the other two pairs of ports are used for creating bidirectional links (chords) to the two Mth neighboring routers. Routers $r_{(k+M) \bmod K}$ and $r_{(k-M) \bmod K}$ are said to be the Mth neighboring routers of router r_k on the ring, where $k = 0, 1, \ldots, K - 1$. Figure 12.9 depicts a meshed ring with $K = 6$ wavelength routers, each connected to its two $M = 2$nd neighboring routers.

Each wavelength router is characterized by a wavelength routing matrix that determines to which output port each wavelength from a given input port is routed. The wavelength routing matrix is chosen such that the average distance between each source–destination pair is minimized with a minimum number of required wavelengths. Figure 12.10 illustrates an example for an optimal set of wavelength paths for $K = 4$, $M = 2$, and $W = 3$.

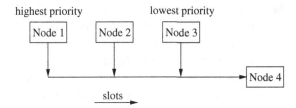

Figure 12.11 Medium access priorities in ring networks.

SMARTNet is able to significantly increase the capacity of a bidirectional ring network with shortest path routing and destination stripping. For uniform traffic it was shown that a meshed ring with $K = 6$ wavelength routers and $M = 2$ increases the network capacity by 720% compared to unidirectional source-stripping rings, at the expense of additional wavelength routers and chords which add to the network costs.

12.6   Fairness control and QoS support

Several of the aforementioned access protocols were extended to achieve fairness and QoS support. In the following, we discuss these protocol extensions in greater detail.

12.6.1   Fairness control

In general, the bandwidth of a network is shared by all nodes and each node ready to send data should have the same opportunity to transmit data. Most of the networks described earlier are based on a unidirectional ring. In this architecture each wavelength can be considered a unidirectional bus terminating at a certain destination, as shown in Fig. 12.11. In an empty-slot access protocol upstream nodes have a better-than-average chance to receive an empty slot for transmission while downstream nodes have a worse-than-average chance. At heavy traffic this can lead to starvation of downstream nodes since they "see" slots which are mostly used by upstream nodes. To avoid starvation, the transmission rate of nodes has to be controlled in order to achieve fairness among all nodes. However, restricting nodes in their transmission decreases the channel utilization. In general, there is a trade-off between fairness and channel utilization.

MMR

Since SRR (see Section 12.2.2) is not able to enforce fairness by itself, a fairness control algorithm has to be superimposed to SRR. To this end, the so-called *Multi-MetaRing* (MMR) fairness algorithm is used (Marsan et al., 1997a). The MMR algorithm adapts a mechanism originally proposed for the MetaRing high-speed electronic metropolitan area network (Cidon and Ofek, 1990, 1993; Chen et al., 1992). Fairness in MetaRing is granted by the circulation of a control message, named SAT (standing for SATisfied). Nodes are assigned a maximum number of packets to be transmitted between two SAT visits (also known as *quota* or *credit*). Each node normally forwards the SAT message on the ring with no delay, unless it is not SATisfied in the sense that it has not transmitted the permitted number of packets since the last time it forwarded the SAT. The SAT is delayed at unSATisfied nodes until SATisfaction is obtained (i.e., either the node's packet buffer is empty or the number of permitted packet transmissions is achieved).

In the so-called MMR Single SAT (MMR-SS) scheme a single SAT message regulates the transmissions of all nodes on all wavelength channels. Each node can transmit at most K packets to each destination since the last SAT visit. Each SATisfied node forwards the SAT to the upstream node. Thus, the SAT logically rotates in the opposite direction with respect to data (although the physical propagation is co-directional). Therefore, the SAT propagation delays are very large since the SAT message has to traverse almost the entire network to reach the upstream node. Alternatively, in the so-called MMR Multiple SAT (MMR-MS) scheme one SAT message for each wavelength can be used. It was shown in Marsan et al. (1997b) that the best extension of the MetaRing fairness control scheme to a WDM ring seems to be the one using one SAT for each wavelength which circulates together with data packets and is addressed to the node upstream of the node that emits the SAT.

M-ATMR

The access protocol of Section 12.3.2 suffers from fairness problems due to destination stripping. In Bengi and van As (2001), an extension of the well-established *asynchronous transfer mode ring* (ATMR) fairness protocol was adopted to the multiple channel WDM ring case, referred to as M-ATMR.

In M-ATMR each node gets a certain number of transmission credits for each destination. When a node has used all its credits or has nothing to send it gets into the inactive state. In order to properly apply the credit reset mechanism, every node has to know which node was the last active node. To achieve this, each active node overwrites a so-called *busy address field* in the header of every incoming slot with its own address (the busy address field may be included into the SCM header of each WDM wavelength channel). Thus, each node receiving a slot with its own busy address assumes that all other nodes are inactive. If the last active node detects inactivity of all other nodes it generates a reset immediately after its own transmission. The reset mechanism causes the nodes to set up their credits to the predefined values. This way, it is guaranteed that every node uses a maximum number of slots between two subsequent reset cycles. It was shown in Bengi and van As (2001) that the M-ATMR fairness protocol applied for best-effort traffic provides throughput and delay fairness for both uniform and client/server traffic scenarios.

DQBR

The so-called *distributed queue bidirectional ring* (DQBR) fairness protocol for the CC-based HORNET of Section 12.3.1 is an adaptation of the distributed queue dual bus (DQDB) protocol (White et al., 2002b).

The DQBR fairness protocol works as follows. In each CC frame a bit stream of length W called the *request bit stream* follows the wavelength-availability information. When a node on the network receives a packet in VOQ w, the node notifies the upstream nodes about the packet by setting bit w in the request bit stream in the CC that travels upstream with respect to the direction the packet will travel. All upstream nodes take note of the requests by incrementing a counter called the *request counter* (RC). Each node maintains a separate RC for each wavelength. Thus, if bit w in the request bit stream is set, RC w is incremented. Each time a packet arrives at VOQ w, the node stamps the value in RC w onto the packet and then clears the RC. The stamp is called a *wait counter* (WC). After the packet reaches the head of the VOQ, if the WC equals n it must allow n frame availabilites to pass by for downstream packets that were generated earlier. When an availability passes by the node on wavelength w, the WC for the packet at the head of VOQ w is decremented (if the WC equals zero, the RC w is decremented). Not until the WC equals zero can the packet be transmitted. The counting system ensures that the packets are sent in the order that they arrived to the network. It was shown in (White et al., 2002b) that with DQBR the throughput is equal for all nodes.

12.6.2   QoS support

Many applications (e.g., multimedia traffic) require QoS with respect to throughput, delay, and jitter. To meet these requirements, networks typically provide different service classes, for example, constant bit rate (CBR) and variable bit rate (VBR). In general, traffic with stringent throughput, delay, and jitter requirements is supported by means of circuit switching via reservation of network resources, resulting in *guaranteed* QoS. On the other hand, to provide QoS to bursty traffic more efficiently, nodes process and forward packets with different priorities while benefitting from statistical multiplexing, leading to *statistical* QoS. In the following, we review different approaches for providing QoS in WDM ring networks.

SR^3

Synchronous round robin with reservations (SR^3) is derived from the SRR (see Section 12.2.2) and the aforementioned MMR protocols and allows nodes to reserve slots, thereby achieving a stronger control on access delays (Marsan et al., 1997c). SR^3 can be used in conjunction with SRR and MMR, requiring a marginal algorithmic complexity increase with no additional signaling messages.

In SR^3 time is subdivided into successive periods called *reservation frames*. Each reservation frame comprises P SRR frames. Each node can reserve up to P slots with a given destination per reservation frame (i.e., at most one slot per destination per SRR frame). Reservations are effective if all network nodes are aware of other nodes' reservations. SAT messages are used to broadcast the reservation information. Each SAT distributes information regarding current reservations on the channel it regulates. Each SAT contains a *reservation field* (SAT-RF), which is subdivided into $N - 1$ subfields; each subfield is assigned to a particular node for reservations. If node i needs to reserve $1 \leq h \leq P$ slots per reservation frame on channel j, it waits until it receives the j-SAT; it then forwards the reservation request by properly setting the ith SAT-RF subfield to the value h. The j-SAT visits all nodes during the next tour of the multiring. By the time node i receives the j-SAT again, all nodes in the network are aware of the request of node i. Node i can thus update its reservation request on channel j every time it releases the j-SAT.

It was shown in Marsan et al. (1997c) that SR[3] guarantees a throughput-fair access to each node. Moreover, the bandwidth left unused by guaranteed services can be shared by best-effort traffic very effectively. Even for the basic best-effort service, that requires no service guarantee, the reservation scheme can be very beneficial; the average and variance of access delays are greatly reduced when slots are reserved, leading to an improved performance and fairness. The reservation scheme can also be extended to multiple service classes. It was shown in Marsan et al. (1999) that in an unbalanced multiclass traffic scenario a very good separation of the different traffic classes is obtained; the performance of higher-priority traffic is largely unaffected by lower-priority traffic, even when the latter one is grown to overload conditions.

Reservation scheme for QoS support

For QoS support in the WDM ring network of Section 12.3.2 a connection-oriented protocol (based on connection set-up and termination) was proposed in Bengi and van As (2002). In order to enable connection-oriented packet transmission for real-time services, the ring is subdivided into so-called *connection frames*. The real-time connections are established by reserving equally spaced slots within successive connection frames such that each destination node can be reached by a certain slot on the corresponding wavelength. Best-effort data traffic is supported by using a slot for packet transmission only when it is unreserved and empty. It was shown that this QoS approach is able to meet the delay requirements almost deterministically. Note that this scheme allows for reserving only one fixed-size slot (i.e., only fixed-size packets are supported).

A similar reservation scheme for providing QoS was presented in Bengi and van As (2001, 2002). In addition to the W normal VOQs, each node has W real-time VOQs. Packets in the real-time VOQs are transmitted via connections in equally spaced, reserved slots. At each wavelength the ring is subdivided into frames each consisting of N/W slots, one slot per destination node receiving at that wavelength. A single reservation slot carries a connection set-up and a connection termination field, each consisting of N bits on the subcarrier. When a node sets a bit in the set-up field the slot to the corresponding destination is reserved in each frame. After one circulation of the reservation slot all nodes are aware of the reservation and the set-up flag is cleared. All nodes keep track of the reservations by maintaining a table that is updated when the reservation slot passes. To free the reserved slots the same set/circulation/reset procedure is performed with the corresponding bit in the termination field.

MTIT – QoS with lightpaths

The MTIT protocol of Section 12.4.1 can be extended to support not only packet switching but also circuit switching with guaranteed QoS (Fumagalli et al., 1998). The proposed solution allows for the all-optical transmission of packets with source stripping and circuits via a tell-and-go establishment of lightpaths (i.e., wavelength routes) with destination stripping. When using the lightpath establishment technique (to be discussed next), a point-to-point connection is established between the source and the destination. The on-off switches at both the source and destination corresponding to the lightpath wavelength are set in the off state. This allows for spatial reuse of the wavelength.

Each node maintains a so-called *local lightpath table* (LLT) for all active lightpaths that is updated each time a token passes. A so-called *token lightpath table* (TLT) is transmitted with each token to broadcast the changes of lightpath deployment on the ring on the wavelength associated with the token. Each token comprises two lists for circuit set-up and tear-down. Specifically, a node holding a token can set up a lightpath to a destination node at the token's wavelength by making an entry in the add list of the token. The path to the destination must not be occupied by another lightpath. A lightpath is torn down by the source by making an entry in the delete list of the token.

13 RINGOSTAR

The aforementioned wavelength division multiplexing (WDM) ring networks appear to be natural candidates to extend existing optical single-channel ring networks (e.g., RPR) to multichannel systems by means of WDM. In WDM rings, optical single-channel rings are multichannel upgraded by exploiting the already existing fiber infrastructure without requiring any additional fiber links and modifications of the ring topology. Clearly, deploying WDM on the existing ring infrastructure saves on fiber requirements. At the downside, however, WDM rings require all ring nodes to be WDM upgraded at the same time (e.g., each ring node is equipped with a transceiver array or wavelength (de)multiplexer). Furthermore, WDM rings are able to survive only a single link or node failure due to their underlying ring topology, similar to their single-channel counterparts.

An alternative approach to multichannel upgrade optical single-channel rings relies on topological modifications of the basic ring architecture. Many ways exist to modify and enhance the topology of ring networks, resulting in so-called *augmented rings* (Aiello et al., 2001). In this chapter, we describe a novel multichannel upgrade of optical single-channel ring networks where the ring network is left untouched and only a subset of ring nodes needs to be WDM upgraded and interconnected by a single-hop star WDM subnetwork in a pay-as-you-grow fashion (Maier and Reisslein, 2006). The resultant hybrid ring–star network, called RINGOSTAR, requires additional fiber links to build the star subnetwork, as opposed to WDM rings. Unlike WDM rings, however, RINGOSTAR does not require all ring nodes to be WDM upgraded at the same time. Instead, RINGOSTAR provides an evolutionary multichannel upgrade path of optical single-channel ring networks in that only a subset of ring nodes need to be WDM upgraded and attached to the star WDM subnetwork according to given traffic demands and/or cost constraints. Hence, RINGOSTAR not only saves on nodal WDM upgrade requirements but also enables operators to provide cautious upgrades of existing ring networks and to realize their survival strategy in a highly competitive environment. Furthermore, the additional star subnetwork and the two novel performance-enhancing techniques *proxy stripping* and *protectoration* make RINGOSTAR survivable against multiple link and node failures and let RINGOSTAR clearly outperform conventional WDM rings in terms of spatial reuse, bandwidth efficiency, and network lifetime, as discussed in greater detail in the following.

13.1   Architecture

To improve the spatial reuse, resilience, and bandwidth efficiency of optical single-channel ring networks such as RPR we propose to augment the bidirectional ring by a single-hop star subnetwork. A subset of the ring nodes is connected to the single-hop star subnetwork, preferably by bidirectional pairs of dark fiber. Note that recently most conventional carriers, a growing number of public utility companies, and new network operators make use of their rights of way especially in metropolitan areas to build and offer so-called dark-fiber networks. These dark-fiber providers have installed a fiber infrastructure that exceeds their current needs. The unlit fibers provide a cost-effective way to build very-high-capacity networks or upgrade the capacity of existing (ring) networks. Buying one's own dark fibers is a promising solution to reduce network costs as opposed to leasing bandwidth which is an ongoing expense. Nodes can be attached to the single-hop star subnetwork one at a time in a pay-as-you-grow manner according to given traffic demands. To guarantee minimum short-cuts in terms of hops, the star subnetwork is assumed to be a single-hop network (i.e., all attached nodes are able to communicate with each other in a single hop).

   The hub of the star subnetwork may be a wavelength-broadcasting passive star coupler (PSC), a wavelength-routing arrayed waveguide grating (AWG), or a combination of both. For more detailed information on various possible AWG- and PSC-based single-hop star network architectures together with medium access control (MAC) protocols the interested reader is referred to Maier and Reisslein (2004). The design of the hub depends on the capacity requirements. To interconnect a small number P of ring nodes with low traffic loads it might be sufficient to use a PSC. If the number P of ring nodes attached to the star subnetwork and/or their traffic loads increase it appears reasonable to use an AWG instead of a PSC. Due to its spatial wavelength reuse capability (i.e., all wavelength channels can be simultaneously used for data transmission at each AWG input port), the AWG provides a much larger number of communication channels and capacity than the PSC, which does not allow for spatial wavelength reuse.

   Figure 13.1 shows the network and node architectures of RINGOSTAR using an AWG as the hub of the star subnetwork (Herzog, Maier, and Wolisz, 2005). RINGOSTAR interconnects N nodes which are subdivided into two groups of $P = D \cdot S$ ring-and-star homed nodes and $N_r = N - D \cdot S$ ring homed nodes, where $D \geq 1$ denotes the number of input/output ports of the AWG and $S \geq 1$ denotes the degree of the combiners and splitters attached to the input and output ports of the AWG, respectively. All N nodes are attached to the bidirectional dual-fiber ring by means of two pairs of fixed-tuned transmitter (FT) and fixed-tuned receiver (FR), one for each single-channel fiber. The placement of the P ring-and-star homed nodes is done according to given traffic patterns. Nodes that generate and/or receive a large amount of traffic (e.g., hot-spot nodes) are best attached to the star subnetwork to benefit from its short-cuts. For uniform traffic, where each node generates and receives the same amount of traffic, the P nodes are best equally spaced among the remaining N_r ring homed nodes on the ring, as shown in Fig. 13.1 for $N = 16$ and $P = 4$ (and $N_r = N - P = 12$). Unlike the ring homed

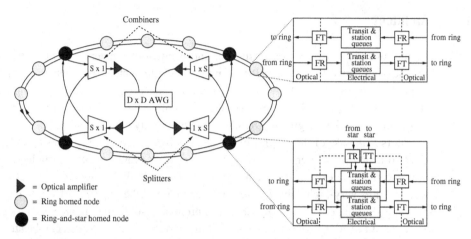

Figure 13.1 RINGOSTAR architecture with $N = 16$ and $D = S = 2$. After Maier and Reisslein (2004). © 2004 IEEE.

nodes, the ring-and-star homed nodes are also attached to the central AWG-based star subnetwork by using an additional tunable transceiver and fiber pair. More precisely, the tunable transmitter (TT) of a given ring-and-star homed node is connected to a combiner input port and its tunable receiver (TR) is located at the opposite splitter output port. Thus, the number of ring-and-star homed nodes is given by $P = D \cdot S \leq N$. Generally, for a given P, different combinations of D and S are possible (e.g., $D = S = 2$ for $P = 4$), as shown in Fig. 13.1. If necessary, an optical amplifier (e.g., Erbium doped fiber amplifier) is placed between each combiner and AWG input port and each AWG output port and splitter to compensate for fiber, splitting, and insertion losses in the star subnetwork. In summary, the structure of each ring homed and ring-and-star homed node is FT^2–FR^2 and FT^2–TT–FR^2–TR, respectively.

Each node performs OEO conversion and has separate electrical transit and station queues for either fiber ring. In case of ring-and-star homed nodes, the station queues are also connected to the star transceiver. Figure 13.2 shows the buffer structure in greater detail for both ring and ring-and-star homed nodes under the assumption that the optical single-channel ring network is an RPR network (RPR was described in Chapter 11). We show only the buffers for one fiber ring. The same buffer structure is replicated at each node for the other counterdirectional fiber. Figure 13.2(a) depicts the architecture of ring homed nodes. Similar to RPR, each ring homed node is equipped with two transit (primary transit queue [PTQ] for high-priority traffic, secondary transit queue [STQ] for low-priority traffic) and two station (receive and transmit) queues. Similarly, ring-and-star homed nodes have the same transit and station queues, as well as two additional station (one transmit and one receive) queues to store data packets going to and coming from the star, as illustrated in Fig. 13.2(b). Nodes attached to the star subnetwork perform proxy stripping, a novel packet stripping technique developed for RINGOSTAR.

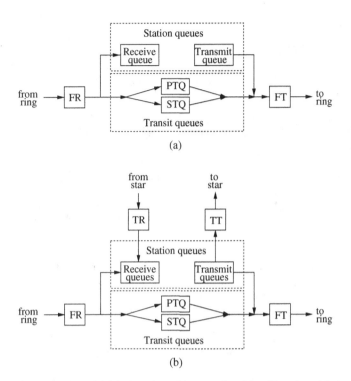

Figure 13.2 RINGOSTAR node architecture for either fiber ring: (a) ring homed node and (b) ring-and-star homed node. After Herzog, Maier, and Wolisz (2005). © 2005 IEEE.

13.2   Proxy stripping

To benefit from the short-cuts provided by the single-hop star subnetwork, each of the P nodes performs *proxy stripping*, as shown in Fig. 13.3(a) (Herzog, Adams, and Maier, 2005). With proxy stripping, each of the P nodes that is neither source nor destination pulls incoming data packets from the ring and sends them across the short-cuts to another proxy stripping node that is either the destination of the data packets or closest to the corresponding destination node on the ring by using the MAC protocol of the given single-hop star subnetwork, as described shortly. In the latter case, the receiving proxy stripping node forwards the data packets on the shortest path by choosing the appropriate fiber ring. The destination node finally takes the data packets from the ring (destination stripping), as illustrated in Fig. 13.3(b) for source-destination pair A–C. Note that packets undergo proxy stripping only if the short-cuts provide a shorter path in terms of hops, where a hop denotes the distance between two adjacent nodes. Otherwise, data packets remain on the ring until they are received by the destination node, as shown in Fig. 13.3(b) for source-destination pair A–B. Practically, this can be done by monitoring each packet's source and destination MAC addresses and making a table lookup at proxy stripping nodes with table entries indicating whether a given data packet has to be proxy stripped or not.

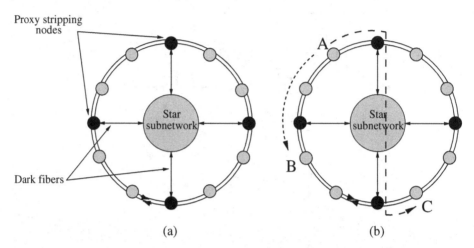

Figure 13.3 Proxy stripping: (a) $N = 12$ ring nodes, where $P = 4$ are interconnected by a dark-fiber star subnetwork, (b) proxy stripping in conjunction with destination stripping and shortest path routing. After Herzog and Maier (2006). © 2006 IEEE

To formally describe proxy stripping let us introduce the following variables for a given pair of source node s and destination node d, where $s, d \in \{0, 1, \ldots, N - 1\}$:

- $h_{rs}(s)$: hop distance between source node s and its closest proxy stripping node.
- $h_{rs}(d)$: hop distance between destination node d and its closest proxy stripping node.
- $h_{\text{ring}}(s, d)$: minimum hop distance between source node s and destination node d on the ring (i.e., without proxy stripping).
- $h_{\text{star}}(s, d)$: minimum hop distance between source node s and destination node d via short-cuts of the single-hop star subnetwork (i.e., with proxy stripping). Note that $h_{\text{star}}(s, d) = h_{rs}(s) + 1 + h_{rs}(d)$.

Generally speaking, if the hop distance on the ring between a given source node s and destination node d is "small enough," the source node sends the data packet(s) on the ring without undergoing proxy stripping. More precisely, if $h_{\text{ring}}(s, d) \leq h_{\text{star}}(s, d)$ then source node s sends the data packet(s) to destination node d along the ring on the shortest path by choosing the appropriate ring. Destination node d takes the transmitted data packet(s) from the ring (destination stripping). Note that in this case there is no proxy stripping (like node pair A–B in Fig. 13.3(b)). Proxy stripping takes place only if $h_{\text{ring}}(s, d) > h_{\text{star}}(s, d)$, that is, if the short-cuts form a shorter path between nodes s and d than either peripheral ring. Specifically, source node s sends the data packet(s) to its closest proxy stripping node. Note that the chosen direction does not necessarily have to be the same as that in shortest-path routing on the original bidirectional ring. This implies that all ring nodes are aware of the presence and location of proxy stripping nodes. As a consequence, packets travel along smaller ring segments resulting in a decreased mean hop distance. Alternatively, data packets could be sent in the same direction as is done in shortest-path-routing bidirectional rings. In doing so, we would obtain a larger mean hop distance. However, nodes would not need to have knowledge about the location of the P proxy stripping nodes. Note that this would allow for a *transparent* proxy stripping dark-fiber upgrade of RPR in which the remaining $N - P$ nodes do not have to be modified at all. With proxy stripping, each of the P nodes takes the corresponding data packet(s) from the ring and sends the data packet(s) across the star subnetwork to the proxy stripping node that is either the destination node or closest to the destination node. A given proxy stripping node pulls only data packets from the ring whose source and destination addresses satisfy the condition $h_{\text{ring}}(s, d) > h_{\text{star}}(s, d)$. After transmitting a given data packet across the single-hop star subnetwork, the corresponding proxy stripping node receives the packet and, if necessary, forwards it on the ring toward destination node d on the shortest path by using the appropriate ring. Destination node d finally takes the data packet from the ring (destination stripping). Besides proxy stripping and forwarding data packets, proxy stripping nodes also generate traffic. Note that in this case $h_{rs}(s) = 0$. Again, if $h_{\text{ring}}(s, d) \leq h_{\text{star}}(s, d)$, then the proxy stripping source node s transmits the data packet on that ring which provides the shortest path to destination node d. Otherwise, if $h_{\text{ring}}(s, d) > h_{\text{star}}(s, d)$, then the proxy stripping source node s sends the data packet across the star subnetwork to the corresponding proxy stripping node which is either the destination itself or forwards the data packet onward to node d via the shortest path ring.

Note that proxy stripping may be applied in both single-queue and dual-queue modes of RPR. In single-queue mode, proxy-stripped packets are put in an additional queue at the corresponding proxy stripping node for transmission across the star subnetwork. After traversing the star subnetwork, proxy-stripped packets are either received by the respective proxy stripping node or, if necessary, are forwarded by putting them into the corresponding ring transit queue. In dual-queue mode, each proxy stripping node has two additional queues for transmission across the star subnetwork. Proxy-stripped packets are put into one of these two queues according to their priority. After transmitting the proxy-stripped packets across the star subnetwork they are either received by the respective proxy stripping node or, if necessary, are forwarded by putting them into the corresponding ring transit queue according to their priority. In doing so, proxy stripping preserves the priority of all packets.

13.3   Access and fairness control

As an evolutionary WDM upgrade, RINGOSTAR builds on the access and fairness control protocols used in the original single-channel ring network (e.g., RPR), which must be adapted to take the short-cuts of the star subnetwork and proxy stripping into account. For data transmission on the bidirectional ring of RINGOSTAR, all N ring homed and ring-and-star homed nodes use the ring protocols without requiring any modifications. For the star subnetwork of RINGOSTAR, the P ring-and-star homed nodes must deploy separate access and fairness protocols to arbitrate data transmission across the AWG of the star subnetwork in a fair manner, as outlined in the following.

13.3.1　Reservation on star subnetwork

Access on the AWG-based star subnetwork can be controlled by using a reservation protocol with pretransmission coordination, which works as follows (Herzog, Maier, and Wolisz, 2005). Prior to transmitting a data packet from the star transmit queue the corresponding ring-and-star homed node broadcasts a control packet on one of the fiber rings by means of source stripping. The control packet consists of three fields: (1) source address of the proxy-stripping ring-and-star homed node; (2) destination address of the ring-and-star homed node that is closest to destination node d; and (3) length of the corresponding data packet. Each ring-and-star homed node receives the broadcast control packet and is thus able to acquire and maintain global knowledge of all P ring-and-star homed nodes' reservation requests for the star subnetwork. Note that the control packets are sent by using RPR's high-priority traffic class service. Hence, the pretransmission coordination of the star subnetwork builds on RPR and requires neither additional hardware nor software. It was shown in Davik et al. (2004) that the latency of high-priority (control) traffic is constant and equal to the round-trip propagation delay of the ring, even under overload conditions. As a result, all ring-and-star homed nodes receive control packets after a deterministic period of time and are able to process control packets and acquire and maintain global knowledge in a synchronized manner. Based on this global knowledge, all P nodes schedule the transmission and reception of the corresponding data packets on the single-hop star subnetwork in a distributed fashion executing a deterministic first-come-first-served and first-fit (FCFS-FF) scheduling algorithm. Note that the chosen scheduling alogrithm is relatively simple. Its low time complexity helps avoid scalability problems in very-high-speed networks where each node has to process broadcast control packets (Ramaswami and Sivarajan, 2001).

Note that the aforementioned reservation with pretransmission coordination suffers from an inefficient use of ring bandwidth and nodal processing resources since each control packet travels along the entire ring. Consequently, each control packet traverses all N nodes and is processed at both ring and ring-and-star homed nodes, even though only the P ring-and-star homed nodes need the control information, resulting in wasted ring bandwidth and nodal processing resources. To mitigate these inefficiencies, control packets may be sent across the star subnetwork instead of the ring subnetwork. The overhead caused by the round trip of the control packet on the ring can be reduced by deploying a wavelength-insensitive PSC in parallel with the AWG and broadcasting the control packet on the star subnetwork via the PSC. We discuss this type of star subnetwork in greater detail in the subsequent section when addressing the fault tolerance of RINGOSTAR.

After transmitting a given data packet across the star, the corresponding ring-and-star homed receiving node puts the data packet into its star receive queue. If necessary, the ring-and-star homed node forwards the data packet on the ring toward the destination node d on the shortest path by using the appropriate fiber ring. Destination node d finally takes the data packet from the ring (destination stripping).

We have seen in Chapter 12 that previously reported multichannel extensions of optical single-channel ring networks deploy WDM on the ring, resulting in meshed or nonmeshed WDM ring networks. Both meshed and non-meshed WDM rings have in common that *all* ring nodes have to be WDM upgraded, be it by arrays of fixed-tuned transceivers, tunable transceivers, wavelength multiplexers, or demultiplexers. Nonmeshed WDM rings yield at most the same spatial reuse factor as their single-channel counterparts. To see this, note that for opaque WDM rings with OEO conversion at each node the spatial reuse factor is the same in both nonmeshed WDM and single-channel rings. In WDM rings with optical bypassing, however, the spatial reuse factor in WDM rings can be smaller than in their single-channel counterparts. This is because optical bypassing makes the wavelength access less flexible and efficient since nodes can be addressed only on certain wavelengths which do not necessarily provide the shortest path. For instance, for uniform traffic in bidirectional rings the mean hop distance equals $N/4$ and the spatial reuse factor is upper bounded by four due to missing alternate physical short-cuts. In RINGOSTAR the spatial reuse factor of all WDM wavelength channels is given by the physical degree of the AWG D, which in principle can be chosen arbitrarily large (practically, for large D a free-space rather than planar AWG has to be used to provide a sufficiently small channel crosstalk). As a consequence, RINGOSTAR potentially achieves a much better WDM upgrade than conventional approaches which deploy WDM on the ring. Note that the parameter D together with S determine not only the number of ring-and-star homed nodes $P = D \cdot S$, but also the degree of spatial wavelength reuse on the ring. By means of proxy stripping, data transmissions are bounded to smaller ring segments, resulting in a decreased mean hop distance, an increased number of simultaneous transmissions on different ring segments, and thus an increased spatial reuse factor on the ring. With properly chosen P, the mean hop distance is smaller than $N/4$ and the spatial reuse factor on the ring can be pushed well beyond that of bidirectional rings (4) for uniform traffic. Clearly, there is a trade-off between ring bandwidth reuse and network costs. For increasing P the spatial reuse factor on the ring improves and the costs of the network increase since more nodes require an additional tunable transceiver attached to the star via a separate fiber pair. It was shown in Herzog, Maier, and Wolisz (2005) and Herzog and Maier (2006) that in RINGOSTAR the overall mean hop distance and network diameter (maximum number of hops between any pair of source and destination nodes) are dramatically decreased and the capacity is significantly increased due to improved spatial wavelength reuse and bandwidth efficiency on both star and ring subnetworks. Specifically, by WDM upgrading and interconnecting only 64 nodes of a 256-node RINGOSTAR network the mean hop distance is less than 5% of that of bidirectional rings with destination stripping and shortest-path routing. In terms of capacity, a 256-node RINGOSTAR network with a single additional (tunable) transceiver at only 64 nodes significantly outperforms unidirectional, bidirectional, and meshed WDM rings in which each of the 256 nodes needs to be WDM upgraded by using an array of 16 (fixed-tuned) transceivers.

Clearly, the gained capacity comes at some expense. For increasing the capacity a subset of nodes must be connected to the star subnetwork. Each of these nodes must be upgraded with tunable transceivers and the access protocol of the star sub-network requires additional processing capacity. Moreover, an AWG, splitters, and combiners must be deployed and interconnected by fibers. Another trade-off between costs and performance concerns the additional delay introduced by the pretransmission coordination required for each transmission over the star subnetwork. The round-trip time of control packets on the peripheral ring could be avoided by deploying a passive star coupler in parallel with the AWG, at the expense of using an additional single passive component in the star subnetwork, as discussed in the subsequent section.

RINGOSTAR is well suited for efficiently transporting uniform traffic typically found in metro core networks with any-to-any traffic demands between central offices. For metro edge networks with their strongly hubbed (hot-spot) traffic patterns, RINGOSTAR with its short-cuts on the star WDM subnetwork provides a promising cost-effective solution in that only the hub node (hot spot) is equipped with multiple transceivers while operating the remaining nodes with a single transceiver. Note that the RINGOSTAR WDM upgrade is not restricted to RPR. RINGOSTAR with its star WDM subnetwork and proxy stripping is generally applicable to asynchronous (e.g., Token ring) and synchronous (e.g., empty slot) ring networks.

13.3.2 Adaptation of DVSR

For fairness control in RINGOSTAR the so-called *distributed virtual-time scheduling in rings* (DVSR) fairness protocol can be extended to incorporate proxy stripping. DVSR was initially proposed to improve fairness control in RPR networks (Gambiroza et al., 2004). Similar to DVSR, to establish RIAS fair transmission rates in RINGOSTAR, packets arriving at the transit queue(s) and station queues are first-in/first-out (FIFO) queued at each node. One fairness control packet circulates upstream on each ring. Each fairness control packet consists of $N + DS/2$ fields. The first N fields contain the fair rates of all ring links and the remaining $DS/2$ fields contain the fair rates of the star links, where one control packet carries the rates of the even-numbered and the other one the rates of the odd-numbered star links. Each node monitors both fairness control packets and writes its locally computed fair rates in the corresponding fields of the fairness control packets. To calculate the fair link rates, each node measures the number of bytes, l_k, arriving from node k, including the station itself, during the time interval T between the previous and the actual arrival of the control packet. Each node performs separate measurements for either ring using two separate time windows. Proxy stripping nodes additionally count the number of bytes arriving from the star for each node and use the time window of the fairness control packet that carries the fair rate of the corresponding proxy stripping node. The fair rate F of a given link is equal to the max-min fair share among all measured link rates l_k/T with respect to the link capacity C currently available for fairness-eligible traffic.

Each node limits the data rate of its $N - 1$ ingress flows by using token buckets whose refill rates are set to the current fair rates of the corresponding destinations. Using the same two time windows as in the calculation of the link fair rates, each node i counts the bytes ρ_{ij} sent to destination j during the time window. Thus, there are two sets of $N - 1$ byte counters, one for each time window. Each time a fairness

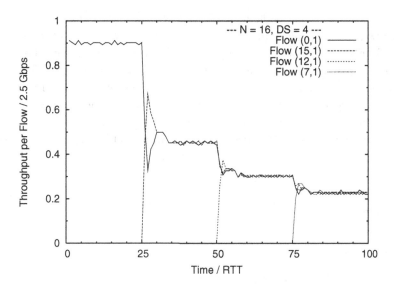

Figure 13.4 Dynamics of adapted DVSR fairness control protocol. After Herzog and Maier (2006). © 2006 IEEE.

control packet arrives, a given node caculates the fair rate of each ingress flow as follows. According to RIAS, the total capacity available to a given node on a certain link equals the fair rate F which is shared among all its ingress flows crossing that link. Based on the measured ingress rates ρ_{ij}/T of these flows and the available capacity F, the max-min fair share f is calculated for each crossed link. The refill rate of each token bucket is set to the minimum fair share f of these links.

The adapted DVSR fairness control protocol was investigated by means of simulation in Herzog and Maier (2006) for uniform self-similar traffic with Hurst parameter 0.75, where each node does not send any traffic to itself and sends a generated data packet to the remaining $N - 1$ nodes with equal probability $1/(N - 1)$. In this study, RPR best-effort traffic class C was considered under the assumption that no bandwidth is reserved for traffic class A and 10% of the ring bandwidth is left for traffic class B, that is, class C traffic must not use more than 90% of the ring bandwidth. Each node was assumed to continuously have data to send on the ring which operates at 2.5 Gbps. The dynamics of the adapted DVSR fairness control are illustrated in Fig. 13.4 for $N = 16$, $D = 4$, and $S = 1$. The figure shows the throughput of four different flows versus time which is given in round-trip time (RTT) of the ring. All four flows cross the ring link (0,1) from node 0 to node 1, where node 0 is assumed to be a proxy stripping node. Initially, only flow (0,1) from source node 0 to destination node 1 is active, achieving a normalized throughput of 0.9. Next, flow (15,1) is activated at 25 RTTs. After a convergence time of approximately 10 RTTs both flows equally share the available bandwidth on link (0,1). Note that before the new fair rates are established, flow (15,1) fills up the transit queue of node 0, resulting in a throttled rate of flow (0,1) and a throughput peak of flow (15,1). After 50 RTTs, flow (12,1) is activated. Flow (12,1) is first sent from node 12 to 0 via the star subnetwork and then uses link (0,1) to reach node 1. We observe that it takes about 10 RTTs to converge to the new fair rates after flow (12,1) has been activated. Finally, flow (7,1) is activated after 75 RTTs. The flow uses the star subnetwork as a short-cut from node 8 to node 0, similarly to flow (12,1). Since the fair rate of link (0,1) is transmitted upstream it takes longer for node 7 to receive changes of the fair rate of link (0,1) than for node 12. Note that some packets collide at node 0 and have to be retransmitted since now two flows use the star subnetwork as a short-cut, resulting in an increased delay. However, the convergence time remains at approximately 10 RTTs. In summary, the sending rates adapt precisely to the theoretically expected rates in about 10 RTTs and do not suffer from severe oscillations afterward.

13.4   Protectoration

13.4.1   Limitations of RPR protection

As a state-of-the-art optical single-channel ring network, RPR aims at improving throughput efficiency and service differentiation in addition to the resilience of optical packet-switched bidirectional dual-fiber ring networks. To achieve a higher level of resilience, two methods of protection are considered in RPR – wrapping and steering – as described in great detail in Chapter 11. With wrapping, upon detection of a link or node failure the two ring nodes adjacent to the failed link or node switch all traffic arriving on the incoming fiber onto the outgoing fiber to reach the destination node going in the opposite direction. Thus, the two ring nodes adjacent to the failure wrap all traffic away from the failed link or node. With steering, after learning that a failure has occurred a given source node injects the traffic in the direction opposite to the link or node failure (i.e., the source node steers the traffic away from the failure). Note that in the case of a link and/or node failure the two protection techniques result in a rather inefficient use of bandwidth resources. Specifically, wrapped traffic travels from the source node to the corresponding wrapping node and then back to the source node, thus consuming bandwidth without getting closer to the destination. After returning to the source node the wrapped traffic continues traveling toward the destination node along a secondary path which is longer than the primary path in terms of hops. Note that the bandwidth inefficiency due to wrapping occurs only during a transient period before the failure notification has propagated to the source node. Steering avoids the wasting of bandwidth due to the round trip of traffic between source node and wrapping node, but it does not eliminate the increased bandwidth consumption incurred on the secondary path. For instance, it was shown in Spadaro et al. (2004) that in the event of a failure the loss of traffic in a 63-node RPR ring network may be as high as 94% due to the increased length of the backup path. We note that the traffic scenario in Spadaro et al. (2004) is rather peculiar but realistic, as the authors indicate. It was primarily used to illustrate possible problematic situations of RPR network applications. Furthermore, note that both fault recovery techniques are able to protect traffic only against a single link or node failure. In case of multiple failures, the full connectivity of RPR is lost (i.e., the ring is divided into two or more disjoint subrings).

RPR as a metro ring network will be deployed often in interconnected ring architectures. Metropolitan interconnected rings are composed of metro core and metro edge rings, where a metro core ring interconnects several metro edge rings, as depicted in Fig. III.1. Metro core rings have to meet several requirements. Aside from configurability, reliability, flexibility, scalability, and large capacity, metro core rings have to be extremely survivable (Saleh and Simmons, 1999). If the metro core ring network fails, all customers are potentially left without service. Thus, survivability in metro core ring networks is crucial. Especially storage networking protocols, being one of the important metro applications without built-in adequate survivability, rely almost entirely on the failure recovery techniques of the optical layer (Gerstel and Ramaswami, 2003).

As mentioned earlier, RPR with its current protection schemes provides only inefficient and limited survivability. It is worthwhile mentioning that pre-standard RPR solutions exist which implement advanced resilience solutions in addition to wrapping and steering. These solutions are able to protect traffic against multiple failures on the same fiber ring. For example, in the so-called *single ring recovery* (SRR) protocol, an extension to the spatial reuse protocol (SRP) of dynamic packet transport (DPT) rings, wrapping and steering take place in the case of a single link/node failure on either ring, similar to RPR. If on one of both rings multiple failures occur, wrapping is not deployed and all ring nodes use only the other failure-free ring. In doing so, the failed ring may have multiple failures without losing full network connectivity. This, however, only holds if the other ring is failure free. If there are failures on both rings full network connectivity is lost and the bidirectional ring is divided into disjoint subrings. Note that SRR affects the entire ring network in that *all* ring nodes must support SRR. If one or more ring nodes do not support SRR, SRR will have no effect.

In the following section, we introduce the *protectoration* technique which affects only a subset of ring nodes, as opposed to SRR. More importantly, protectoration is able to guarantee full network connectivity also in the presence of multiple failures on both fiber rings, as opposed to SRR (Maier et al., 2005).

13.4.2 Protectoration

In this section, we outline the underlying architecture and operation of the protectoration technique and elaborate on its impact on RPR.

Architecture

Figure 13.5 shows the network architecture, whose star subnetwork differs from that shown in Fig. 13.1. The star subnetwork is based on a central hub which consists of a $D \times D$ AWG in parallel with a $D \times D$ PSC, where $D \geq 1$. Each ring-and-star homed node i, $i = 1, \ldots, P$, has a home channel λ_i on the PSC (i.e., a unique wavelength channel λ_i on which node i receives data transmitted over the PSC). In addition, there is a control wavelength channel λ_c on the PSC. Consequently, there are $\Lambda_{\mathrm{PSC}} = P + 1 = D \cdot S + 1$ wavelength channels on the PSC, which make up the PSC waveband. The AWG waveband consists of $\Lambda_{\mathrm{AWG}} = D \cdot R$ contiguous data wavelength channels, where

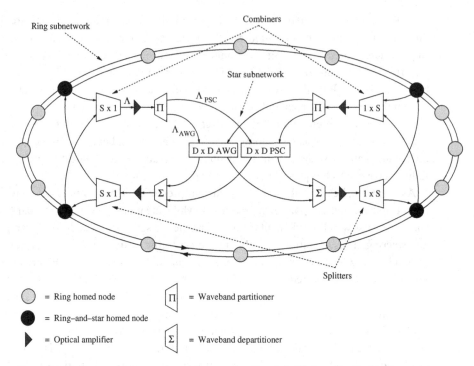

Figure 13.5 Protectoration network architecture for $N = 16$ and $P = D \cdot S = 2 \cdot 2 = 4$. After Maier et al. (2005). © 2005 IEEE.

$R \geq 1$ denotes the number of used free spectral ranges (FSRs) of the underlying $D \times D$ AWG. A total of $\Lambda = \Lambda_{\text{AWG}} + \Lambda_{\text{PSC}}$ contiguous wavelength channels are operated in the star subnetwork, as further detailed shortly.

The signals from S ring-and-star homed nodes on the Λ wavelength channels are transmitted on S distinct fibers to a $S \times 1$ combiner, which combines the signals onto the Λ wavelength channels of one fiber leading to a waveband partitioner. The waveband partitioner partitions the set of Λ wavelengths into the AWG and PSC wavebands, which are fed into an AWG and PSC input port, respectively. The signals from the opposite AWG and PSC output ports are collected by a waveband departitioner and then equally distributed to the S ring-and-star homed nodes by a $1 \times S$ splitter. If necessary, optical amplifiers are used between combiner and partitioner as well as splitter and departitioner to compensate for attenuation and insertion losses of the star subnetwork. A total of D of these arrangements, each consisting of combiner, amplifier, waveband partitioner, waveband departitioner, amplifier, and splitter, are used to connect all $P = D \cdot S$ ring-and-star homed nodes to the central hub.

The architecture of ring homed nodes is identical to that of RPR nodes depicted in Fig. 13.2(a). Figure 13.6 depicts the architecture of a ring-and-star homed node with PSC data channel λ_i, where $\lambda_i \in \{1, 2, \ldots, D \cdot S\}$. Each ring-and-star homed node has the same number and type of transceivers and queues as a ring homed node for transmission and reception on both rings. In addition, each ring-and-star homed node has several

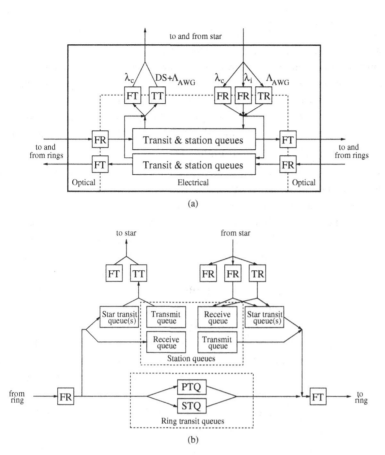

Figure 13.6 Protectoration architecture of ring-and-star homed node with home channel $\lambda_i \in \{1, 2, \ldots, D \cdot S\}$: (a) node architecture for both rings and (b) buffer structure for either ring. After Maier et al. (2005). © 2005 IEEE.

transceivers which are attached to the star subnetwork, by means of a pair of outgoing and incoming fibers. The outgoing fiber is connected to a combiner and the incoming fiber is connected to the splitter which is attached to the opposite AWG-PSC input ports.

As shown in Fig. 13.6(a), for control transmission on the star subnetwork, each ring-and-star homed node is equipped with a transmitter (FT) fixed tuned to the control wavelength channel λ_c of the PSC waveband, which consists of $\Lambda_{\text{PSC}} = 1 + D \cdot S$ wavelength channels. The remaining $D \cdot S$ wavelength channels of the PSC waveband and all $\Lambda_{\text{AWG}} = D \cdot R$ wavelength channels of the AWG waveband are accessed for data transmission by a tunable transmitter (TT) whose tuning range equals $D \cdot S + \Lambda_{AWG} = D(S + R)$. Similarly, for control reception on the star subnetwork each ring-and-star homed node is equipped with a receiver (FR) fixed tuned to the control wavelength channel λ_c of the PSC waveband. For data reception on the PSC each ring-and-star homed node has a separate fixed-tuned receiver (FR) operating at its own dedicated *home channel* $\lambda_i \in \{1, 2, \ldots, D \cdot S\}$. Each data wavelength channel of the PSC waveband is dedicated to a different ring-and-star homed node for reception. Data packets transmitted on PSC data wavelength channels do not suffer from receiver collisions, also referred to as destination conflicts (a receiver collision occurs when the receiver of the intended destination node is not tuned to the wavelength channel on which the data packet was sent by the corresponding source node). Moreover, on the wavelength channels of the AWG waveband, data packets are received by a tunable receiver (TR) whose tuning range equals $\Lambda_{AWG} = D \cdot R$. All transceivers of the star subnetwork are connected to the station queues. Note that the required tuning range of the tunable receiver (Λ_{AWG}) is smaller than that of the tunable transmitter ($D \cdot S + \Lambda_{AWG}$). These requirements take into account the current state of the art of tunable transceivers. While fast tunable transmitters with a relatively large tuning range have been shown to be feasible, tunable receivers are less mature in terms of tuning time and/or tuning range.

Figure 13.6(b) depicts the buffer structure of a ring-and-star homed node in greater detail for one ring direction. Each ring-and-star homed node has for each ring direction the following queues: one or two ring transit queues (depending on the operation mode), one ring transmit queue, and one ring receive queue (the additional queue for control is not shown). To send locally generated traffic to and receive traffic destined for itself from the star subnetwork each ring-and-star homed node has an additional star transmit queue and star receive queue. Furthermore, packets that are pulled from the ring (coming in from both directions of the ring) and forwarded onto the star subnetwork are placed in an additional ring-to-star transit queue (single-queue mode) or one of two additional ring-to-star transit queues according to their priority (dual-queue mode). Traffic that is received from the star subnetwork and needs to be forwarded on either ring is placed in the additional star-to-ring transit queue (single-queue mode) or one of two additional star-to-ring transit queues according to its priority (dual-queue mode) of the corresponding fiber ring. (The additional queue for sending control on the star subnetwork is not shown.) The service among the transmit and transit queues of the star subnetwork is arbitrated similar to that of their counterparts of the ring subnetwork. We arbitrate the transmissions from the ring transit queue and the star-to-ring transit queue of a given priority level in a given ring direction using a round-robin policy.

Operation

Next, let us describe and discuss the protectoration technique in greater detail. We first describe the normal operation of the network (i.e., without any failures). We then explain the network operation in the presence of link, node, and other failures.

Operation without failures

1. **Wavelength Channel Allocation Scheme in Star Subnetwork:** Figure 13.7 illustrates how the Λ contiguous wavelength channels of the star subnetwork are used for control and data transmission. Time is divided into frames which are repeated periodically. Each frame consists of $F \geq D \cdot S$ slots, where one slot is equal to the transmission time of a control packet (function and format of a control packet are defined shortly). As shown in Fig. 13.7, all $D \cdot S$ home channels of the PSC waveband

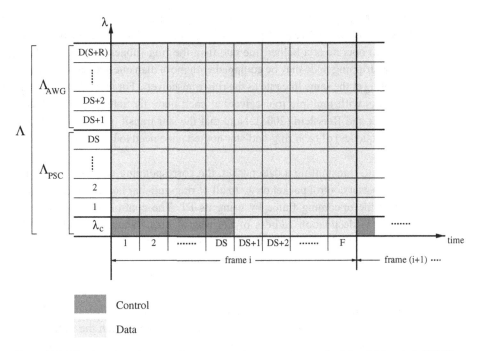

Figure 13.7 Wavelength assignment in protectoration star subnetwork. After Maier et al. (2005). © 2005 IEEE.

and all wavelength channels of the AWG waveband are used for data transmission. All these data wavelength channels are not statically assigned to nodes. Instead, access to these wavelength channels is arbitrated by broadcasting control packets on the control wavelength channel λ_c of the PSC prior to transmitting data packets, as explained in greater detail in the following subsection. Control packets are allowed to be sent on λ_c during the first $D \cdot S$ slots of each frame. More precisely, each of these $D \cdot S$ slots is dedicated to a different ring-and-star homed node such that channel collisions of control packets are avoided. The remaining $(F - D \cdot S)$ slots of each frame can be used for data transmission on λ_c. Note that data packets sent during these slots on λ_c are received by all ring-and-star homed nodes by using their receiver fixed tuned to λ_c. Thus, these slots allow for broadcasting in the star subnetwork.

2. **Wavelength Access:** A ring-and-star homed node puts data packet(s) pulled from the ring in one of the two corresponding star transit queues belonging to its TT that is attached to the star subnetwork. The star transit queue of the TT is chosen according to the priority of the pulled data packet(s). The service among these two star transit queues that store in-transit traffic coming from the ring and the transmit queue that stores locally generated traffic is arbitrated by applying the same scheduling algorithms as used on the ring. That is, ring-to-star in-transit traffic is given priority over star traffic locally generated by the proxy-stripping node. Similar to the transit queues on the ring, the star transit queues thus provide a lossless path for in-transit traffic. Depending on the traffic pattern as well as the number and location of the

proxy-stripping nodes the amount of proxy-stripped traffic may become large. To provide lossless delivery of proxy-stripped packets the star subnetwork in general needs to operate at a higher line rate than the ring subnetwork. Alternatively, each proxy-stripping node may be equipped with more than one star data transceiver, each operating at the same line rate as the ring transceivers. For more details on star WDM networks with multiple transceivers at each node the interested reader is referred to Maier and Reisslein (2004). Note that the star transit queues (as well as the star station queues) of each ring-and-star homed node need to be added to the RPR MAC layer.

Prior to transmitting a data packet, the corresponding ring-and-star homed node broadcasts a control packet on λ_c to all P ring-and-star homed nodes in its assigned slot of the upcoming frame by using its FT. The control packet consists of three fields: (1) destination address of the ring-and-star homed node that is closest to destination node d, (2) length of the corresponding data packet, and (3) priority of the corresponding data packet. After announcing the data packet in its assigned control slot, the ring-and-star homed node transmits the corresponding data packet on the home channel λ_i of the addressed ring-and-star homed node in the subsequent L slots by using its TT, where $\lambda_i \in \{1, 2, \ldots, D \cdot S\}$ and L denotes the length of the data packet in number of slots. Data packets are sent within the same frame as the corresponding control packet and have a maximum length of $(F - D \cdot S)$ slots (i.e., $1 \leq L \leq F - D \cdot S$). We note that due to this assumption a small fraction of each home channel λ_i is not used at the beginning of each frame. However, this could be easily avoided by letting nodes send data packets across the boundary of adjacent frames. After an end-to-end propagation delay of the PSC of the star subnetwork all ring-and-star homed nodes receive the broadcast control packet by using their FRs fixed tuned to λ_c. The corresponding data packet is successfully received at the addressed ring-and-star homed node by using its FR fixed tuned to λ_i, unless one or more other ring-and-star homed nodes have transmitted data packets on λ_i in at least one of the aforementioned L slots. In the latter case, all involved data packets are assumed to be corrupted due to (channel) collision and have to be retransmitted. Collided data packets are kept in the queues until the transmission is successful. Note that due to the fact that control packets are sent collision-free all ring-and-star homed nodes are aware of the original order of the corresponding data packets. As a consequence, even though collided data packets need to be retransmitted, the receiving ring-and-star homed nodes are able to restore the original sequence of data packets and thus maintain in-order packet delivery.

The retransmission of collided data packets works as follows. Due to the dedicated access control of the control channel λ_c, collisions of control packets are prevented. Therefore, for collided data packets no control packets have to be retransmitted. Instead, each ring-and-star homed node is able to find out which transmitted data packets have experienced channel collision by processing the previously (success-fully) transmitted control packets. More precisely, the index j, $1 \leq j \leq D \cdot S$, of the used control slot and the destination and length fields of the control packet enable each ring-and-star homed node to determine whether the corresponding data packet has collided or not. Collided data packets are not retransmitted on the home channels of the PSC but across the AWG by using one of the Λ_{AWG} wavelength channels. Given the index j of the control slot, which uniquely identifies not only the given source ring-and-star homed node but more importantly also the input port of the AWG to which it is attached, together with the destination field of the corresponding control packet all ring-and-star homed nodes are able to determine the wavelength in each FSR of the AWG which provides a single-hop connection between the corresponding pair of source and destination ring-and-star homed nodes. The actual retransmissions on the chosen wavelength channels are scheduled in a distributed fashion by all ring-and-star homed nodes. The scheduling starts at the beginning of frame $i + 1$ upon receiving the control packets in frame i after one end-to-end propagation delay of the PSC of the star subnetwork. At the end of every frame each ring-and-star homed node collects all control packets belonging to collided data packets. By using each control packet's priority field, each ring-and-star homed node first processes all high-priority control packets and then all low-priority control packets. Control packets of the same priority class are randomly chosen for scheduling. All ring-and-star homed nodes deploy the same random algorithm and same seed and therefore build the same schedule. Note that randomizing the scheduling counteracts the static control slot assignment, resulting in an improved fairness among the ring-and-star homed nodes. Otherwise, source nodes with a smaller index j would be more successful in the scheduling than nodes with a larger index. The corresponding data packets of the selected control packets are scheduled on a first-fit basis starting from the lowest possible wavelength channel at the earliest possible time. The data packet is retransmitted on the corresponding AWG wavelength channel at the scheduled time. After the successful retransmission of a given data packet across the AWG, the corresponding ring-and-star homed receiving node puts the data packet in the star subnetwork receive queue belonging to its TR if the data packet is destined for itself. Otherwise, the ring-and-star homed node forwards the data packet on the ring toward the destination node d on the shortest path by using the appropriate fiber ring and placing the data packet in the corresponding star-to-ring transit queue. Destination node d finally takes the data packet from the ring. We note that the aggregated length of the collided packets can be larger than $F - D \cdot S$. This fact does not pose any problems since the retransmission takes place over the AWG where transmissions are permitted to be scheduled across frame boundaries.

Operation with failures

Aside from link and node failures, other network elements may fail. In the star subnetwork, splitters, amplifiers, combiners, waveband partitioners/departitioners, PSC, or AWG may go down. Note that the various failures affect the network differently. For instance, while a fiber cut between a given ring-and-star homed node and the attached combiner disconnects only a single node from the star subnetwork, the entire star subnetwork goes down if the central hub fails (i.e., if both AWG and PSC fail). In the following, we assume that each node is able to detect any type of failure in both ring and star subnetworks. For a more detailed discussion on available techniques for fault

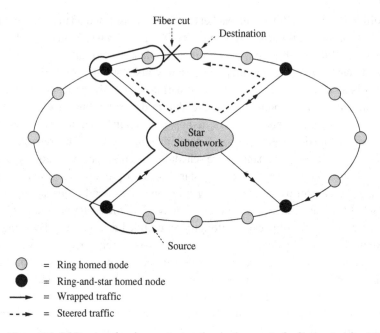

Fiber cut

Destination

Star
Subnetwork

Source

⊘ = Ring homed node
⬤ = Ring-and-star homed node
──▶ = Wrapped traffic
--▶ = Steered traffic

Figure 13.8 RPR network using protectoration in the event of a fiber cut. After Maier et al. (2005). © 2005 IEEE

detection in the ring and star subnetwork we refer the interested reader to Suwala and Swallow (2004) and Fan et al. (2004), respectively.

The protectoration technique builds on the wrapping and steering techniques of RPR and thus provides an evolutionary upgrade of RPR. Moreover, protectoration makes RPR resilient against multiple link and node failures in an efficient manner, as we shall see shortly. In the following, we first consider link and node failures only in the ring subnetwork while the star subnetwork is assumed to work properly. Subsequently, we also take failures in the star subnetwork into account.

1. **Failures only in Ring Subnetwork:** Figure 13.8 depicts a bidirectional RPR network with $N = 16$ nodes, where $P = 4$ ring-and-star homed nodes are interconnected via the star subnetwork. For illustration we consider a pair of source and destination nodes and a single fiber cut, as shown in Fig. 13.8. The source node sends all data packets intended for the destination node to its closest proxy-stripping node, which in turn forwards the data packets across the single-hop star subnetwork to the proxy-stripping node that is closest to the destination node. Upon detection of the fiber cut, the data packets are wrapped and return to the proxy-stripping node closest to the destination node. Now, instead of forwarding the wrapped traffic to the source node on the counterclockwise ring (as done in conventional RPR), the corresponding proxy stripping node sends the wrapped data packets across the single-hop star subnetwork to that proxy stripping node which is on the other side of the fiber cut. The latter proxy-stripping node receives the data packets from the star subnetwork and forwards them to the destination node on the counterclockwise ring. Note that

the former proxy-stripping node (on the left-hand side of the link failure) deploys not only proxy stripping but also steering of wrapped traffic. Thus, wrapped traffic does not have to go back all the way to the source node but is sent along a single-hop path to another proxy stripping node. In the event of a single fiber cut, wrapped traffic neither makes a round-trip between source and destination nodes nor takes any long secondary path. As a result, wrapped traffic consumes significantly fewer bandwidth resources on the ring network, resulting in a more efficient use of bandwidth. After learning about the fiber cut, the source node – and its closest reachable ring-and-star homed node, in case proxy stripping results in an intact source-to-destination route with a smaller hop count – transmits the data packets along a different path. In our example, after learning that the fiber cut has occurred, the source node sends the data packets intended for the destination node to its closest proxy-stripping node, which in turn sends the proxy-stripped packets on the star subnetwork directly to the proxy-stripping node on the right-hand side of the destination node. The latter proxy-stripping node finally forwards the data packets to the destination node on the counterclockwise ring. Hence, protectoration requires significantly fewer bandwidth resources on the ring than conventional steering which would use the peripheral counter-clockwise ring that is significantly longer than the short-cuts of the star subnetwork in terms of hops.

Note that a ring-and-star homed node is able to determine which packets have to be proxy stripped by using the source and destination addresses available in RPR's MAC address fields of each packet. These MAC addresses, together with the direction a given packet comes from, enable each ring-and-star homed node to determine whether a given data packet has undergone wrapping or not. If wrapping has taken place, the corresponding ring-and-star homed node recomputes the shortest path taking the link failure into account and sends a given wrapped packet along the updated shortest path. Depending on the updated shortest path, the corresponding ring-and-star homed node forwards a given wrapped data packet either on the ring or star subnetwork. Note that each node computes the shortest path by using its topology database. Each node maintains and updates its topology database by means of RPR's built-in topology discovery protocol, as explained in Chapter 11.

Clearly, single-failure scenarios also include node failures apart from link failures. While a single failed ring homed node triggers the same procedure as that mentioned earlier, special attention has to be paid to a failed ring-and-star homed node. If a ring-and-star homed node goes down it is not further available for proxy stripping traffic from the ring subnetwork and forwarding traffic coming from the star subnetwork. In this case, the two ring homed nodes adjacent to the failed proxy-stripping node detect the failure and inform the remaining nodes by sending control packets. After learning about the failed proxy-stripping node the remaining nodes do not send traffic to the failed ring-and-star homed node. Instead, the neighboring proxy-stripping nodes of the failed proxy-stripping node take over its role of proxy-stripping regular traffic and steering wrapped traffic.

Next, let us consider multiple failures in the ring subnetwork. If there are multiple link and/or node failures on the ring subnetwork, nodes can use the star subnetwork to bypass the failures. Thus, with an intact star subnetwork multiple link and/or node failures on the ring subnetwork may occur simultaneously without losing full connectivity. Note, however, that full connectivity in the event of multiple failures is only guaranteed if no more than one link or node failure occurs between each pair of ring-and-star homed nodes. Otherwise, one or more nodes between a given pair of ring-and-star homed nodes are disconnected from the network.

2. **Failures in Both Ring and Star Subnetworks:** Failures in the star subnetwork include fiber cuts and nonfunctional network devices such as failed combiners/splitters, waveband (de)partitioners, AWG, PSC, and amplifiers. Depending on the failure, only one, a subset, or all ring-and-star homed nodes are disconnected from the star subnetwork. More precisely, a fiber cut between a given ring-and-star homed node and the combiner/splitter port to which it is attached disconnects only the ring-and-star homed node from the star subnetwork. If a given combiner/splitter, amplifier, waveband (de)partitioner, or any fiber between these devices goes down, all S corresponding ring-and-star homed nodes are disconnected from the star subnetwork. If the central hub (AWG and PSC) goes down, the connectivity of the star subnetwork is entirely lost, reducing the network to a conventional bidirectional RPR ring network. If a given ring-and-star homed node detects that it is disconnected from the star subnetwork, it is unable to send and receive traffic to and from the star subnetwork. After detecting disconnection, the affected ring-and-star homed node informs all remaining nodes by broadcasting a control packet on either ring and acts subsequently as a conventional ring homed node. Failures in the ring subnetwork are handled as described earlier.

Impact on RPR

The bidirectional RPR ring network with its two protection techniques, wrapping and steering, is able to guarantee full connectivity only if no more than one link or node failure occurs. Full connectivity in the event of multiple link and/or node failures can also be achieved by interconnecting several ring nodes via a star subnetwork. In doing so, the ring is divided into several segments, each comprising the nodes between two adjacent ring-and-star homed nodes. Each segment is able to recover from a single link or node failure without losing full connectivity of the network. Thus, the number of fully recoverable link and/or node failures is identical to the number of ring-and-star homed nodes, provided that there is no more than one failure in each segment.

The proposed multiple-failure recovery technique combines the fast recovery time of protection (wrapping) and the bandwidth efficiency of restoration (steering together with proxy stripping). Accordingly, we call this hybrid approach *protectoration*. Protectoration does not change the scheduling algorithms of RPR. The service among the station and transit queues of the star subnetwork is arbitrated as that of their counterparts of the ring subnetwork. On the ring, in-transit traffic has priority over station traffic. Similarly, proxy-stripped traffic which is pulled from the ring is stored in the ring-to-star transit queue(s) of the corresponding ring-and-star homed node and is sent with higher priority across the star subnetwork than locally generated traffic. Traffic which arrives from the star subnetwork and needs to be forwarded on the ring is put in the star-to-ring transit queue(s) and is forwarded together with ring in-transit traffic in a round-robin fashion with higher priority than traffic locally generated by the corresponding ring-and-star homed node. Hence, in-transit traffic is given priority over station traffic on both ring and star subnetworks. As a result, with proxy stripping, in-transit packets are not lost due to buffer overflow, thus maintaining the lossless property of RPR. Note that this holds only if the amount of both ring in-transit traffic and star-to-ring in-transit traffic remain below a certain threshold. Since at each ring-and-star homed node two queues (star-to-ring and ring transit queues) compete for the bandwidth on the outlink to the downstream ring node, packets may be lost due to buffer overflow, unless congestion control is applied. We explained in Section 13.3.2 how RPR fairness control algorithms are able to alleviate the congestion and thus prevent packet loss in both ring and star-to-ring transit queues.

The protectoration technique supports the service differentiation of dual-queue-mode RPR by storing packets that are sent on or received from the star subnetwork in one of the ring-to-star or star-to-ring transit queues, respectively, according to their priority. Also note that on the star subnetwork reservation control packets carry the priority of the data packets which are subsequently sent across the star. This not only enables the receiving ring-and-star homed node to put an incoming data packet in the corresponding star-to-ring transit queue but also allows for a collided data packet to be retransmitted according to its priority indicated in the corresponding control packet, which has been sent collision-free due to the dedicated time slot assignment on the control channel.

Under stable operational conditions packets sent between a given pair of source and destination nodes are delivered in order along the shortest path in the hybrid ring–star network by using proxy stripping. Note that proxy stripping does not require any modifications of the RPR protection techniques wrapping and steering, which are used in the event of network failures. Like in RPR, packets marked eligible for wrapping are wrapped by the nodes adjacent to a given failure, which are subsequently proxy stripped by the next ring-and-star homed node and sent along the updated shortest path on the star subnetwork, thus bypassing the failed ring link or node. Steering does not need to be modified either. Proxy stripping can be used for both the *strict* packet mode, which is the default packet mode in IEEE 802 protocols, and the *relaxed* packet mode of RPR. These two packet modes affect the steering of traffic by the source node after learning about a failure. In brief, in the strict packet mode, all ring nodes stop adding packets and discard all in-transit packets until their topology database is updated and provides a stable and consistent topology image. In the relaxed mode, ring nodes may steer packets immediately after a failure without waiting for their topology image to become stable and consistent. Note that steering is done by source nodes, whereas proxy stripping is done by intermediate nodes between a given pair of source and destination nodes. To guarantee in-order packet delivery in the hybrid ring-and-star network, source nodes must apply strict-mode steering while intermediate nodes perform proxy stripping of both wrapped traffic and steered traffic, as discussed in detail earlier.

In summary, the proposed protectoration technique used in RINGOSTAR provides full recovery from multiple link and node failures without requiring any major modifications of the basic RPR protocols and mechanisms and appears to be a promising solution to improve the resilience of RPR dramatically. It was shown in Maier et al. (2005) that protectoration without any failures achieves a significantly higher throughput-delay performance than a failure-free RPR network. For a large number of failures on the star subnetwork, the throughput-delay performance of the protectoration network degenerates to the performance of a conventional RPR network. The impact of the failures on the ring subnetwork on the throughput-delay performance depends largely on the position of the failure (i.e., the distance from the ring-and-star homed node where the ring subnetwork is connected to the star subnetwork) and is largely independent from the number of failures.

13.5   Network lifetime

Traffic modeling is of great importance for the accurate design and performance evaluation of telecommunications networks. Various models have been proposed to describe traffic, study network performance, and evaluate traffic engineering schemes (Frost and Melamed, 1994; Adas, 1997; Dwivedi and Wagner, 2000). Apart from traffic modeling, *traffic forecasting* is an important means which allows network operators to predict when and where in the network bottlenecks emerge and capacity upgrades are needed to accomodate the expected traffic growth. Recently, a methodology for forecasting Internet backbone traffic was presented which allows prediction of required capacity upgrades over several months into the future within reasonable error bounds. The obtained results are very useful for network operators to know when and where link upgrades and/or link additions have to take place. However, the proposed methodology does not take into account events that are hard to predict (e.g., breaking news, flash crowd events, denial-of-service attacks, or link failures). Those events are in general hard to predict and may be typically short lived, but they occur frequently and have a significant impact on the traffic load of the network, as measurements show (Papagiannaki et al., 2005).

The amount of unpredictable traffic is expected to increase due to the aforementioned short-lived events and the fact that server farms that provide popular Internet services may be located anywhere on the network. If a server farm provides a service that becomes popular the capacity demand increases in that part of the network. As a result, changes and shifts in the traffic load are expected to occur more frequently and more suddenly than in the past (Maxemchuk et al., 2005). To cope with these traffic scenarios efficiently, the ability of a network to sustain unexpected changes and shifts in traffic load is becoming increasingly important. Recently, a novel quantitative measure called *network lifetime* was proposed and investigated in order to predict the ability of a network to survive unpredicted traffic changes, isolate the capacity bottlenecks of a network, and to compare different network topologies (Maxemchuk et al., 2005). As we will see shortly, the network lifetime is a measure of the growth and shifts in the traffic load that a network can sustain. The results in Maxemchuk et al. (2005) show that the network lifetime of a given network largely depends on the assumed initial traffic matrices.

Furthermore, it was shown that bidirectional rings, together with hierarchical networks, have the poorest performance in terms of network lifetime, while so-called chordal rings perform best not only for uniform traffic growth but also for traffic demand perturbations (to be defined shortly). A chordal ring is a ring where so-called chords directly connect pairs of ring nodes that are multiple hops apart on the ring, each chord having a length of one hop. An example for chordal rings are the so-called meshed rings, which were discussed in Section 12.5.

The network lifetime is a quantitative measure of the maximum uniform growth and shifts, also referred to as perturbations, in a given initial traffic matrix that a network can sustain without exceeding its link capacities and violating its routing constraints before links must be added. Specifically, let $\Psi^*(U)$ denote the network lifetime. The network lifetime is a function of U, where $U \geq 0$ denotes the so-called unexpected traffic growth in a traffic distribution. More precisely, the amount of traffic between node pair (i, j) or at a single node i in a given initial traffic matrix increases by U while the total amount of traffic remains constant, where $1 \leq i, j \leq N$ and N denotes the number of network nodes. That is, with increasing U, more traffic is moved to node pair (i, j) or a single node i from the remaining nodes. For $U = U_{\max}$ all traffic has been moved to node pair (i, j) or a single node i with no traffic left at the remaining nodes, where the computation of $U_{\max}$ is given in Maxemchuk et al. (2005). For any value of $U \in [0, U_{\max}]$, the network lifetime $\Psi^*(U)$ denotes the maximum uniform growth factor by which the traffic load between all node pairs (i, j) and at a single node i, $1 \leq i, j \leq N$, of a given initial traffic matrix can be multiplied and supported by the network without exceeding its link capacities and violating its routing constraints.

The network lifetime of optical bidirectional, chordal, meshed rings, and RINGOSTAR was examined in Maier (2006) and Maier and Herzog (2007) for realistic (a)symmetric uniform and hot-spot initial traffic matrices, as typically found in metro edge and metro core ring networks. The obtained results show that under realistic traffic scenarios chordal rings are less sensitive to small to medium traffic perturbations than bidirectional rings, especially under uniform traffic. However, chordal rings do not necessarily provide an improved network lifetime. They may be superior or inferior to bidirectional rings in terms of network lifetime, depending on the type of chords in use. The findings further show that RINGOSTAR dramatically improves the network lifetime of bidirectional ring networks and that RINGOSTAR is superior to chordal and meshed rings for uniform and in particular hot-spot traffic.

Part IV

Optical access and local area networks

Overview

Future broadband optical access networks not only have to unleash the economic potential and societal benefit by opening up the first/last mile bandwidth bottleneck between bandwidth-hungry end users and high-speed backbone networks but they also must enable the support of a wide range of new and emerging services and applications (e.g., triple play, video on demand, video conferencing, peer-to-peer [P2P] audio/video file sharing, multichannel high-definition television [HDTV], multimedia/multiparty online gaming, telemedicine, telecommuting, and surveillance) to get back on the road to prosperity. Due to their longevity, low attenuation, and huge bandwidth, asynchronous transfer mode (ATM) or Ethernet-based passive optical networks (PONs) are already widely deployed in today's operational access networks (e.g., fiber-to-the-premises [FTTP] and fiber-to-the-home [FTTH] networks) (Abrams et al., 2005). Typically, these PONs are time division multiplexing (TDM) single-channel systems, where the fiber infrastructure carries a single upstream wavelength channel and a single downstream wavelength channel. To support the aforementioned emerging services and applications in a cost-effective and future-proof manner and to unleash the full potential of FTTX networks, PONs need to evolve by addressing the following three tasks (Shinohara, 2005):

1. **Cost Reduction:** Cost is key in access networks due to the small number of cost-sharing subscribers compared to that of metro and wide area networks. Devices and components that can be mass produced and widely applied to different types of equipment and situations must be developed. Importantly, installation costs which largely contribute to the overall costs must be reduced. A promising example for cutting installation costs is NTT's envisioned do-it-yourself (DIY) installation which deploys a user-friendly hole-assisted fiber which exhibits negligible loss increase and sufficient reliability, even when it is bent at right angles, clinched, or knotted, and can be produced economically.
2. **Colorless ONU:** The next target is to make the optical network unit (ONU), which connects one or more subscribers to the PON, colorless (i.e., wavelength independent). Colorless ONUs require either no light source at all, or only a broadband light source, resulting in decreased costs, simplified maintenance, and reduced stock inventory issues.
3. **WDM PON:** The third and final target is to increase the number of wavelength channels by means of wavelength division multiplexing (WDM). The use of WDM technologies allows access network operators to respond to user requests for service

upgrades and network evolution. Deploying WDM adds a new dimension to current TDM PONs. The benefits of the new wavelength dimension are manifold. Among others, it may be exploited (1) to increase the network capacity, (2) to improve the network scalability by accommodating more end users, (3) to separate services, or (4) to separate service providers.

All three tasks are currently addressed by various research groups worldwide (Koonen, 2006). For example, in the United States, FTTH costs per connected home have dropped by a factor of almost 5 between the years 1993 and 2004 (Green, 2006). A low-cost WDM light source for a multiservice wavelength-routing WDM PON was demonstrated in Park et al. (2004). A two-stage PON architecture that serves more end users and provides a longer access reach was investigated in Shami et al. (2005).

PONs come in a number of flavors, for example, ATM-based, broadband PON (BPON), and gigabit PON (GPON). In the following, we focus on Ethernet PONs (EPONs), standardized by the IEEE 802.3ah Ethernet in the First Mile (EFM) Task Force. EPONs aim at converging the low-cost equipment and simplicity of Ethernet and the low-cost infrastructure of PONs. Given the fact that 95% of local area networks (LANs) use Ethernet, EPONs and their WDM upgraded descendants are likely to become increasingly the norm due to their capability of natively carrying variable-size IP packets in a simpler and more efficient way than their ATM-based counterparts, which suffer from a severe cell tax and 125-μs framing overhead (Kramer, 2005).

Chapter 14

PONs have been considered attractive due to their longevity, low operational costs, and huge bandwidth and are already widely deployed in the first/last mile of today's operational access networks. In particular, EPONs aim at converging the low-cost equipment and simplicity of Ethernet and the low-cost fiber infrastructure of PONs. In Chapter 14, we will provide an introduction to the EPON architecture and multipoint control protocol (MPCP). After presenting a taxonomy of dynamic bandwidth allocation (DBA) algorithms, we provide an in-depth discussion of various state-of-the-art DBA algorithms proposed for EPON to date. DBA algorithms are used in EPON to arbitrate the upstream transmissions of the attached subscribers to the central office.

Chapter 15

To cope with the steadily increasing number of users and bandwidth-hungry applications, single-channel EPON will need to be upgraded in order to satisfy the growing traffic demands in the future. Clearly, one way to upgrade the capacity of a single-channel EPON is to increase the line rate. Such an upgrade would affect all ONUs in that all installed

ONU transceivers had to be replaced with higher-speed transceivers. In Chapter 15, we outline an alternative upgrade approach of single-channel EPON that requires only a subset of ONUs to be WDM upgraded in a pay-as-you-grow fashion according to given traffic growth, state-of-the-art transceiver technology, operator preferences, and/or cost constraints. After reviewing the state of the art of WDM EPON, we address the requirements of a smooth migration path from current single-channel TDM EPON to WDM EPON. We then elaborate on required WDM extensions to MPCP, taking backward compatibility with MPCP and future-proofness against arbitrary WDM ONU structures into account. Finally, we briefly describe dynamic wavelength allocation (DWA) algorithms for WDM EPON.

Chapter 16

WDM EPONs are becoming rapidly mature due to the fact that the major tasks of cost reduction, design of so-called colorless (wavelength-independent) ONUs, and capacity upgrade by means of WDM are expected to be addressed successfully in the near term. The next evolutionary step toward unleashing the potential of WDM EPONs is their all-optical interconnection and integration with Ethernet-based metro networks such as Resilient Packet Ring (RPR). In Chapter 16, we will elaborate on the traffic characteristis of emerging applications (e.g., online gaming and P2P file sharing) and discuss their impact on future optically integrated access metro WDM network architectures and services. Specifically, we will describe a novel optically integrated access metro network, termed STARGATE, which is able to provide all-optical circuits at the wavelength and subwavelength granularity on demand to interconnect ONUs residing in different WDM EPONs.

Chapter 17

Ethernet is a low-cost networking technology originally designed for LANs. Since its first specification in 1980, Ethernet has come a long way and is widely deployed in 95% of today's LANs. In Chapter 17, we will briefly review the evolution of the original 10-Mbps Ethernet to Gigabit Ethernet (GbE) and 10-Gigabit Ethernet (10GbE) standards. The media access control and physical layers specified in the GbE and 10GbE standards will be explained in greater detail, thereby highlighting the different duplex modes, performance enhancements, and variety of supported physical transmission media. Finally, we elaborate on the interoperability of 10GbE not only with Ethernet-based LANs but also with SONET/SDH-based metropolitan area networks (MANs) and wide area networks (WANs).

Chapter 18

Chapter 18 deals with the interface between optical (wired) networks and wireless access networks. We will elaborate on so-called radio-over-fiber (RoF) networks. RoF networks are deployed to carry radio frequencies over optical fiber links in support of a variety of wireless applications (e.g., global system for mobile communications [GSM], universal mobile telecommunications systems [UMTS], and wireless LAN [WLAN]). In this chapter, we describe some recently proposed and experimentally demonstrated RoF network architectures and their integration with FTTH, WDM PON, and rail track networks.

14　EPON

Access networks connect business and residential subscribers to the central offices (COs) of service providers, which in turn are connected to metropolitan area networks (MANs) or wide area networks (WANs). Access networks are commonly referred to as the *last mile* or *first mile*, where the latter term emphasizes their importance to subscribers. In today's access networks, telephone companies deploy digital subscriber line (xDSL) technologies and cable companies deploy cable modems. Typically, these access networks are hybrid fiber coax (HFC) systems with an optical fiber–based feeder network between CO and remote node and an electrical distribution network between remote node and subscribers. These access technologies are unable to provide enough bandwidth to current high-speed Gigabit Ethernet local area networks (LANs) and evolving services and applications (e.g., distributed gaming or video on demand). Future first-mile solutions not only have to provide more bandwidth but also have to meet the cost-sensitivity constraints of access networks arising from the small number of cost-sharing subscribers.

In so-called FTTX access networks the copper-based distribution part of access networks is replaced with optical fiber (e.g., fiber to the curb [FTTC] or fiber to the home [FTTH]). In doing so, the capacity of access networks is sufficiently increased to provide broadband services to subscribers. Due to the cost sensitivity of access networks, these all-optical FTTX systems are typically unpowered and consist of passive optical components (e.g., splitters and couplers). Accordingly, they are called passive optical networks (PONs). PONs had attracted a great deal of attention well before the Internet spurred bandwidth growth. The *full service access network* (FSAN) group was the driving force behind the ITU-T G.983 broadband PON (BPON) specification which defines a PON with asynchronous transfer mode (ATM) as its native protocol data unit (PDU). ATM suffers from several shortcomings. Due to the segmentation of variable-size IP packets into fixed-size cells, ATM imposes a cell tax overhead. Moreover, a single corrupted or dropped ATM cell requires the retransmission of the entire IP packet even though the remaining ATM cells belonging to the corresponding IP packet are received correctly, resulting in wasted bandwidth and processing resources. And finally, ATM equipment (switches, network interface cards [NICs]) is quite expensive.

Recently, Ethernet PONs (EPONs) have attracted a great amount of interest both in industry and academia as a promising cost-effective solution of next-generation broadband access networks, as illustrated by the formation of several fora and working groups, including the *EPON forum*, the *Ethernet in the first mile (EFM) alliance*, and the

IEEE 802.3ah working group. EPONs carry data encapsulated in Ethernet frames, which makes it easy to carry IP packets and eases the interoperability with installed Ethernet LANs. EPONs represent the convergence of low-cost Ethernet equipment (switches, NICs) and low-cost fiber architectures. Furthermore, given the fact that more than 90% of today's data traffic originates from and terminates in Ethernet LANs, EPONs appear to be a natural candidate for future first-mile solutions. The main standardization body behind EPON is the IEEE 802.3ah Task Force. This task force developed the so-called multipoint control protocol (MPCP) which arbitrates the channel access between CO and subscribers. MPCP is used for dynamically assigning the upstream bandwidth (subscriber-to-service provider), which is the key challenge in the access protocol design for EPONs. Note that MPCP does not specify any particular dynamic bandwidth allocation (DBA) algorithm. Instead, it is intended to facilitate the implementation of DBA algorithms.

To understand the importance of DBA in EPONs, note that the traffic on the individual links in the access network is quite bursty. This is in contrast to MANs or WANs, where the bandwidth requirements are relatively smooth due to the aggregation of many traffic sources. In an access network, each link represents a single or small set of subscribers with very bursty traffic conditions, due to a small number of ON/OFF sources. Because of this bursty nature the bandwidth requirements vary widely with time. Therefore, the static allocation of bandwidth to the individual subscribers (or sets of subscribers) in an EPON is very inefficient. Employing a DBA algorithm that adapts to instantaneous bandwidth requirements is much more efficient, by capitalizing on the benefits of statistical multiplexing. Hence, DBA is a critical feature for EPON design (McGarry et al., 2004; Kramer, 2005).

In the following sections, we first describe the EPON architecture and highlight the major operational functions of MPCP. We then provide an overview of state-of-the-art DBA algorithms proposed for EPON.

14.1   Architecture

Typically, EPONs (and PONs in general) have a physical tree topology with the CO located at the root and the subscribers connected to the leaf nodes of the tree, as illustrated in Fig. 14.1. At the root of the tree is an optical line terminal (OLT) which is the service provider equipment residing at the CO. The EPON connects the OLT to multiple optical network units (ONUs) through a 1:N optical splitter/combiner. An ONU can serve a single residential or business subscriber, referred to as fiber to the home/business (FTTH/B), or multiple subscribers, referred to as fiber to the curb (FTTC). Each ONU buffers data received from the attached subscriber(s). In general, the round-trip time (RTT) between OLT and each ONU is different. For instance, in Fig. 14.1 the OLT is connected to five ONUs, each with a different RTT. Due to the directional properties of the optical splitter/combiner the OLT is able to broadcast data to all ONUs in the downstream direction. In the upstream direction, however, ONUs cannot communicate

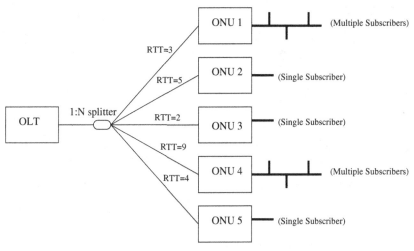

RTT: Round Trip Time

Figure 14.1 EPON architecture.

directly with one another. Instead, each ONU is able to send data only to the OLT. Thus, in the downstream direction an EPON may be viewed as a point-to-multipoint network and in the upstream direction, an EPON may be viewed as a multipoint-to-point network (Kramer et al., 2003). Due to this fact, the original Ethernet media access control (MAC) protocol does not operate properly since it relies on a broadcast medium. Instead, the MPCP arbitration mechanism is deployed, as discussed in the subsequent section.

In the upstream direction, all ONUs share the transmission medium. To avoid collisions, several approaches can be used. Wavelength division multiplexing (WDM) is currently considered cost prohibitive since the OLT would require a tunable receiver or a receiver array to receive data on multiple wavelength channels and each ONU would need to be equipped with a wavelength-specific transceiver. At present, time division multiplexing (TDM) is considered a more cost-effective solution. With TDM a single transceiver is required at the OLT and there is just one type of ONU equipment (Kramer et al., 2001). Note that this does not prevent EPONs from being upgraded to multiple wavelength channels (WDM) in the future. Given the aforementioned different connectivity in upstream and downstream direction of EPONs, the OLT appears to be the best-suited node to arbitrate the time sharing of the channel, as discussed next.

14.2   Multipoint control protocol (MPCP)

To increase the upstream bandwidth utilization the OLT dynamically allocates a variable time slot to each ONU based on the instantaneous bandwidth demands of the ONUs, best by means of *polling* (Zheng and Mouftah, 2005). To facilitate DBA and arbitrating the upstream transmissions of multiple ONUs the MPCP, specified in IEEE standard

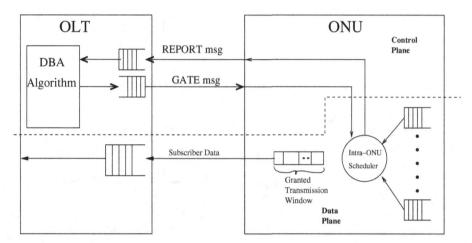

Figure 14.2 Operation of multipoint control protocol (MPCP). After McGarry et al. (2004).
© 2004 IEEE.

802.3ah, is deployed in EPON. Besides autodiscovery, registration, and ranging (RTT computation) operations for newly added ONUs, MPCP provides the signaling infrastructure (control plane) for coordinating the data transmissions from the ONUs to the OLT.

As shown in Fig. 14.2, MPCP uses two types of messages to facilitate arbitration: REPORT and GATE. Each ONU has a set of queues, possibly prioritized, holding Ethernet frames ready for upstream transmission to the OLT. The REPORT message is used by an ONU to report bandwidth requirements (typically in the form of queue occupancies) to the OLT. A REPORT message can support the reporting of up to eight queue occupancies of the corresponding ONU. Upon receiving a REPORT message, the OLT passes it to the DBA algorithm module. The DBA module calculates the upstream transmission schedule of all ONUs such that channel collisions are avoided. Scheduling can be done in two ways: inter-ONU scheduling and intra-ONU scheduling. Inter-ONU scheduling arbitrates the transmissions of different ONUs while intra-ONU scheduling arbitrates the transmissions of different priority queues in each ONU. There are two possible implementations. Either inter-ONU scheduling is implemented at the OLT and each ONU performs its own intra-ONU scheduling or both inter-ONU scheduling and intra-ONU scheduling are implemented at the OLT. After executing the DBA algorithm, the OLT transmits GATE messages to issue transmission grants. Each GATE message can support up to four transmission grants. Each transmission grant contains the transmission start time and transmission length of the corresponding ONU. Each ONU updates its local clock using the timestamp contained in each received transmission grant. Thus, each ONU is able to acquire and maintain global synchronization. Each ONU sends backlogged Ethernet frames during its granted transmission window according to the corresponding intra-ONU scheduling. The transmission window may comprise multiple Ethernet frames; packet fragmentation is not allowed. As a consequence, if the next

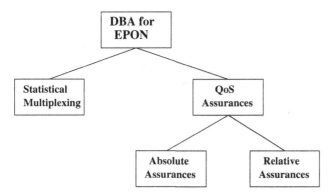

Figure 14.3 Classification of dynamic bandwidth allocation (DBA) algorithms for EPON. After McGarry et al. (2004). © 2004 IEEE.

frame does not fit into the current transmission window it has to be deferred to the next granted transmission window.

Note that MPCP does not specify any particular DBA algorithm. MPCP simply provides a framework for the implementation of various DBA algorithms, which are described in greater detail next.

14.3 Dynamic bandwidth allocation (DBA)

According to McGarry et al. (2004), DBA algorithms for EPONs can be classified into algorithms with statistical multiplexing and algorithms with quality of service (QoS) assurances. The latter ones are further subdivided into algorithms with absolute and relative QoS assurances, as shown in Fig. 14.3. In the following, we discuss the DBA algorithms of each class in greater detail. Finally, we briefly touch on *decentralized* DBA algorithms which have been attracting some attention recently.

14.3.1 Statistical multiplexing methods

Interleaved polling with adaptive cycle time (IPACT)
In the interleaved polling with adaptive cycle time (IPACT) approach, the OLT polls the ONUs individually and issues transmission grants to them in a round-robin fashion (Kramer et al., 2002a). The grant window size of each ONU's first grant, $G(1)$, is set to some arbitrary value. After n cycles, the backlog (in bytes) in each ONU's transmission buffer, $Q(n)$ (reported queue size), is piggybacked to the current data transmission from the corresponding ONU to the OLT during its grant window $G(n)$. The backlog $Q(n)$ is measured at the instant when the ONU generates the request message, which is piggy-backed to the data transmission in cycle n. This backlog $Q(n)$ is used to determine the grant window size of the next grant $G(n + 1)$ of the ONU. In doing so, bandwidth is dynamically assigned to ONUs according to their queue occupancies. If a given ONU's queue is empty, the OLT still grants a transmission window of zero byte to that ONU such

that the ONU is able to report its queue occupancy for the next grant. IPACT deploys in-band signaling of bandwidth requests by using escape characters within Ethernet frames instead of sacrificing an entire Ethernet frame for control (as in MPCP), resulting in a reduced signaling overhead. The OLT keeps track of the round-trip times of all ONUs. As a result, the OLT can send out a grant to the next ONU in order to achieve a very tight guard band between consecutive upstream transmissions, resulting in an improved bandwidth utilization. The guard band between two consecutive upstream transmissions is needed to compensate for RTT fluctuations and to give the OLT enough time to adjust its receiver to the transmission power level of the next ONU.

In IPACT, each ONU is served once per round-robin polling cycle. The cycle length is not static but adapts to the instantaneous bandwidth requirements of the ONUs. By using a maximum transmission window (MTW), ONUs with high traffic volumes are prevented from monopolizing the bandwidth. The OLT allocates the upstream bandwidth to ONUs in one of the following ways:

- **Fixed Service:** This DBA algorithm ignores the requested window size and always grants the MTW size. As a result, the cycle time is constant.
- **Limited Service:** This DBA algorithm grants the requested number of bytes, but no more than the MTW.
- **Credit Service:** This DBA algorithm grants the requested window plus either a constant credit or a credit that is proportional to the requested window.
- **Elastic Service:** This DBA algorithm attempts to overcome the limitation of assigning at most one fixed MTW to an ONU in a round. The maximum window granted to an ONU is such that the accumulated size of the last N grants does not exceed N MTWs, where N denotes the number of ONUs. Thus, if only one ONU is backlogged, it may get a grant of up to N MTWs.

The simulation results reported in Kramer et al. (2002a) indicate that both the average packet delays and the average queue lengths with the IPACT method with the limited, credit, or elastic service DBA were almost two orders of magnitude smaller compared with the fixed service DBA (fixed service is a static bandwidth allocation) under light traffic loads. Under heavy loads, the average packet delays and average queue lengths for all four types of service were similar. Generally, limited, credit, and elastic service DBA all provided very similar average packet delays and average queue lengths.

In summary, IPACT improves the channel utilization efficiency by reducing the overhead arising from walk times (propagation delay) in a polling system. This is achieved by overlapping multiple polling requests in time. As opposed to static TDM systems, IPACT allows for statistical multiplexing and dynamically allocates upstream bandwidth according to the traffic demands of the ONUs within adaptive polling cycles. Furthermore, IPACT deploys an efficient in-band signaling approach that avoids using extra Ethernet frames for control. By using an MTW, throughput fairness among the ONUs is achieved. On the downside, this original design for IPACT does not support QoS assurances or service differentiation by means of reservation or prioritization of bandwidth assignment. An IPACT extension to support multiple service classes was developed in Kramer et al. (2002a), which we discuss in Section 14.3.3.

Control theoretic extension of IPACT

In IPACT, the ONU requests (reports) the amount of backlogged traffic $Q(n)$ as grant for the next cycle. One drawback of this approach is that the request does not take into consideration the amount of traffic arriving to the ONU between the generation of the request message in cycle n and the arrival of the grant $G(n + 1)$ for the next cycle at the ONU. As a consequence, the traffic arriving after the generation of a request message is only taken into consideration in the next request message and hence experiences typically a queueing delay of one cycle in the ONU.

To overcome this queueing delay, a control theoretic extension to IPACT was proposed in Byun et al. (2003). In this extension the amount of traffic arriving to the ONU between two successive requests is estimated and this estimate is incorporated into the grant to the ONU. More specifically, the estimation works as follows. Recall that $Q(n - 1)$ denotes the amount of backlogged traffic in the ONU at the instant when the request of cycle $n - 1$, which is used by the OLT to calculate the grant $G(n)$, is generated. Let $A(n - 1)$ denote the amount of traffic arriving to the ONU between generating the request for cycle $n - 1$ and receiving the grant for cycle n. With these definitions, the difference between the grant for cycle n and the amount of traffic backlogged in the ONU when the grant arrives is approximately $D(n) = G(n) - [Q(n - 1) + A(n - 1)]$. The OLT allocates bandwidth based on the size of the previous grant and the scaled version of the difference reported by the ONUs. More specifically, the grant for cycle $n + 1$ is calculated as $G(n + 1) = G(n) - \alpha \cdot D(n)$, where α is the gain factor. Using control theoretic arguments it is shown in Byun et al. (2003) that for piecewise constant traffic with infrequent jumps the system is asymptotically stable for $0 < \alpha < 2$.

Note that this refinement to IPACT essentially views the bandwidth assignment as an automatic control system with the goal to keep the difference $D(n)$ close to zero. A proportional (P) control is proposed for this system with the control gain α. The advantage of this control theoretic approach is that the grant size is typically closer to the size of the backlog at the instant of receiving the grant at the ONU. This in turn results in lower queueing delays. On the downside, the control system may require careful tuning to achieve a prompt response to changes in the traffic load without creating oscillations in the system. This may be a challenging problem if the traffic load is highly variable.

14.3.2   Absolute QoS assurances

Bandwidth guaranteed polling

The bandwidth guaranteed polling (BGP) method proposed in Ma et al. (2003) divides ONUs into the two disjoint sets of bandwidth guaranteed ONUs and best-effort ONUs. Bandwidth guaranteed nodes are characterized by their service-level agreement (SLA) with the network provider. The SLA specifies the bandwidth this node is to be guaranteed.

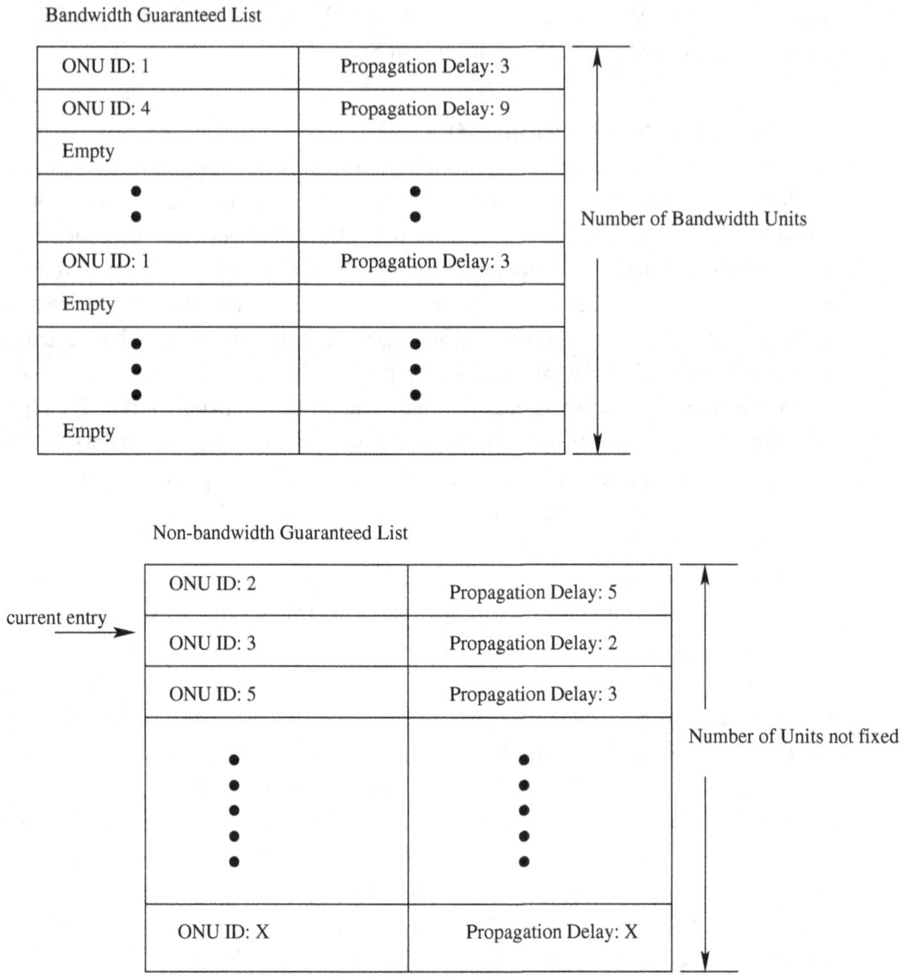

Bandwidth Guaranteed List

ONU ID: 1	Propagation Delay: 3
ONU ID: 4	Propagation Delay: 9
Empty	
• •	• •
ONU ID: 1	Propagation Delay: 3
Empty	
• • •	• • •
Empty	

Number of Bandwidth Units

Non-bandwidth Guaranteed List

current entry →

ONU ID: 2	Propagation Delay: 5
ONU ID: 3	Propagation Delay: 2
ONU ID: 5	Propagation Delay: 3
• • • • •	• • • •
ONU ID: X	Propagation Delay: X

Number of Units not fixed

Figure 14.4 Bandwidth guaranteed polling (BGP) tables. After McGarry et al. (2004). © 2004 IEEE.

The total upstream bandwidth is divided into equivalent bandwidth units, whereby the bandwidth unit is chosen such that the total upstream bandwidth in terms of the bandwidth unit is larger than the number of ONUs. For instance, for a network with 64 ONUs and an upstream bandwidth of 1 Gbps, the equivalent bandwidth unit may be chosen as 10 Mbps (i.e., the total upstream bandwidth corresponds to 100 bandwidth units). The OLT maintains 2 Entry Tables, one for bandwidth guaranteed ONUs (ONUs with IDs 1 and 4 in the example in Fig. 14.4) and one for best-effort ONUs (ONUs with IDs 2, 3, and 5 in the example in Fig. 14.4). Each table entry (row) has two fields, namely ONU ID and propagation delay from ONU to OLT. The table for bandwidth guaranteed ONUs has as many rows (entries) as there are bandwidth units in the total upstream bandwidth. In the preceding example, the bandwidth guaranteed ONU table has 100 rows. The table for best-effort nodes is not fixed in size. Entries in the bandwidth guaranteed ONU table are established for each bandwidth guaranteed ONU based on its SLA. If an ONU requires more than one bandwidth unit, then these units are spread evenly through the table as illustrated in Fig. 14.4 for ONU with ID 1, which is guaranteed a bandwidth of two bandwidth units (i.e., 20 Mbps in the example). Rows in the guaranteed bandwidth ONU table that are not occupied can be dynamically assigned to best-effort nodes. The OLT polls the best-effort ONUs during the rows that are not used by the bandwidth guaranteed ONUs in the order they are listed in the best-effort table.

The OLT begins polling ONUs using the information in the two tables. The OLT polls an ONU by sending a Grant message to grant a window of size G, which is initially set to one bandwidth unit. The ONU decides based on the size of its output buffer if it has enough data to fully utilize the granted transmission window. The ONU sends a reply to the OLT with the amount of the window it intends to utilize B and then transmits this amount of data. The OLT, upon receiving a reply from an ONU, checks the amount of the granted window the currently polled ONU intends to use. If B is zero, the OLT immediately polls the next ONU in the table. (Note that this wastes the bandwidth during the RTT to that next ONU, whereas the polling to the first ONU can be interleaved with the preceding data transmission to avoid wasting bandwidth.) If B is between zero and some threshold G_{reuse}, whereby $G - G_{reuse}$ specifies the minimum portion of the bandwidth unit that can be effectively shared, the OLT polls the next best-effort ONU ready for transmission and grants it a transmission window $G - B$. Lastly, if B is larger than the threshold G_{reuse}, the OLT will not poll the next ONU until the current grant has passed.

The simulation results reported in Ma et al. (2003) indicate that for bandwidth guaranteed ONUs with 4 or more entries, the delays were an order of magnitude smaller than with IPACT. However, for bandwidth guaranteed ONUs with only 1 entry as well as non-bandwidth guaranteed ONUs, the delays were orders of magnitude larger than with IPACT under light loads and almost an order of magnitude larger than with IPACT under heavy loads. On the other hand, for bandwidth guaranteed ONUs with 4 or more entries, the queue lengths were similar to IPACT for light loads and were orders of magnitude shorter than with IPACT under heavy loads. However, for bandwidth guaranteed ONUs, with only 1 entry as well as non-bandwidth guaranteed ONUs, the queue lengths were orders of magnitude larger than with IPACT under light loads and similar to IPACT under heavy loads. It was also found that the throughput with BGP tends to be lower than the throughput with IPACT, especially at heavy loads.

Overall, the advantage of the BGP approach is that it ensures that an ONU receives the bandwidth specified by its SLA and that the spacing between transmission grants corresponding to SLAs has a fixed bound. The approach also allows for the statistical multiplexing of traffic into unreserved bandwidth units as well as unused portions of a guaranteed bandwidth unit (i.e., if an ONU does not have enough traffic to use all the bandwidth specified in its SLA). One drawback of the table-driven upstream transmission grants of fixed bandwidth units is that the upstream transmission tends to become fragmented, with each fragment requiring a guard band, which tends to reduce the throughput and bandwidth utilization.

Deterministic effective bandwidth

In Zhang et al. (2003), a system in which ONUs and OLT employ deterministic effective bandwidth (DEB) admission control and resource allocation in conjunction with Generalized Processor Sharing (GPS) scheduling is developed. In this system, a given ONU maintains several queues, typically one for each traffic source or each class of traffic sources. A given queue is categorized as either a QoS queue or a best-effort queue, depending on the requirements of the corresponding traffic source (class). A given traffic source feeding into a QoS queue is characterized by leaky bucket parameters. The leaky bucket parameters are traffic descriptors widely used in QoS networking and give the peak rate of the source, the maximum burst that the source can send at the peak rate, as well as the long run average rate of the source. A source also specifies the maximum delay it can tolerate. The leaky bucket traffic characterization together with the delay limit of the source (class) are used to determine whether the system can support the traffic in the QoS queues at all ONUs without violating delay bounds (and also without dropping any traffic at a QoS queue) using techniques derived from the general theory of deterministic effective bandwidths.

During the operation of the network, the OLT assigns grants to a given ONU based on the aggregate effective bandwidth of the traffic of the QoS queues at the ONU. Roughly speaking, a given ONU is assigned grants proportional to the ratio of the aggregate effective bandwidth of the traffic of the ONU to the total aggregate effective bandwidth of the traffic of all ONUs supported by the OLT. In turn, a given ONU uses the grants that it receives to serve its QoS queues in proportion to the ratio of the effective bandwidth of the traffic of a queue to the aggregate effective bandwidth of the traffic of the QoS queues supported by the ONU. A given ONU uses the grants not utilized by QoS queues to transmit from best-effort queues.

The advantage of the deterministic effective bandwidth approach is that individual flows (or classes of flows) are provided with deterministic QoS guarantees, ensuring lossless, bounded-delay service. In addition, best-effort traffic flows can utilize bandwidth not needed by QoS traffic flows. One main drawback of the DEB approach is that it requires increased complexity to conduct admission control and update proportions of effective bandwidths of ongoing flows. In particular, conducting admission control and allocating grant resources may result in significant overhead for short-lived traffic flows (or classes of traffic).

14.3.3   Relative QoS assurances

DBA for multimedia

In dynamic bandwidth allocation for multimedia (Choi and Huh, 2002), traffic in each ONU is placed into one of three priority queues (high, medium, and low). These priorities are then used by the DBA algorithm to assign bandwidth. The sizes of the three priority queues in each ONU are reported to the OLT. The OLT based on the priority queue sizes issues grants separately for each of the priorities in each of the ONUs. In particular, bandwidth is first handed out to the high-priority queues, satisfying all the requests of the high-priority flows. The DBA algorithm then considers the requests from the medium-priority flows. If it can satisfy all of the medium-priority requests with what is left over from the high-priority requests it does so. Otherwise it divides the remaining bandwidth between all medium-priority flows, where the fraction of the bandwidth granted to each medium-priority flow is related to the fraction requested by each flow to the total of all medium-priority requests. Finally, if there is any leftover bandwidth after satisfying the high- and medium-priority requests, this leftover bandwidth is distributed between the low-priority flows in a manner identical to the case where all the medium-priority flow requests cannot be fully satisfied.

Note that in the DBA for multimedia approach, bandwidth is essentially allocated using strict priority based on the requirements of each priority traffic class of the entire PON (all the ONUs connected to a single OLT). One feature of this approach is that the OLT controls the scheduling within the ONU. This comes at the expense of reporting the occupancies of the individual priority queues and issuing multiple grants to each ONU per cycle. Also, the OLT has the additional burden of deciding on the scheduling among the queues in the ONU. Note that the strict priority scheduling based on the traffic classes at the PON level may result in starvation of ONUs that have only low priority traffic.

DBA for QoS

The DBA for QoS (Assi et al., 2003) is a method of providing per-flow QoS in an EPON using differentiated services. Within each ONU, priority packet queuing and scheduling is employed per the differentiated services framework. This is similar to the DBA for multimedia approach, but recall that in DBA for multimedia the priority scheduling was performed at the PON level (all the ONUs connected to a single OLT). In contrast, in DBA for QoS, the priority scheduling is performed at the ONU level.

Before we proceed to DBA for QoS (Assi et al., 2003), we review the IPACT extension to multiple service classes (Kramer et al., 2002b), which may be viewed as a precursor to DBA for QoS. In Kramer et al. (2002b), a simulation study is conducted of supporting differentiated service to three classes of traffic with strict priority scheduling inside the ONU. The authors noticed an interesting phenomenon they dubbed "light-load penalty." What they noticed was that, under light loading, the lower-priority class experienced a significant average packet delay increase; the maximum packet delays for the higher priorities also exhibited similar behavior. This appears to be caused by the fact that the queue reporting occurs at some time before the strict priority scheduling is performed, thus allowing higher-priority traffic arriving after the queue reporting but before the transmission grant to preempt the lower-priority traffic that arrived before the queue reporting. It appears this problem is exacerbated under light loading. The authors discuss two methods for dealing with the light-load penalty. The first method involves scheduling the packets when the REPORT message is transmitted and placing them in a second stage queue. This second stage queue is the queue that will be emptied out first into the timeslot provided through a grant in a GATE message. The second method involves predicting the number of high-priority packets arriving between the queue reporting and the grant window so that the grant window will be large enough to accommodate the newly arriving high-priority packets. This second method inherently lowers the delay experienced by higher priority traffic compared to the two-stage queueing approach.

In DBA for QoS (Assi et al., 2003) the authors incorporate a method similar to the aforementioned two-stage queueing approach. Specifically, in the DBA for QoS method the packet scheduler in the ONU employs priority scheduling only on the packets that arrive before some t_{request}, which is the time at which the REPORT message is sent to the OLT. This avoids the problem of having the ONU packet scheduler request bandwidth based on buffer occupancies at time t_{request} and then actually schedule packets at time t_{grant} to fill the granted transmission window. If this mechanism is not employed, lower priority queues can be starved more severely because higher priority traffic arriving between t_{request} and t_{grant} would tend to take away transmission capacity from the lower-priority queues. Note that this problem only arises with strict priority scheduling which schedules lower-priority packets only when the higher-priority packet queues are empty. With weighted fair queueing (WFQ), which serves the different priority queues in proportion to fixed weights, this problem would not arise.

In DBA for QoS, each ONU is assigned guaranteed bandwidth in proportion to its SLA. More specifically, let B_{total} denote the total upstream bandwidth. Let w_i denote the weighing factor for ONU i. The weighing factors are set in proportion to the SLA of ONU i, such that the weighing factors of all ONUs supported by the OLT sum to one (i.e., $\sum_i w_i = 1$). ONU i is then assigned the guaranteed bandwidth $B_i = B_{\text{total}} \cdot w_i$. Note that the sum of all the guaranteed bandwidths equals the total available bandwidth. In other words, the total upstream bandwidth is divided up among the ONUs in proportion to their SLAs.

For every transmission grant cycle, each of the ONUs requests bandwidth corresponding to its total backlog. If the requested bandwidth is smaller than the guaranteed bandwidth, the difference (i.e., the excess bandwidth) is pooled together with the excess bandwidth from all other lightly loaded ONUs (ONUs whose requested bandwidth is less than their guaranteed bandwidth). This pooled excess bandwidth is then distributed to each of the highly loaded ONUs (ONUs whose requested bandwidth is larger than their guaranteed bandwidth) in a manner that weighs the excess assigned in proportion to the size of their request. Note that this proportional scheduling approach is in contrast to the strict priority scheduling of DBA for multimedia which does not allocate any bandwidth to lower-priority traffic classes until the bandwidth demands of all higher priority traffic classes are met.

We note that DBA for QoS allows for the option of sending the individual priority queue occupancies to the OLT via REPORT messages (a REPORT message supports reporting queue sizes of up to eight queues) and having the OLT generate transmission windows for each individual priority queue (the GATE message supports sending up to four transmission grants). This option puts the priority scheduling that would otherwise be handled by the ONU under the control of the OLT.

DBA for QoS (Assi et al., 2003) also considers the option of reporting queue size using an estimator for the occupancy of the high-priority queue. The estimator makes a one-step prediction of the traffic arriving to the high-priority queue between the time of the report and the time of the grant. In particular, the amount of traffic arriving to the high-priority queue between report and grant in a cycle $n - 1$ is used to estimate the arrival in cycle n. The ONU then reports in cycle n the actual backlog at the time of request plus the estimated new arrivals until the time of the grant.

The simulations reported in Assi et al. (2003) compare the average and maximum delays for the proposed DBA for QoS scheme for the service classes best effort, assured forwarding, and expedited forwarding with the delays achieved with a static bandwidth allocation to the individual ONUs. It was found that the proposed DBA for QoS scheme achieves significantly smaller delays, especially at high loads. This is primarily due to the statistical multiplexing between the different ONUs permitted by DBA for QoS. It was also found that the proposed DBA for QoS scheme is quite effective in differentiating the delays for the different service classes, with the highest-priority expedited forwarding class achieving the smallest delays. The simulations in Assi et al. (2003) also considered the average utilization of the upstream bandwidth and found that the proposed DBA for QoS schemes achieve around 90% utilization compared to around 50% with static bandwidth allocation.

14.3.4   Decentralized DBA algorithms

Note that all of the aforementioned DBA algorithms are centralized schemes. The OLT acts as the central control unit by performing inter-ONU scheduling or both inter-ONU and intra-ONU scheduling. Recently, research on *decentralized* DBA algorithms and distributed scheduling has begun (Foh et al., 2004; Sherif et al., 2004). To enable distributed scheduling, however, the original EPON architecture has to be modified such that each ONU's upstream transmission is echoed at the splitter to all ONUs, each equipped with an additional receiver to receive the echoed transmissions. In doing so, all ONUs are able to monitor the transmission of every ONU and to arbitrate upstream channel access in a distributed manner, similar to Ethernet LANs. Note that in such alternate EPON solutions both inter-ONU and intra-ONU scheduling take place at the ONUs without participation of the OLT. The reported performance results show that such decentralized EPONs and DBA algorithms are able to provide high bandwidth utilization. We refer the interested reader to Foh et al. (2004) and Sherif et al. (2004) for further details on decentralized DBA algorithms.

15 WDM EPON

Current Ethernet passive optical networks (EPONs) are single-channel systems; that is, the fiber infrastructure carries a single downstream wavelength channel and a single upstream wavelength channel, which are typically separated by means of coarse wavelength division multiplexing (CWDM). In the upstream direction (from subscriber to network), the wavelength channel bandwidth is shared by the EPON nodes by means of time division multiplexing (TDM). In doing so, only one common type of single-channel transceiver is used network wide, resulting in simplified network operation and maintenance. At present, single-channel TDM EPONs appear to be an attractive solution to provide more bandwidth in a cost-effective manner.

Given the steadily increasing number of users and bandwidth-hungry applications, current single-channel TDM EPONs are likely to be upgraded in order to satisfy the growing traffic demands in the future. Clearly, one approach is to increase the line rate of TDM EPONs. Note, however, that such an approach implies that *all* EPON nodes need to be upgraded by replacing the installed transceivers with higher-speed transceivers, resulting in a rather costly upgrade. Alternatively, single-channel TDM EPONs may be upgraded by deploying multiple wavelength channels in the installed fiber infrastructure in the upstream and/or downstream directions, resulting in wavelength division multiplexing (WDM) EPONs. As opposed to the higher-speed TDM approach, WDM EPONs provide a cautious upgrade path in that wavelength channels can be added one at a time, each possibly operating at a different line rate. More importantly, only EPON nodes with higher traffic demands may be WDM upgraded by deploying multiple fixed-tuned and/or tunable transceivers while EPON nodes with lower traffic demands remain unaffected. Thus, using WDM enables network operators to upgrade single-channel TDM EPONs in a pay-as-you-grow manner where only a *subset* of EPON nodes may be upgraded gradually.

15.1 State of the art

The design and feasibility of cost-effective WDM structures for optical line terminal (OLT) and optical network units (ONUs) were addressed in Kani et al. (2003). The proposed WDM OLT structure consists of a multicarrier generator and supplies hundreds of optical carriers, thus greatly reducing the number of required laser diodes at the WDM OLT. Each ONU is assigned a separate pair of dedicated upstream and downstream wavelength channels. To decrease costs ONUs deploy no light source, but simply modulate the optical carriers supplied by the OLT for upstream transmission. This *remote modulation* scheme realizes wavelength-independent (colorless) ONUs, resulting in reduced costs and simplified operation and maintenance. Note that in the proposed architecture each pair of wavelength channels is dedicated to a different ONU. Thus, upstream wavelength channels are not shared among ONUs and no WDM dynamic bandwidth allocation (DBA) is performed.

A WDM PON in which each upstream wavelength channel can be shared among multiple ONUs by means of TDM was examined in Shin et al. (2005). The work focuses on the design and feasibility study of cost-effective burst-mode ONU transmitters which can operate on any wavelength channel without requiring wavelength tuning. Such so-called wavelength-selection-free transmitters require neither wavelength stability circuits nor network operators to stock spare transmitters for each wavelength channel, resulting in reduced costs. WDM DBA algorithms were not discussed in greater detail.

The interconnection of multiple PONs of arbitrary topology was investigated in Hsueh et al. (2005b). The proposed PON interconnection is highly scalable and provides an evolutionary upgrade path from single-channel TDM PONs to WDM PONs where each ONU is assigned a separate pair of dedicated upstream and downstream wavelength channels and existing field-deployed PON infrastructures remain intact. The transmitter(s) at the OLT may be shared among all interconnected PONs. For upstream transmission, each ONU may have a different node structure. For instance, one ONU may deploy a single tunable transmitter while another ONU may deploy an array of fixed-tuned transmitters. In contrast, for downstream transmission ONU node structures are less flexible. More precisely, the receiver of each ONU has to operate on a different wavelength band. Thus, each ONU has to have a different wavelength-band-selective receiver and each wavelength band must not be used by more than a single node at any given time. Different ONU node structures which provide a platform for deploying WDM DBA were discussed and examined. However, no specific WDM DBA algorithm was presented.

A WDM-DBA algorithm with online scheduling for WDM EPONs was described and investigated in Kwong et al. (2004). In the WDM EPON under consideration all ONUs are equipped identically with an array of fixed-tuned transceivers, one for each upstream/downstream wavelength channel. The proposed WDM IPACT with a single polling table (WDM IPACT-ST) dynamic wavelength allocation (DWA) algorithm is a multichannel extension of IPACT, which we have discussed in Section 14.3.1. In WDM IPACT-ST, transmission windows are assigned to ONUs in a round-robin fashion allowing them to transmit on the first available upstream wavelength channel. It was shown that the resultant WDM IPACT-ST EPON outperforms a single-channel TDM IPACT EPON in terms of delay. This is due to the fact that in TDM EPONs the polling cycle time increases linearly with the number of attached ONUs, as opposed to WDM EPONs which use multiple wavelength channels simultaneously to accommodate an increasing number of ONUs while maintaining a short polling cycle.

The WDM EPON presented in Xiao et al. (2005) aims at integrating both APON and EPON. The proposed so-called byte size clock (BSC) protocol is scalable in bandwidth assignment since heavy users may be assigned a single wavelength whereas light users may share a single wavelength. In BSC, time is divided into periodically recurring time frames. Each frame consists of dedicated reservation minislots, one for each ONU, and data slots, which are assigned on demand. Users send request packets in their assigned minislots at start of frame i. ONUs then transmit their respective data packets in accordance with the OLT grants received in frame $i - 1$. Once the request packets of frame i are received by the OLT, it computes the grants and broadcasts them back to the ONUs in frame $i + 1$. Note that a data packet has to go through a delay of at least one frame due to the reservation. It was shown that this delay can be reduced by *preallocating* a minimum number of dedicated data slots to certain ONUs which can increase in subsequent frames if the ONUs make reservations in their minislots. As a result, part of the data packets can be sent without reservation, leading to an improved throughput-delay performance and a decreased queue length of the corresponding ONUs. The performance of the preallocation BSC protocol can be further improved by means of *delta compression*. By using delta compression to compute the delta (difference) between packets and transmitting only the delta instead of the original packets, the need to reserve a large number of data slots is avoided as well as the number of preallocated time slots is kept to a minimum. On the downside, in BSC all nodes need to be synchronized and the resultant TDM frame time structure does not comply with IEEE 802.3ah.

15.2   TDM to WDM EPON migration

The WDM upgrade of single-channel TDM EPONs will very likely occur over long periods of time on a pay-as-you-grow manner. Hence, the type of WDM ONU node structures in a given EPON can differ as current technology and economic constraints as well as service provider preferences dictate.

Given that the main driver for supporting WDM on an EPON is the expansion of the bandwidth available on the EPON, it is not advantageous to this goal to have a tunable transceiver in the OLT. By having a tunable transceiver, only a single wavelength channel can be used at any given time. This would not expand the bandwidth currently available with a single fixed-tuned transceiver. In fact, it would actually provide less bandwidth because of the dead time imposed every time there is a wavelength switch due to the nonzero transceiver tuning time. Therefore, it appears reasonable that any WDM upgrade of an EPON has an array of fixed-tuned transceivers in the OLT, one for each operating wavelength channel.

The envisioned goal for managing WDM in EPON in an evolutionary manner should be to manage the different wavelength channels to increase the available bandwidth of the EPON without imposing a particular WDM architecture on the ONU, thereby allowing the ONUs to take on whatever architecture is preferred at the time when they are upgraded, possibly using transceivers with different tuning time and tuning range. ONUs should also be able to be upgraded incrementally as needed (e.g., adding new fixed-tuned and/or tunable transceivers incrementally). The evolutionary WDM upgrade of EPON should not impose any particular WDM ONU architecture, thus allowing these decisions to be dictated by economics, state of the art of transceiver manufacturing technology, and service provider preferences. Note that such an evolutionary and flexible WDM upgrade path not only increases the capacity but also meets key requirements of PONs (Frigo, 1997). Specifically, the evolutionary WDM upgrade allows for cautious pay-as-you-grow upgrades and thus helps operators realize their survival strategy for highly cost-sensitive access networks.

Note that although the evolutionary WDM upgrade allows for arbitrary ONU WDM node structures, it is reasonable to maintain a common channel that all ONUs support for transmission and reception, whereby the legacy EPON channel appears to be the best candidate for the common channel. Such a common channel for reception allows the OLT to forward broadcast frames to all ONUs with a single transmission on one wavelength channel.

15.3   WDM extensions to MPCP

To guarantee compliance of the aforementioned evolutionary WDM upgrade of EPON with IEEE 802.3ah, the multipoint control protocol (MPCP) has to be extended accordingly. In McGarry et al. (2006), several WDM extensions to MPCP were recommended that enable the OLT to schedule transmissions to and receptions from ONUs on any wavelength channel(s) supported by the OLT and the respective ONU. Figure 15.1 shows the recommended WDM extensions to MPCP, which are discussed in greater detail next.

15.3.1   Discovery and registration

For backward compatibility, the discovery and registration of ONUs take place on the original wavelength channel of TDM EPONs. During the registration process a discovered ONU conveys the following information about its WDM architecture to the OLT:

- **TX_type and RX_type:** Two bits each indicating the transmitter and receiver type, respectively, by using the following assigned values: $0 =$ no WDM, $1 =$ fixed-tuned, $2 =$ tunable, and $3 =$ reserved.
- **TX_tuning_time and RX_tuning_time:** Sixteen bits each indicating the tuning time of the transmitter and receiver, respectively, as an integer multiple of unit time (e.g., microsecond) (if TX_type/RX_type $= 2$).
- **Wavelength_id_type:** One bit indicating the encoding scheme of the supported wavelengths by using the following assigned values: $0 =$ two-level hierarchical encoding scheme (waveband identifier/bitmap of supported wavelengths within waveband), $1 =$ flat encoding scheme (bitmap of supported wavelengths).
- **TX_waveband and RX_waveband:** Four bits each indicating the identifier of the supported waveband of the transmitter and receiver, respectively. Wavebands are defined in compliance with the WDM channel spacings specified in ITU-T G.694.1 (if Wavelength_id_type $= 0$).

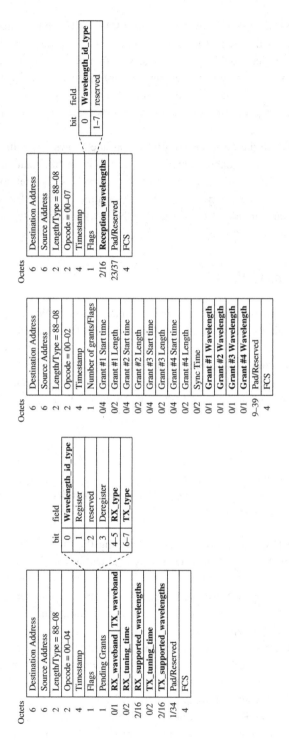

Octets — (a) REGISTER_REQ MPCPDU

Octets	field
6	Destination Address
6	Source Address
2	Length/Type = 88-08
2	Opcode = 00-04
4	Timestamp
1	Flags
1	Pending Grants
0/1	**RX_waveband | TX_waveband**
0/2	**RX_tuning_time**
2/16	**RX_supported_wavelengths**
0/2	**TX_tuning_time**
2/16	**TX_supported_wavelengths**
1/34	Pad/Reserved
4	FCS

bit	field
0	**Wavelength_id_type**
1	Register
2	reserved
3	Deregister
4-5	**RX_type**
6-7	**TX_type**

(a) REGISTER_REQ MPCPDU

(b) GATE MPCPDU

Octets	field
6	Destination Address
6	Source Address
2	Length/Type = 88-08
2	Opcode = 00-02
4	Timestamp
1	Number of grants/Flags
0/4	Grant #1 Start time
0/2	Grant #1 Length
0/4	Grant #2 Start time
0/2	Grant #2 Length
0/4	Grant #3 Start time
0/2	Grant #3 Length
0/4	Grant #4 Start time
0/2	Grant #4 Length
0/2	Sync Time
0/1	**Grant #1 Wavelength**
0/1	**Grant #2 Wavelength**
0/1	**Grant #3 Wavelength**
0/1	**Grant #4 Wavelength**
9–39	Pad/Reserved
4	FCS

(b) GATE MPCPDU

(c) RX_CONFIG MPCPDU

Octets	field
6	Destination Address
6	Source Address
2	Length/Type = 88-08
2	Opcode = 00-07
4	Timestamp
1	Flags
2/16	**Reception_wavelengths**
23/37	Pad/Reserved
4	FCS

bit	field
0	**Wavelength_id_type**
1-7	reserved

(c) RX_CONFIG MPCPDU

Figure 15.1 WDM extensions to MPCP protocol data units (PDUs): (a) REGISTER_REQ, (b) GATE, and (c) the proposed RX_CONFIG (extensions are shown bold). After McGarry et al. (2006). © 2006 IEEE.

- **TX_supported_wavelengths and RX_supported_wavelengths:** Sixteen bits each (Wavelength_id_type = 0) or 128 bits each (Wavelength_id_type = 1) indicating the bitmap of the supported wavelengths of the transmitter and receiver, respectively.

This information is mapped into the reserved fields of the REGISTER_REQ MPCP protocol data unit (PDU), as shown in Fig. 15.1(a).

15.3.2   Upstream coordination

To facilitate the need for the OLT to assign a specific wavelength channel for the upstream transmission from a given ONU to the OLT, an 8-bit wavelength identifier (allowing for the support of up to 256 unique wavelengths) is issued along with every transmission grant by the OLT by using the reserved fields of the GATE MPCPDU, as depicted in Fig. 15.1(b).

15.3.3   Downstream coordination

To let the OLT (re)configure the receiving wavelength(s) of a given ONU currently no appropriate MPCPDU exists. Two new MPCPDUs, RX_CONFIG (Opcode = 00-07) and RX_CONFIG_ACK (Opcode = 00-08), were proposed in McGarry et al. (2006). The OLT sends the RX_CONFIG MPCPDU to (re)configure a given ONU's receiver(s). The ONU acknowledges the (re)configuration by sending the RX_CONFIG_ACK MPCPDU to the OLT. As shown in Fig. 15.1(c), the Reception_wavelengths field of the RX_CONFIG MPCPDU consists of 16 or 128 bits, depending on the applied encoding scheme given in the respective Flags field. The RX_CONFIG_ACK MPCPDU (not shown in the figure) consists of the echoed Flags and Reception_wavelengths fields.

Note that the aforementioned WDM extensions to IEEE 802.3ah EPON maintain compliance with the IEEE 802.1d bridging by assigning one *logical link ID* (LLID) for each ONU, irrespective of the number of physical wavelength channels supported by the ONU. The IEEE 802.1d bridging uses the LLID (which is associated with a logical port number) to keep an entry in a filtering database that specifies to which logical port a frame with a particular destination address is forwarded to. Assigning one LLID for each ONU allows for a unique entry in the filtering database; the multiple physical wavelength channels to an ONU are simply used as another dimension (in addition to the time dimension) for dynamic bandwidth allocation, as discussed in the next section.

15.4   Dynamic wavelength allocation (DWA)

While in a conventional single-channel TDM EPON the DBA problem is limited to scheduling the upstream transmissions on the single wavelength channel, in a WDM EPON the DBA problem is expanded to scheduling the upstream transmissions on the different upstream wavelengths supported by the ONUs. In other words, in a WDM EPON decisions not only on when and for how long to grant an ONU upstream transmission but also on which wavelength channel to grant the upstream transmission are required, giving rise to the DWA problem. DWA algorithms for dynamically allocating grants to ONUs for upstream transmissions on the different upstream wavelengths in a WDM EPON can be classified into *online scheduling* and *offline scheduling* algorithms (McGarry et al., 2006).

15.4.1   Online scheduling

In an online scheduling DWA algorithm, a given ONU is scheduled for upstream transmission as soon as the OLT receives the REPORT message from the ONU. In other words, the OLT makes scheduling decisions based on individual requests and without global knowledge of the current bandwidth requirements of the other ONUs. For instance, the OLT may schedule the upstream transmission of an ONU on the wavelength channel that is available the earliest among the upstream wavelength channels supported by the ONU. The amount of the bandwidth (i.e., the length of the granted transmission) allocated to an ONU can be determined according to any of the existing DBA algorithms for single-channel TDM EPON.

15.4.2   Offline scheduling

In an offline scheduling DWA algorithm, the ONUs are scheduled for transmission once the OLT has received the current MPCP REPORT messages from *all* ONUs, allowing the OLT to take into consideration in the scheduling the current bandwidth requirements of all ONUs. Since an offline scheduler makes scheduling decisions for all ONUs at once, all of the REPORTs, which are usually appended to the end of the data stream of a gated transmission window, from the previous cycle must be received. This requires that the scheduling algorithm is executed after the OLT receives the end of the last ONU's gated transmission window. Due to this, a report-to-scheduling delay is introduced, as opposed to online scheduling DWA algorithms.

The results obtained in McGarry et al. (2006) indicate that at medium to high traffic loads online scheduling DWA algorithms tend to result in lower packet delays compared to their offline counterparts.

16 STARGATE

We have seen in Chapter 15 that wavelength division multiplexing (WDM) upgraded Ethernet passive optical networks (EPONs) are expected to become mature in the near term. In this chapter, we consider WDM EPONs and, arguing that the key tasks of cost reduction and design of colorless ONUs will be addressed successfully in the near term, elaborate on the question "WDM EPON – what's next?" Our focus will be on evolutionary upgrades and further cost reductions of WDM EPONs and the all-optical integration of Ethernet-based WDM EPON and WDM upgraded RPR networks. The resultant Ethernet-based optical access-metro area network, called STARGATE, was recently proposed in Maier et al. (2007) and will be described at length in the following.

Research on the interconnection of multiple (E)PONs has begun only very recently. In Hsueh et al. (2005a), multiple PONs of arbitrary topology are connected to the same central office (CO) whose transmitters may be shared for downstream transmission among all attached PONs. In An et al. (2005), a common fiber collector ring network interconnects multiple PONs with the CO whose transmitters are used not only for downstream from CO to subscribers but also for upstream transmission from subscribers to CO by means of remote modulation. Note that in both proposed PON interconnection models, any traffic sent between end users residing in different PONs has to undergo OEO conversion at the common CO (i.e., PONs are not interconnected all-optically).

RPR can easily bridge to Ethernet networks such as EPON and may also span into metropolitan area networks (MANs) and wide area networks (WANs). This makes it possible to perform layer 2 switching from access networks far into backbone networks (Davik et al., 2004). It remains to be seen whether end-to-end Ethernet networks turn out to be practical. From an all-optical integration point of view, however, end-to-end optical *islands of transparency* are not feasible and are expected to be of limited geographical coverage due to physical transmission impairments as well as other issues such as management, jurisdiction, and billing issues. As a matter of fact, islands of transparency with optical bypassing capability are key in MANs in order to support not only various legacy but also future services in an easy and cost-effective manner (Saleh and Simmons, 1999).

The proposed STARGATE architecture all-optically integrates Ethernet based access and metro networks. The rationale behind STARGATE is based on the following three principles:

- **Evolutionary Downstream SDM Upgrades:** Eventually, when per-user bandwidth needs grow, incrementally upgrading existent EPON tree networks with additional fibers may prove attractive or become even mandatory. In fact, some providers are already finding this option attractive since long runs of multifiber cable are almost as economical in both material and installation costs as the same lengths of cables with one or a few fibers. Interestingly, the standard IEEE 802.3ah supports not only the commonly used point-to-multipoint (P2MP) topology but also a hybrid EPON topology consisting of point-to-point (P2P) links in conjunction with P2MP links. In STARGATE, we explore the merits of deploying an additional P2P or P2MP fiber link in EPON tree networks to connect the OLT with a subset of one or more ONUs in an evolutionary fashion according to given traffic demands and/or cost constraints. It is important to note that STARGATE requires an additional P2P or P2MP fiber link only in the downstream direction from OLT to ONU(s) and none in the upstream direction. Thus, STARGATE makes use of evolutionary downstream space division multiplexing (SDM) upgrades of WDM/time division multiplexing (TDM) EPONs.

- **Optical Bypassing:** The problem with using SDM in EPONs is the increased electro-optic port count at the optical line terminal (OLT). To avoid this, STARGATE makes use of optical bypassing. Specifically, all wavelengths on the aforementioned additional P2P or P2MP downstream fiber link coming from the metro edge ring are not terminated at the OLT, thus avoiding the need for OEO conversion and additional transceivers at the OLT, as explained in greater detail shortly. Note that OEO conversion usually represents the major part of today's optical networking infrastructure costs. Due to the small to moderate distances of STARGATE's access-metro networks, optical bypassing and the resultant transparency can be easily implemented, thereby avoiding OEO conversion and resulting in major cost savings (Noirie, 2003).

- **Passive Optical Networking:** Finally, the last design principle of STARGATE is based on the idea of letting low-cost PON technologies follow low-cost Ethernet technologies from access networks into metro networks. In doing so, not only PONs but also MANs benefit from passivity, which is a powerful tool to build low-cost high-performance optical networks (Green, 2002). As we will see shortly, STARGATE makes use of an athermal (temperature-insensitive) arrayed waveguide grating (AWG) wavelength router, which eliminates the need for temperature control and monitoring the wavelength shift of the AWG and thus leads to a simplified network management and to reduced costs (Park et al., 2004).

We note that passive optical networking in all-optical wavelength-routing WDM networks has recently begun to gain momentum within the so-called *time-domain wave-length interleaved networking* (TWIN) concept, that enables cost-effective and flexible optical networks to be built using readily available components (Widjaja et al., 2003). In TWIN, fast TDM switching and packet switching in the passive optical wavelength-selective WDM network core are emulated through the use of emerging fast-tunable lasers at the optical network edge, thus avoiding the need for fast optical switching and optical buffering. The original TWIN did not scale well since the number of nodes, N, was limited by the number of available wavelengths, W (i.e., $N = W$). Very recently, the so-called TWIN with wavelength reuse (TWIN-WR) was proposed, where the number of nodes is independent of the number of wavelengths (i.e., $N > W$) (Nuzman and Widjaja, 2006). Both TWIN and TWIN-WR require the network-wide scheduling of transmissions in order to avoid channel collisions. Unlike TWIN, a source node in TWIN-WR may not be able to send traffic directly to any destination node in an optical single hop, resulting in multihopping via intermediate electrical gateways. As we will see shortly, STARGATE differs from TWIN and TWIN-WR in a number of ways. First, STARGATE supports extensive spatial wavelength reuse while providing optical *single-hop* communication among all ONUs. Second, STARGATE requires only *local* scheduling of transmissions within each separate EPON in order to completely avoid channel collisions throughout the network, thus avoiding the need for network-wide scheduling. Third, STARGATE does not require any time-of-day synchronization. And finally, while TWIN and TWIN-WR are designed to support WANs of arbitrary topology, STARGATE targets access and metro networks, whose *regular* topologies (tree, ring, star) help simplify scheduling significantly.

In the following, the STARGATE architecture and operation are described in greater detail.

16.1   Architecture

We first describe the network architecture of STARGATE, followed by a description of the various node architectures.

16.1.1   Network architecture

The network architecture of STARGATE is shown in Fig. 16.1. STARGATE consists of an RPR metro edge ring that interconnects multiple WDM EPON tree networks among each other as well as to the Internet and server farms (RPR was described at length in Chapter 11). More specifically, the RPR network consists of P COs and N_r RPR ring nodes. The CO in the upper right corner of the figure is assumed to be attached to the Internet and a number of servers via a common router. We refer to this CO as the *hot-spot CO*. The P COs are interconnected via a single-hop WDM star subnetwork whose hub is based on a wavelength-broadcasting $P \times P$ passive star coupler (PSC) in parallel with an athermal wavelength-routing $P \times P$ AWG. Each CO is attached to a separate input/output port of the AWG and PSC by means of two pairs of counterdirectional fiber links. Each fiber going to and coming from the AWG carries $\Lambda_{AWG} = P \cdot R$ wavelength channels, where $R \geq 1$ denotes the number of used free spectral ranges (FSRs) of the underlying AWG. Each fiber going to and coming from the PSC carries $\Lambda_{PSC} = 1 + H + (P - 1)$ wavelength channels, consisting of one control channel λ_c, $1 \leq H \leq P - 1$ dedicated home channels for the hot-spot CO, and $P - 1$ dedicated home channels, one for each of the remaining $P - 1$ COs. All COs, except the hot-spot CO, are collocated with a separate OLT of an attached WDM EPON. Let N_c denote the number of ONUs in a given WDM EPON c, which may be different in each WDM EPON. Moreover, let Λ_{OLT}^c denote the number of used wavelengths in WDM EPON c in both directions between the N_c ONUs and the corresponding OLT (i.e., there are Λ_{OLT}^c

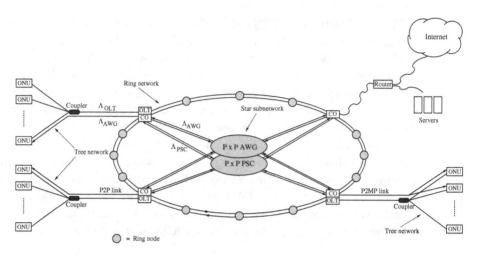

Figure 16.1 STARGATE network architecture comprising $P = 4$ central offices (COs) and $N_r = 12$ RPR ring nodes. After Maier et al. (2007). © 2007 IEEE.

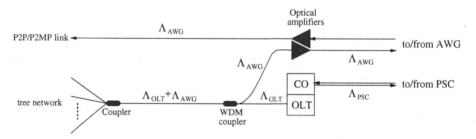

Figure 16.2 Optical bypassing of optical line terminal (OLT) and central office (CO). After Maier et al. (2007). © 2007 IEEE.

upstream and Λ_{OLT}^c downstream wavelength channels in WDM EPON c). Furthermore, each WDM EPON deploys an additional P2P or P2MP downstream fiber link from the CO to a single ONU or multiple ONUs, respectively. Each downstream fiber link carries the Λ_{AWG} wavelength channels coming from the AWG of the star subnetwork.

Figure 16.2 depicts the interconnection of a given WDM EPON and the star subnetwork in greater detail, illustrating the optical bypassing of the collocated OLT and CO. Note that in the figure Λ_{OLT} comprises both upstream and downstream wavelength channels which run in opposite directions on the tree network to and from the OLT, respectively. In contrast, the Λ_{AWG} wavelength channels are carried on the tree network only in the upstream direction, while in the downstream direction they are carried on the separate P2P or P2MP downstream fiber link. As shown in Fig. 16.2, a WDM coupler is used on the tree network in front of the OLT to separate the Λ_{AWG} wavelength channels from the Λ_{OLT} wavelength channels and to guide them directly onward to the AWG of the star subnetwork, possibly optically amplified if necessary. In doing so, the Λ_{AWG} wavelength channels are able to optically bypass the CO and OLT. Similarly, the Λ_{AWG}

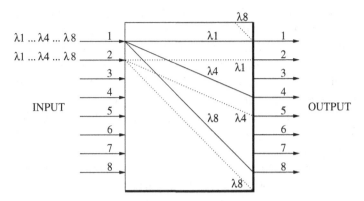

Figure 16.3 Wavelength routing of an 8×8 arrayed waveguide grating (AWG) using $R = 1$ free spectral range (FSR).

wavelength channels coming from the AWG optically bypass both CO and OLT and directly travel on the P2P or P2MP link onward to the subset of attached ONUs. As a result, the ONU(s) as well as the hot-spot CO that send and receive on any of the Λ_{AWG} wavelength channels are able to communicate all-optically with each other in a single hop across the AWG of the star subnetwork. In other words, the *star* forms a *gate* for all-optically interconnecting multiple WDM EPONs. Accordingly, we term the network STARGATE.

The wavelength-routing AWG allows for the spatial reuse of all Λ_{AWG} wavelength channels at each AWG port, as shown in Fig. 16.3 for an 8×8 AWG ($P = 8$) and $\Lambda_{AWG} = P \cdot R = 8$ wavelengths ($R = 1$). Observe that any wavelength can be simultaneously deployed at two (and more) AWG input ports without resulting channel collisions at the AWG output ports. This also holds if multiple FSRs ($R > 1$) are used, where each FSR provides a separate wavelength channel between every pair of AWG input and output ports. The wavelength-routing characteristics of the AWG have the following two implications: First, due to the fact that the AWG routes wavelengths arriving at a given input port independently from all other AWG input ports, no network-wide scheduling but only local scheduling at each AWG input port is necessary to avoid channel collisions on the AWG. Second, note that in Fig. 16.3 each AWG input port reaches a given AWG output port on a different wavelength channel. Consequently, under full spatial wavelength reuse, Λ_{AWG} different wavelengths arrive at each AWG output port simultaneously. To avoid receiver collisions (destination conflicts), each AWG output port must be equipped with a receiver operating on all Λ_{AWG} wavelengths.

16.1.2   Node architecture

Similar to an RPR node, each CO is equipped with two separate fixed-tuned transceivers, one for each direction of the dual-fiber ring. In addition, each CO has one transceiver fixed tuned to the control channel λ_c of the star subnetwork. For data reception on its PSC home channel, each CO (except the hot-spot CO) has a single fixed-tuned receiver. The hot-spot CO is equipped with $1 \leq H \leq P - 1$ fixed-tuned receivers. For data transmission on the PSC, each CO (except the hot-spot CO) deploys a single transmitter which can be tuned over the $(P - 1) + H$ home channels of the COs. The hot-spot CO deploys H tunable transmitters whose tuning range covers the home channels of the remaining $P - 1$ COs as well as the Λ_{AWG} wavelengths. Unlike the remaining COs, the hot-spot CO is equipped with an additional multiwavelength receiver operating on Λ_{AWG}. In WDM EPON c, the OLT is equipped with an array of c fixed-tuned transmitters and c fixed-tuned receivers, operating at the Λ_{OLT}^{c} downstream and Λ_{OLT}^{c} upstream wavelength channels, respectively. Note that STARGATE does not impose any particular WDM node structure on the ONUs except for ONUs which receive data over the AWG. Those ONUs must be equipped with a multiwavelength receiver operating on the Λ_{AWG} wavelengths in order to avoid receiver collisions, as explained in Section 16.1.1.

16.2   Operation

In this section, we explain the operation of the STARGATE network. In STARGATE, we use the WDM extensions to EPON's multipoint control protocol (MPCP) REPORT and GATE messages recommended in McGarry et al. (2006) (see Chapter 15), which enable each OLT to schedule transmissions to and receptions from its attached WDM-enhanced ONUs on any wavelength channels supported by both the OLT and the respective ONU, as explained in the following.

16.2.1   Discovery and registration

In each WDM EPON, the REGISTER_REQ MPCP message with WDM extensions described in McGarry et al. (2006) is deployed for the discovery and registration of ONUs. The REGISTER_REQ message is sent from each ONU to its OLT and carries the MAC address as well as detailed information about the WDM node structure of the ONU. In doing so, the OLT of each WDM EPON learns about the MAC address and WDM node structure of each of its attached ONUs. After registration, the OLTs exchange via the PSC (to be described shortly) the MAC addresses of their attached ONUs that are able to receive data over the AWG. As a result, the OLTs know not only which MAC addresses can be reached via the AWG but also to which AWG output ports the corresponding ONUs are attached and thus on which of the Λ_{AWG} wavelengths they can be reached from a given AWG input port.

16.2.2   Piggyback REPORT MPCP message

The IEEE 802.3ah REPORT MPCP message can carry one or more queue sets, each set comprising up to eight queues, as shown in Fig. 16.4. In STARGATE, we let ONUs use the first queue set to report bandwidth requirements on the Λ_{OLT} upstream wavelengths for sending data to the OLT (e.g., conventional triple-play traffic [data, voice, video]).

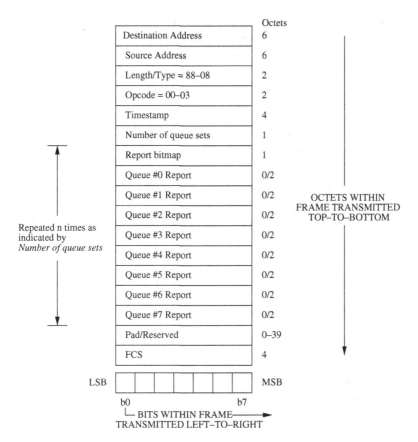

	Octets
Destination Address	6
Source Address	6
Length/Type = 88–08	2
Opcode = 00–03	2
Timestamp	4
Number of queue sets	1
Report bitmap	1
Queue #0 Report	0/2
Queue #1 Report	0/2
Queue #2 Report	0/2
Queue #3 Report	0/2
Queue #4 Report	0/2
Queue #5 Report	0/2
Queue #6 Report	0/2
Queue #7 Report	0/2
Pad/Reserved	0–39
FCS	4

Repeated n times as indicated by *Number of queue sets*

OCTETS WITHIN FRAME TRANSMITTED TOP-TO-BOTTOM

LSB   MSB

b0   b7

└ BITS WITHIN FRAME
TRANSMITTED LEFT–TO–RIGHT

Figure 16.4 REPORT MPCP message.

To report bandwidth requirements on any of the Λ_{AWG} wavelength channels to ONUs located in different EPONs, a given ONU uses one or more additional queue sets and writes the MAC addresses of the destination ONUs in the reserved field of the REPORT MPCP message and sends it to the OLT. Thus, the bandwidth requirements on Λ_{AWG} ride piggyback on those on Λ_{OLT} within the same REPORT message. In Section 16.3, we will discuss an illustrative example of how the piggyback REPORT MPCP message can be used for online gaming and peer-to-peer file sharing.

16.2.3   STARGATE MPCP message

The WDM extended GATE message in McGarry et al. (2006) is used to coordinate the upstream transmission on the Λ_{OLT} wavelengths within each WDM EPON and also to coordinate the all-optical transmission on any of the Λ_{AWG} wavelengths across the star subnetwork between two ONUs residing in different WDM EPONs, provided both ONUs support the Λ_{AWG} wavelengths. Based on the MAC addresses of the destination ONUs carried piggyback in the REPORT message, the OLT of the source WDM EPON uses the extended GATE MPCP message, which we term STARGATE message, to grant the source ONUs a time window on the wavelengths which the AWG routes to the destinations according to the DBA algorithm in use at the OLT.

16.2.4 STARGATING service

Similar to EPON, the STARGATE network is not restricted to any specific dynamic bandwidth allocation (DBA) algorithm. However, DBA algorithms for STARGATE should be able to dynamically set up transparent all-optical circuits across the AWG at the wavelength and subwavelength granularity with predictable quality of service (QoS) in terms of bounded delay and guaranteed bandwidth between ONUs of different WDM EPONs. Each OLT uses its DBA module to provide *gated* service across the AWG-based *star* network, a service which we correspondingly term STARGATING. This gated service enables the dynamic set-up of low-latency circuits on any of the Λ_{AWG} wavelengths in support of applications such as P2P file sharing and periodic game traffic, as discussed in greater detail shortly.

16.2.5 Access control on ring and PSC

ONUs unable to send and receive data across the AWG as well as RPR ring nodes send their data on the tree, ring, and/or PSC along the shortest path in terms of hops. Channel access on the dual-fiber ring is governed by RPR protocols, described in Chapter 11. On the PSC, time is divided into periodically recurring frames. On the control channel λ_c, each frame consists of P control slots, each dedicated to a different CO. Each CO stores data packets to be forwarded on the PSC in a single first-in/first-out (FIFO) queue with look-ahead capability to avoid head-of-line (HOL) blocking. For each stored data packet the CO broadcasts a control packet in its assigned control slot to all COs. A control packet consists of two fields: (1) destination address and (2) length of the corresponding data packet. All COs receive the control packet and build a common distributed transmission schedule for the collision-free transmission of the corresponding data packet on the home channel of the destination CO at the earliest possible time. The destination CO forwards the received data packet toward the final destination node.

16.3 Applications

It is important to note that providing end users with advanced broadband access and a growing body of content and applications has a significant impact on their everyday lives. According to the Pew Internet & American Life Project, a project that examines the social impact of the Internet, subscribers of advanced access networks increasingly use the Internet as a "destination resort," a place to go just to have fun or spend their free time (Pew Internet & American Life Project, 2006). Let us focus on two applications which become increasingly popular in optical access networks among subscribers spending their free time: *online gaming* and *P2P file sharing*, whereby the latter one is already the predominant traffic type in today's operational access networks (Garcia et al., 2004). In the following, we deploy a top-down approach. We first outline the traffic characteristics of both applications and discuss their impact on the architecture and services of optical access networks. We then discuss how STARGATE is well suited to support both applications in an efficient, cost-effective, and future-proof manner.

16.3.1   Online gaming

Traffic characteristics

Most of the popular online games are based on the client–server paradigm, where the server keeps track of the global state of the game. Online gaming traffic primarily consists of information sent periodically back and forth between all clients (players) and server. Online gaming traffic is very different than web traffic (Feng et al., 2005). Specifically, online gaming requires low-latency point-to-point upstream communication from each client to the server as well as low-latency directed broadcast downstream communication from the server to all clients. To facilitate the synchronous game logic, packets are small since the application requires extremely low latencies which makes message aggregation and message retransmission impractical. The workload of online games consists of large, highly periodic bursts of very small packets with predictable long-term rates, exhibiting packet bursts every 50 or 100 ms where the payload of almost all upstream packets is smaller than 60 bytes and that of downstream packets is spread between 0 and 300 bytes. Upstream packets have an extremely narrow length distribution centered around the mean size of 40 bytes while downstream packets have a much wider length distribution around a significantly larger mean.

Impact

The unique characteristics and requirements of online games pose significant challenges on current network infrastructures (Feng et al., 2005). Game vendors increasingly deploy large server farms in a single location to control the experience of players. The server farms must be provided with a means to efficiently realize directed broadcasting. More importantly, online gaming introduces a significant downward shift in packet size which makes the electronic processing the bottleneck versus the link speed. Networking devices that are designed for handling larger packets will suffer from packet loss or persistent packet delay and jitter. One solution to alleviate this bottleneck is to optically bypass electronic access network devices. The high predictability of online gaming traffic can be exploited for efficient access network resource management by setting up subwavelength circuits.

In online games, low latency and good scalability are the two most important network design aspects (Jiang et al., 2005). If not handled properly, network latency leads to poor game quality, sluggish responsiveness, and inconsistency. Besides latency, scalability is another key aspect of networked online games. The currently predominant client–server model of commercial online games faces scalability issues since the central server may suffer from CPU resource and bandwidth bottlenecks. Recently, the peer-to-peer design of scalable game architectures has begun to increasingly receive attention, where clients' computing resources are utilized to improve the latency and scalability of networked online games (Jiang et al., 2005).

16.3.2   Peer-to-peer file sharing

The use of P2P applications for sharing large audio/video files as well as software has been growing dramatically. P2P traffic represents now by far the largest amount of data traffic in today's operational access networks, clearly surpassing web traffic (Garcia et al., 2004). In P2P file sharing applications, the process of obtaining a file can be divided into two phases: (1) signaling and (2) data transfer (Sen and Wang, 2004). In the signaling phase, by using a specific (proprietary) P2P protocol a host identifies one or more target hosts from which to download the file. In the data transfer phase, the requesting host downloads the file from a selected target host.

Traffic characteristics

According to Garcia et al. (2004), the major P2P application, which represents more than 50% of the entire access network upstream data traffic, exhibits a nearly constant traffic pattern over time, independent of the number of connected subscribers. Moreover, the vast majority of upstream traffic is generated by a small number of hosts, for both weekday and weekend. Specifically, the top 1–2% of IP addresses account for more than 50% and the top 10% of IP addresses account for more than 90% of upstream traffic. Interestingly, note that this behavior, where a few *hot-spot servers* with popular content originate most of the P2P upstream traffic, resembles that of networks which are based on the conventional client–server paradigm. Similar observations were made for P2P downstream traffic, where a few *heavy hitters* are responsible for a high percentage of downstream traffic. Also, heavy hitters tend to have long on-times. Finally, it is worthwhile mentioning that the most popular P2P system in terms of both number of hosts and traffic volume is able to resolve most queries locally by finding nearby peers (Sen and Wang, 2004).

Impact

The high volume and good stability properties of P2P traffic give rise to the use of simple yet highly effective capacity planning and traffic engineering techniques as a promising solution to manage P2P file sharing. Additionally, the fact that individual hot-spot servers and heavy hitters with long on-times generate huge traffic volumes and most of the queries can be resolved locally can be exploited at the architecture and protocol levels of future P2P-friendly optical access networks that must be designed to meet the requirements of P2P applications in a resource-efficient and cost-effective manner.

16.3.3   Discussion

STARGATE is well suited to meet the aforementioned requirements of online gaming and P2P file sharing. The all-optical subwavelength circuits may be used to carry periodic low-latency game traffic and high volumes of stable P2P traffic. Directed broadcasting can be realized by letting the hot-spot CO transmit packets on different wavelengths of Λ_{AWG}, whereby each packet sent on a given wavelength is locally broadcast to all N_c ONUs attached to the $1:N_c$ coupler in EPON c. Individual ONUs which send or receive large amounts of traffic may deploy additional transceivers. STARGATE also scales well in that additional EPON tree networks may be attached to the N_r RPR ring nodes, which in turn might be later connected to the WDM star subnetwork, and additional FSRs of the AWG may be used to increase the number Λ_{AWG} of used wavelengths on the AWG, as needed. Finally, STARGATE provides a high degree of connectivity which not only improves the network resilience and bandwidth efficiency but also decreases the number of required hops between nearby file-sharing peers.

Finally, we note that F-TCP, a light-weight TCP with asynchronous loss recovery mechanism for the transfer of large files in high-speed networks, can be used at the transport layer of STARGATE to achieve reliable P2P file sharing (Chuh et al., 2005).

17   Gigabit Ethernet

Ethernet networks have come a long way and are widely deployed nowadays. In fact, 95% of today's local area networks (LANs) use Ethernet. Ethernet's transmission rate was originally set at 10 megabits per second (10 Mbps) in 1980 and evolved to higher speed versions ever since. A 100-Mbps version, also known as *Fast Ethernet*, was approved as IEEE standard 802.3u in 1995. In order to save time and standards development resources, physical signaling methods previously developed and standardized for Fiber Distributed Data Interface (FDDI) networks were reused in the IEEE standard 802.3u (Thompson, 1997). Fast Ethernet was immediately accepted by customers and its success prompted the development of an Ethernet standard for operation at 1000 Mbps (1 Gbps), leading to *Gigabit Ethernet* (GbE). The standard for Gigabit Ethernet, IEEE standard 802.3z, was formally approved in 1998. At present, 10-Gigabit Ethernet (10GbE) is the fastest of the Ethernet standards. The standardization of 10GbE began in March of 1999 and led to the 10GbE standard IEEE 802.3ae, which was formally approved in 2002.

In this chapter, we highlight the salient features of both 1 and 10 Gbps Ethernet. While 10GbE is the fastest existing Ethernet standard at the time of writing, it is worthwhile to mention that 10GbE does not represent the end of the development of ever-increasing higher-speed Ethernet networks. The standardization of 100-Gigabit Ethernet (100GbE) is currently under development by the IEEE 802.3 Higher Speed Study Group (HSSG). The HSSG was formed in 2006 and aims at providing a standard for 100GbE by the end of 2009.

17.1   Gigabit Ethernet (GbE)

The GbE standard IEEE 802.3z addresses the two lowest layers of the International Standards Organisation (ISO) Open Systems Interconnection (OSI) reference model, i.e., the data link layer (layer 2) and the physical layer (layer 1). More precisely, the standard specifies the media access control (MAC) sublayer of layer 2 besides the physical layer (PHY), as shown in Fig. 17.1. The specifications of the GbE MAC and PHY layers of Fig. 17.1 are described in greater detail in the following. For further information on GbE we refer the interested reader to Frazier (1998) and Frazier and Johnson (1999).

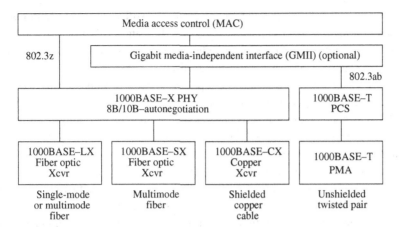

Figure 17.1 Gigabit Ethernet (GbE) MAC and PHY layers diagram. After Frazier (1998). © 1998 IEEE.

17.1.1   Media access control (MAC) layer

GbE supports two types of media access: (1) *shared* access, and (2) *dedicated* acess.

Shared access (half-duplex mode)

With shared access, all stations share the Ethernet medium whose access is governed by the legacy carrier sense multiple access with collision detection (CSMA/CD) protocol. The CSMA/CD protocol allows each Ethernet station to sense the medium for carrier signals prior to transmitting data frames and to detect collisions due to simultaneous transmissions of more than one data frame. The CSMA/CD protocol enables half-duplex communication among stations (i.e., a given station can either transmit or receive data at any given time but is unable to transmit and receive data simultaneously). To efficiently work at 1 Gbps, the 802.3z standard extends the original CSMA/CD protocol by adding the following two enhancements (Tan, 2000):

- **Carrier extension:** The collision detection algorithm of the original CSMA/CD protocol mandates that the round-trip propagation delay between any pair of stations must not exceed the transmission time of the smallest data frame. Due to the fact that the minimum frame size was set to 512 bits (64 bytes) in the original Ethernet, GbE networks operating at a higher line rate than its predecessors could have only a rather limited network diameter. To increase the diameter of GbE networks, a *carrier extension* was introduced to overcome the inherent scalability limitation of CSMA/CD. With carrier extension, the minimum frame size is extended from 512 bits to 512 bytes (4096 bits) by appending a set of special symbols to the end of frames smaller than 512 bytes such that the resulting extended frame is at least 512 bytes long. Frames longer than 512 bytes are not extended. Note that carrier extension is able to maintain backward compatibility with the widely installed base of 10 Mbps and 100 Mbps

Ethernet networks in that the minimum frame size is still 64 bytes (512 bits). At the downside, however, carrier extension decreases the bandwidth efficiency for small frames. To overcome this problem, a method called *frame bursting* may be used to enhance the original CSMA/CD protocol, as discussed next.

- **Frame bursting:** Frame bursting is an optional feature to improve the bandwidth efficiency and throughput of GbE networks that operate in half-duplex mode using the CSMA/CD protocol. Frame bursting enables stations to transmit multiple frames back to back without having to contend for the medium since all the other stations defer their transmission. Specifically, with frame bursting the first transmitted frame is extended by means of the aforementioned carrier extension, if necessary. Subsequently, the transmitting station may send further frames without extension up to a specified limit, referred to as *burstLimit*, which is set to 8192 bytes (65,536 bits). Thus, frame bursting allows a station to transmit multiple short frames, consecutively avoiding the carrier extension technique for all frames but the first one, resulting in an increased bandwidth efficiency and network throughput.

Dedicated access (full-duplex mode)

With dedicated access, Gigabit Ethernet's CSMA/CD protocol together with the aforementioned carrier extension and frame bursting enhancements are disabled and the GbE network operates in full-duplex mode (i.e., a station can send and receive frames simultaneously without interference and contention for a shared medium). The full-duplex mode was defined previously by the IEEE 802.3x specification and began to be widely deployed in switched Ethernet LANs in the early 1990s. Note that the full-duplex mode is not only easier to implement than the half-duplex CSMA/CD-based mode but also provides an increased aggregate network throughput of 2 Gbps (the reduced implementational complexity is likely the reason why the vast majority of commercial GbE equipment operates in full-duplex mode). Furthermore, due to the fact that the CSMA/CD protocol is disabled, there is no round-trip delay constraint and each network link can be of arbitrary length subject to physical transmission impairment limits. The full-duplex Ethernet standard 802.3x also specifies a mechanism for link-level flow control between two adjacent stations. Specifically, the so-called pause protocol enables a receiving station to inhibit (pause) the transmission of a sending station by sending a *pause frame* that indicates the sending station to stop its transmission for a certain period of time. More precisely, The pause frame contains a time value given as an integer multiple of 512 bit transmission times which indicates the length of the transmission pause. If the receiving station becomes uncongested before the pause time expires, the receiving node may send another pause frame to the transmitting node with a time value set to zero. Upon receiving the second pause frame, the transmitting station may resume transmitting data frames immediately. It is worthwhile to mention that the pause protocol can be used only between two adjacent stations located at the end points of a point-to-point link. Pause frames cannot be forwarded through intermediate bridges, switches, or routers, thus preventing the pause protocol from operating on an end-to-end level.

17.1.2   Gigabit-media independent interface (GMII)

To let the MAC layer interoperate with a variety of different physical media, Ethernet has always been using a *media-independent interface* (MII). The optional *Gigabit MII* (GMII) of GbE provides a means to develop further physical media and attach them to the MAC. For instance, the GMII may be used to support Category 5 unshielded twisted pair (UTP) cabling, which is widely installed in many organizations. The IEEE standard 802.3ab defines the specifications of GbE transmission over Category 5 UTP cabling, which is also known as 1000BASE-T. This standard enables organizations to leverage on their already installed copper cabling infrastructure in order to easily set up GbE in a cost-effective manner.

17.1.3   Physical (PHY) layer

The physical (PHY) layer converts the data coming from the MAC layer into optical or electrical signals and sends it across the physical transmission medium. The GbE PHY layer is subdivided into the so-called *physical coding sublayer* (PCS) and *physical medium attachment* (PMA). The PCS and PMA in conjunction with a link configuration protocol, termed *autonegotiation*, are collectively referred to as 1000BASE-X PHY. In the following, the PCS, PMA, and autonegotiation and the various functions carried out by them are explained in greater detail.

- **Physical coding sublayer (PCS):** The PCS made use of the 8B/10B encoding of the ANSI Fibre Channel standard to reduce the time-to-market of GbE equipment, whereby 8-bit data bytes coming from the MAC layer are encoded into 10-bit code groups. The 8B/10B code exhibits excellent properties such as error robustness, transition density, and direct current balance. Apart from 8B/10B encoding, the PCS encapsulates MAC data frames by adding special codes to the start and end of each frame. Owing to the 8B/10B encoding and encapsulation together with the 32-bit cyclic redundancy check at the trailer of each frame, GbE provides 100% detection of all single-, double-, and triple-bit errors in a frame and excellent robustness against undetected errors (Frazier and Johnson, 1999).
- **Physical medium attachment (PMA):** The 10-bit code groups generated by the PCS are transmitted serially by the PMA by means of non-return-to-zero (NRZ) line coding. Note that due to the 8B/10B encoding GbE needs to operate at a clock rate of 1.25 Gbps in order to provide an effective data rate of 1 Gbps.
- **Autonegotiation:** The autonegotiation protocol is used for link initialization and configuration. It was defined for Fast Ethernet to automatically select between 10 and 100 Mbps. In Gigabit Ethernet, the autonegotiation protocol is primarily deployed to determine the duplex mode, i.e., select between half-duplex and full-duplex modes, and to decide whether the pause protocol is used for link-level flow control. Autonegotiation is completed roughly 40 milliseconds after the cables are plugged in or the equipment is turned on (Frazier and Johnson, 1999).

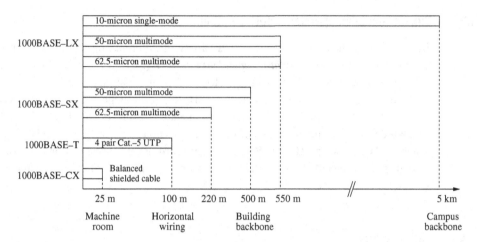

Figure 17.2 Gigabit Ethernet (GbE) supported link distances. After Frazier and Johnson (1999). © 1999 IEEE.

GbE runs over optical fiber, shielded copper cable, and UTP cable, as shown in Fig. 17.1. More specifically, GbE supports two types of optical transceivers, 1000BASE-LX and 1000BASE-SX. The LX transceiver works on single-mode and multimode fibers and is intended for longer distances, whereas the SX transceiver works only on multimode fiber and targets short-distance applications (e.g., desktop links). Both optical transceivers use two separate fibers for transmission and reception. The SX transceiver is less expensive than the LX transceiver. However, the LX transceiver supports significantly longer link distances. Figure 17.2 depicts the link distances supported by both optical transceivers for three widely deployed types of fiber: 10-micron single-mode, 50-micron multimode, and 62.5-micron multimode fibers. The 10-micron single-mode fiber is predominantly used in campus backbone networks with a reach of up to 5 kilometers, whereas the 62.5-micron multimode fiber is widely used in building backbone networks and also in horizontal wiring deployments.

Figure 17.2 also shows the supported link distances of 1000BASE-T and 1000BASE-CX transceivers. The 1000BASE-T transceiver operates on four pairs of Category 5 UTP cables and supports a maximum link distance of 100 m, thus being suitable for horizontal wiring. The 1000BASE-CX transceiver supports shielded copper links of up to 25 m. The CX transceiver is an economically attractive choice for short-distance interconnections (e.g., between devices located within the same rack or within a computer room or telephone closet) (Frazier, 1998).

17.2   10-Gigabit Ethernet (10GbE)

Besides the increased line rate, 10GbE differs from GbE and earlier Ethernet standards primarily in that it operates only over fiber and only in full-duplex mode. Furthermore, 10GbE provides interoperability not only with Ethernet but also with SONET/SDH. As a result, 10GbE can be used in LANs as well as in SONET/SDH-based metropolitan area networks (MANs) and wide area networks (WANs). Various evaluations showed that 10GbE fits seamlessly into existing infrastructures and works equally well in LAN, MAN, and WAN environments (Hurwitz and Feng, 2004).

The 10GbE standard IEEE 802.3ae specifies a total of seven port types (Cunningham, 2001). More specifically, the standard defines four LAN PHY port types for native Ethernet applications and three WAN PHY port types for connection to 10Gbps OC-192 SONET/SDH networks. All three WAN PHY port types and three of the four LAN PHY port types deploy bit-serial transmission across single-mode or multimode fibers by means of direct modulation of laser light sources. The remaining LAN PHY port type sends data across four different wavelengths by means of coarse wavelength division multiplexing (CWDM), which is referred to as wide wavelength division multiplexing (WWDM) in the standard. Similarly to GbE, the WWDM LAN PHY port type deploys 8B/10B encoding on each of the four wavelengths. Each wavelength operates at a line rate of 3.125 Gbps, which translates into a data rate of 2.5 Gbps. In contrast, the other three bit-serial LAN PHY port types use a more efficient 64B/66B encoding instead of the 8B/10B encoding, resulting in a line rate of 10.3125 Gbps. The new 64B/66B encoding allows for a more cost-effective bit-serial implementation of directly modulated lasers than the conventional 8B/10B encoding. Due to the fact that OC-192 SONET/SDH equipment operates at a data rate slightly lower than 10Gbps, the data rate of the WAN PHY port types must be throttled. To achieve this, the interframe spacing is dynamically increased in order to match the data rate of OC-192 SONET/SDH equipment, at the expense of a slightly decreased 10GbE network throughput.

18 Radio-over-fiber networks

In the preceding chapters, we witnessed that optical fiber is widely used as the transmission medium of choice in wide, metropolitan, access, and local area (wired) networks. Passive optical networks (PONs) might be viewed as the final frontier of optical wired networks where they interface with a number of wireless access technologies. One interesting approach to integrate optical fiber networks and wireless networks are so-called *radio-over-fiber* (RoF) networks. In RoF networks, radiofrequencies (RFs) are carried over optical fiber links to support a variety of wireless applications. In this chapter, we describe some recently investigated RoF network architectures and their support of various wireless applications. After reviewing the use of optical fiber links for building distributed antenna systems in fiber-optic microcellular radio networks, we elaborate on the various types of RoF networks and their integration with fiber to the home (FTTH), WDM PON, and rail track networks.

18.1 Fiber-optic microcellular radio

18.1.1 Distributed antenna system

To increase frequency reuse and thereby support a growing number of mobile users in cellular radio networks, cells may be subdivided into smaller units referred to as *microcells*. The introduction of microcells not only copes with the increasing bandwidth demands of mobile users but also reduces the power consumption and size of handset devices. Instead of using a base station antenna with high-power radiation, a distributed antenna system connected to the base station via optical fibers was proposed in Chu and Gans (1991). In the proposed *fiber optic microcellular radio* system, the radio signals in each microcell are transmitted and received to and from mobile users by using a separate small canister that is attached to the base station via optical fiber, as shown in Fig. 18.1. Each canister is equipped with appropriate optical-to-RF and RF-to-optical converters and transceivers for transmitting to and receiving from mobile users, respectively (RF stands for radiofrequency). The base station and each canister are equipped with a laser and optical receiver for transmission and reception on the optical fiber, respectively. For transmission, subcarrier multiplexed radio signals directly modulate the laser. For reception, the radio signals are recovered from the optical signal by means of direct detection. The small canisters can be placed on public property

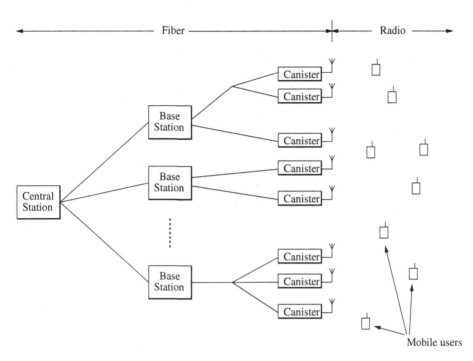

Figure 18.1 Fiber optic microcellular radio system based on canisters connected to base stations via fiber links.

(e.g., streetlight posts or building walls), thus avoiding the need for purchasing or leasing expensive private property to install tall base station antennas. Furthermore, radio transmission from streetlight-level canisters allows for much lower power radiation and thus helps overcome objections about high-power radiation and safety issues. Apart from increasing frequency reuse, canisters might also be deployed to fill in "dead spots" in the coverage area of a base station antenna signal (e.g., indoor microcells in shopping malls). It is worthwhile to mention that the proposed system is identical to the existing cellular radio system except that fibers and canisters need to be placed between the base station and the mobile users. Hand-offs of mobile users between base stations can be done as in the existing cellular radio system.

18.1.2   Dynamic channel assignment

To efficiently support time-varying traffic between the central station of Fig. 18.1 and its attached base stations, a centralized dynamic channel assignment method can be applied at the central station (Ohmoto et al., 1993). The dynamic channel assignment method, referred to as spectrum delivery scheme (SDS), installed at the central station dynamically assigns one or more of the subcarriers used on the fiber links between central station and base stations to any base station according to current traffic demands. SDS helps improve the flexibility of fiber optic microcellular radio networks by assigning more subcarrier channels to a base station serving a heavy traffic zone and fewer subcarrier channels to base stations that cover light traffic zones. In doing so, the

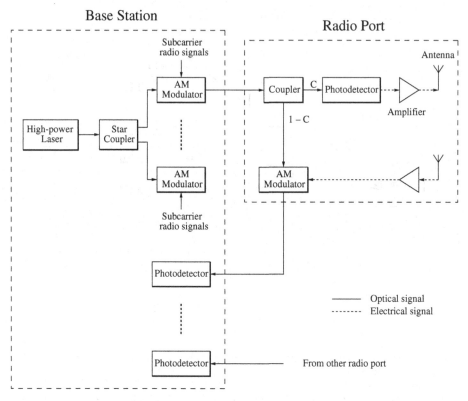

Figure 18.2 Remote modulation at the radio port of a fiber optic microcellular radio network. After Wu et al. (1994). © 1994 IEEE.

dynamic channel assignment using SDS at the central station effectively reduces the call blocking probability in fiber optic microcellular radio networks whose traffic loads vary over time.

18.1.3   Remote modulation

To avoid having to equip each radio port in a fiber optic microcellular radio network with a laser and its associated circuit to control the laser parameters such as temperature, output power, and linearity, a cost-effective radio port architecture deploying *remote modulation* was examined in Wu et al. (1994, 1998). Figure 18.2 depicts the block diagram of the proposed remote modulation technique. A single high-power laser located at the base station is shared among many microcells. The optical output signal of the laser is split by a star coupler and is then externally modulated by subcarrier radio signals using a separate amplitude (AM) modulator for each microcell and its corresponding radio port. At the radio port of each microcell, an optical coupler is placed at the input of the radio port to tap off some portion $1 - C$ of the incoming optical signal for uplink transmission. This portion of the optical signal is remodulated by amplified radio signals received from the antenna using an AM modulator and detected at the

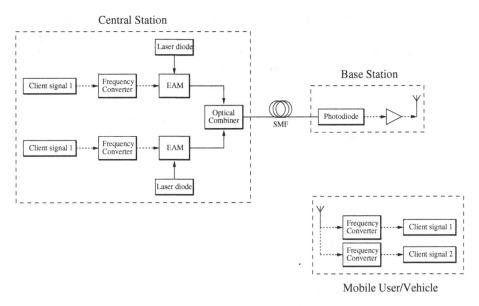

Figure 18.3 Radio-over-SMF network downlink using electroabsorption modulators (EAMs) for different radio client signals. After Tang et al. (2004). © 2004 IEEE.

base station using a separate photodetector for each uplink originating from different microcells. The remaining portion C of the downlink optical signal is detected at the radio port by a photodetector and is subsequently amplified before radiating through the antenna. Remote modulation reduces the power consumption and complexity of radio ports by replacing costly lasers with passive external modulators, giving rise to fiber optic microcellular network architectures with low-cost radio ports.

18.2   RoF networks

18.2.1   Radio-over-SMF

Apart from realizing low-cost microcellular radio networks, optical fibers can also be used to support a wide variety of other radio signals, leading to so-called RoF networks. RoF networks are attractive since they provide transparency against modulation techniques and are able to support various digital formats and wireless standards in a cost-effective manner. It was experimentally demonstrated in Tang et al. (2004) that RoF networks are well suited to simultaneously transmit the following four wireless standards using a single antenna: (1) wideband code division multiple access (WCDMA), (2) IEEE 802.11 wireless local area network (WLAN), (3) personal handyphone system (PHS), and (4) global system for mobile communications (GSM). Toward this end, a method is needed to combine the various radio signals onto a single fiber. Figure 18.3 illustrates the method investigated in Tang et al. (2004) for two different radio client signals transmitted by the central station on the RoF network downlink to a base station and onward to a mobile user or vehicle. As shown in the figure, at the central station both radio client signals are first upconverted to a higher frequency by using a frequency converter. Then the two RF signals go into two different electroabsorption modulators (EAMs) and modulate the optical carrier wavelength emitted by two separate laser diodes. An optical combiner combines the two optical signals onto a common single-mode fiber (SMF) link that interconnects the central station with the base station, resulting in a *radio-over-SMF* network. At the base station, a photodiode converts the incoming optical signal to the electrical domain and radiates the amplified signal through an antenna to a mobile user or vehicle which uses two separate frequency converters to retrieve the two different radio client signals.

18.2.2   Radio-over-MMF

While SMFs are typically found in outdoor optical networks, many buildings have pre-installed multimode fiber (MMF) cables. Cost-effective MMF-based networks can be realized by deploying low-cost vertical-cavity surface-emitting lasers (VCSELs) operating in the 850-nm transmission window. In Lethien et al. (2005), different kinds of MMF in conjunction with commercial off-the-shelf (COTS) components were experimentally tested to demonstrate the feasibility of indoor radio-over-MMF networks for the in-building coverage of second-generation (GSM) and third-generation cellular radio networks (universal mobile telecommunications system (UMTS)) as well as IEEE 802.11 WLAN and digital enhanced cordless telecommunication packet radio service (DECT PRS).

18.3   WDM RoF networks

The introduction of wavelength division multiplexing (WDM) in RoF networks not only increases their capacity but also, and more importantly, increases the number of base stations serviced by a single central station. In WDM RoF networks, newly added base stations can be supported by allocating different wavelength channels to them (Bakaul et al., 2005).

A WDM RoF ring network based on reconfigurable optical add-drop multiplexers (ROADMs) was proposed and experimentally investigated in Lin et al. (2004) and Lin (2005). Specifically, the optical ring network consists of a single WDM fiber loop that connects multiple remote nodes with the central office. Each remote node deploys an array of tunable fiber Bragg gratings (FBGs). To drop a single wavelength, a remote node tunes all its FBGs to the same wavelength such that only the specific wavelength is dropped locally. By tuning its FBGs to different wavelengths, a remote node is able to locally drop more than one wavelength. A number of so-called radio access units (RAUs) is attached to each remote node, whereby each RAU may serve one or more mobile users. Clearly, using ROADMs at remote nodes adds to the flexibility of the WDM RoF network since add-drop wavelengths can be dynamically assigned to remote nodes and the attached RAUs according to given traffic loads that might vary over time.

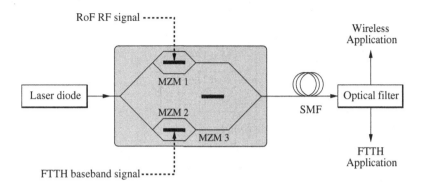

Figure 18.4 Simultaneous modulation and transmission of FTTH baseband signal and RoF RF signal using an external integrated modulator. After Lin et al. (2007). © 2007 IEEE.

18.4   RoF and FTTH networks

To realize future multiservice access networks, it is important to integrate RoF systems with existing optical access networks (e.g., FTTH networks). One of the key concerns of such hybrid optical access networks is to enable RoF and FTTH systems to transmit both wireless RF and wired-line baseband signals on a single wavelength over a single fiber. In Lin et al. (2007), a novel approach for simultaneous modulation and transmission of both RoF RF and FTTH baseband signals using a single external integrated modulator was experimentally demonstrated. Figure 18.4 depicts the hybrid optical access network system based on an external integrated modulator that consists of three different Mach-Zehnder modulators (MZMs) 1, 2, and 3. As shown in the figure, MZM 1 and MZM 2 are embedded in the two arms of MZM 3. The RoF RF and FTTH baseband signals independently modulate the optical carrier generated by a common laser diode. Specifically, the optical RoF RF signal is generated at MZM 1, while the FTTH baseband signal is generated at MZM 2. Subsequently, the optical wireless RF and wired-line baseband signals are combined at MZM 3. After propagation over a standard SMF, an optical filter (e.g., fiber grating) is used to separate the two signals and forward them to the wireless and FTTH application, respectively. The presented experimental results demonstrated that a 1.25 Gb/s baseband signal and a 20-GHz 622 Mb/s RF signal can be simultaneously modulated and transmitted over 50 km standard SMF with acceptable performance penalties.

18.5   RoF and WDM PON networks

We have seen in Chapter 15 that WDM upgraded EPONs in particular and WDM PONs in general become rapidly mature. Therefore, it is desirable to integrate WDM (E)PON transport systems with RoF systems. A seamless integration approach of eight 2.5 Gb/s WDM signals with an RoF system was experimentally demonstrated in Yu et al. (2005). In the proposed integration approach, the simultaneous frequency upconversion of the

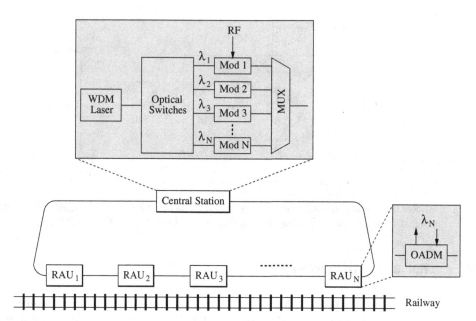

Figure 18.5 Moving cell-based RoF network architecture for train passengers. After Lannoo et al. (2007). © 2007 IEEE.

eight WDM signals was done all-optically by means of four-wave mixing (FWM). Due to the fact that FWM is independent of the signal bit rate and modulation format, it can be used for simultaneous frequency upconversion of different optical WDM signals.

18.6   RoF and rail track networks

Cellular networks used for fast-moving users (e.g., train passengers) suffer from frequent hand-overs when hopping from one base station to another one. The frequent hand-overs cause numerous packet losses, resulting in a significantly decreased network throughput. An interesting approach to solve this problem for train passengers is the use of an RoF network installed along the rail tracks in combination with the so-called *moving cell* concept (Lannoo et al., 2007). The proposed solution provides high-capacity wireless services to high-speed-train passengers using a hierarchical approach that consists of a wireless link between the railway and the train on the one hand and a separate wireless link between the train and the users on the other hand. In each train carriage, one or more WLAN access points are used to provide Internet connection.

Figure 18.5 depicts the moving cell-based RoF network architecture for train passengers. Several RAUs are located along the rail tracks. An optical fiber WDM ring interconnects the RAUs with the central station, where all processing is performed. Each RAU deploys an optical add-drop multiplexer (OADM) fixed tuned to a separate wavelength channel. That is, each RAU is allocated a separate dedicated wavelength channel for transmission and reception to and from the central station. At the central station, a WDM laser generates the desired wavelengths in order to reach the corresponding RAUs. The generated wavelengths are optically switched and passed to an array of RF modulators, one for each RAU. The modulated wavelengths are multiplexed onto the optical fiber ring and received by each addressed RAU on its assigned wavelength. An RAU retrieves the RF signal and transmits it to the antennas of a passing train. In the upstream direction, the RAUs receive all RF signals and send them to the central station for processing. By processing the received RF signals, the central station is able to keep track of the train location and identify the RAU closest to the moving train.

In conventional cellular radio networks, a hand-over would take place whenever the train crosses the cell boundary between two neighboring RAUs. To avoid hand-overs, the applied concept of moving cells lets a cell pattern move together with the passing train such that the train can communicate on the same RFs during the whole connection without requiring hand-overs. The central station implements the moving cells by subsequently sending the RFs used by the train to the next RAU following in the direction the train is moving. Based on the received upstream RF signals, the central station is able to track the location of the train and assign downstream RF signals to the corresponding RAU closest to the train such that the train and moving cells move along in a synchronous fashion.

Part V

Testbeds

19 What worked and what didn't

A variety of optical networking technologies and architectures have been developed and examined over the past decades. Up to date, however, only a few of them led to commercial adoption and revenue generation. According to Ramaswami (2006), Erbium doped fiber amplifiers (EDFAs), reconfigurable optical add-drop multiplexers (ROADMs), wavelength cross-connects (WXCs), and tunable lasers are good examples of devices successfully deployed in today's optical networks. In contrast, other technologies and techniques such as wavelength conversion, optical code division multiple access (OCDMA), optical packet switching (OPS), and optical burst switching (OBS) face significant challenges toward widespread deployment.

Crucial to the commercial success of any proposed networking technology and architecture is not only its performance evaluation by means of analysis or simulation but also a thorough feasibility study of its practical aspects. Toward this end, proof-of-concept demonstrators, testbeds, and field trials play a key role.

In this part, we provide an up-to-date survey of testbed activities on the latest switching techniques proposed for next-generation optical networks. A number of different optical switching techniques have been studied over the last few years. In our survey, we outline current testbed activities of the following major optical switching techniques: generalized multiprotocol label switching (GMPLS), waveband switching (WBS), photonic slot routing (PSR), optical flow switching (OFS), optical burst switching (OBS), and optical packet switching (OPS), which were explained at length in previous chapters. We note that our survey is targeted to networks rather than stand-alone components and devices. Furthermore, we note that regional overviews of optical networking testbeds in Europe and China were recently reported in Fabianek (2006) and Lin and Wu (2006), respectively. In contrast, the following survey provides an overview of worldwide testbed activities rather than focusing on any specific region. Finally, we note that we limited our survey to the most prominent optical switching networking testbeds. Other interesting optical networking testbed activities not addressed in the following include CANARIE, DataTAG, Louisiana Optical Network Initiative, National LambdaRail, OTE-WAVE, SURFnet, UKLight, and the light-trail testbed of Iowa State University.

20 Testbed activities

20.1 GMPLS

20.1.1 LION

The European IST project Layers Interworking in Optical Networks (LION) is a multi-layer, multivendor, and multidomain managed IP/MPLS over automatic switched optical network (ASON) with a GMPLS-based control plane (Cavazzoni et al., 2003). The ASON framework facilitates the set-up, modification, reconfiguration, and release of both *switched* and *soft-permanent* optical connections (lightpaths). Switched connections are controlled by clients as opposed to soft-permanent connections whose set-up and teardown are initiated by the network management system (NMS). An ASON consists of one or more domains, each belonging to a different network operator, administrator, or vendor platform. The points of interaction between different domains are called reference points. Figure 5.1 depicts the ASON reference points between various optical networks and client networks which are connected via lightpaths. Specifically, the reference point between a client network and an administrative domain of an optical network is called user-network interface (UNI). The reference point between the administrative domains of two different optical networks is called external network-network interface (E-NNI). The reference point between two domains (e.g., routing areas) within the same administrative domain of an optical network is called internal network-network interface (I-NNI). The LION testbed comprises three domains consisting of optical add-drop multiplexers (OADMs) and optical cross-connects (OXCs) from different vendors. For video-over-IP (VoIP) and computer-aided design (CAD) applications, the set-up and tear-down of soft-permanent connections through different domains using GMPLS signaling and interworking NMSs was experimentally validated. Furthermore, multilayer resilience tests were successfully carried out demonstrating MPLS fast reroute combined with optical restoration using a holdoff timer at the IP/MPLS layer.

20.1.2 GSN/GSN+

Deutsche Telekom's Global Seamless Network (GSN) field testbed comprises an ASON/GMPLS-based backbone network domain and several key client networks such as IP networks, carrier-grade Ethernet metro and access networks, and storage area networks (SANs) as well as broadband video applications (Foisel et al., 2005). The

ASON/GMPLS backbone network consists of four SDH crossconnects and an ultra-long-haul wavelength division multiplexing (WDM) link. The GMPLS-based control plane provides clients with dynamic transport services via an Optical Internetworking Forum (OIF) UNI1.0, including automatic neighbor and service discovery. The ASON/GMPLS testbed was embedded in the OIF World Interoperability Tests and Demonstration in 2004, which involved seven major carrier laboratories across Asia, Europe, and the United States. UNI and E-NNI interoperability tests were carried out between client networks and multiple transport network domains in a multivendor environment. The extended GSN+ demonstrator focuses on the efficient transport of client traffic over ASON/GMPLS networks and interconnects GSN with other testbeds such as MUPBED, which is described next.

20.1.3   MUPBED

The IST project Multi-Partner European Testbeds for Research Networking (MUPBED) aims at demonstrating a pan-European ASON/GMPLS-based research network in support of advanced applications and collaborative systems (e.g., Grids, disaster recovery, and business continuity) (Szegedi et al., 2006). The MUPBED reference testbed architecture interconnects local testbeds using a layered multidomain approach based on the separation of applications and network services. Different interaction models between Grid applications and network services are studied within MUPBED. Among others, in the GUNI model, a Grid user-network interface (GUNI) between Grid network sites and the underlying ASON/GMPLS network is implemented according to the overlay interconnection model, whereas in the API model, an application programming interface (API) with several enhanced interaction requirements between the Grid middleware and the GMPLS-based control plane is deployed.

20.1.4   DRAGON

The Dynamic Resource Allocation in GMPLS Optical Networks (DRAGON) testbed, located in the Washington, DC, metropolitan area, connects multiple academic, government, and research institutions via two interlocking WDM rings comprising five ROADMs (Lehman et al., 2006). DRAGON develops technologies that allow Grid computing and e-science applications to dynamically require deterministic network services (e.g., dedicated lightpaths) on an interdomain basis. The DRAGON control plane utilizes GMPLS as a basic building block and includes several extensions related to interdomain routing, interdomain signaling, as well as integration of features for authentication, authorization, and accounting (AAA). As shown in Fig. 20.1, each domain has a network-aware resource broker (NARB) which serves as a path computation engine from which end systems can learn about the availability of traffic-engineered LSPs. NARBs peer across domains for interdomain routing to enable interdomain path computation and LSP provisioning. In DRAGON, routing protocols not only exchange traffic engineering (TE)-related information but also associate AAA and scheduling information with

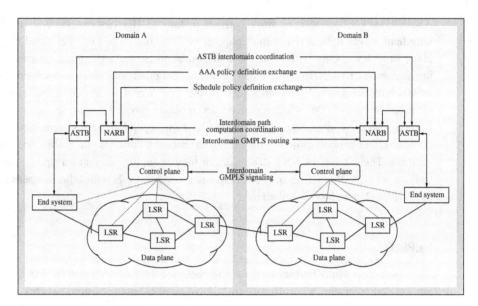

Figure 20.1 DRAGON control plane architecture. After Lehman et al. (2006). © 2006 IEEE.

network resources, giving rise to (1) 3D path computation with constraints for TE, AAA, and scheduling, and (2) policy-based resource allocation. Finally, in DRAGON an end system can request the set-up of an application-specific topology that consists of a set of LSPs from the application-specific topology builder (ASTB) using the services of its attached NARB.

20.1.5   ONFIG

The Optical Network Focused Interest Group (ONFIG) program testbed in Singapore is a GMPLS-controlled bidirectional four-node WDM ring network with the main focus on dynamic set-up of lightpaths with 1:1 protection for distributed storage service applications (Zhou et al., 2006). GMPLS-capable edge IP routers use a web-based connection management system to set up/tear down lightpaths and perform traffic grooming and aggregation to fill established lightpaths. HyperSCSI, a lightweight transport protocol that sends data blocks in Ethernet frames over lightpaths without any TCP/IP overhead, was used to demonstrate efficient storage area networks (SANs).

20.1.6   KDDI

KDDI demonstrated an IPv6 provider edge router approach, referred to as 6PE, for seamless IPv6 integration into GMPLS-controlled optical wavelength switching back-bone networks (Tatipamula et al., 2005). The 6PE method relies on multiprotocol border gateway protocol (MP-BGP) extensions for 6PE routers to exchange IPv6 reachability information. The 6PE BGP routers are dual stack (IPv6 and IPv4), using an IPv4 address for peer communication through the IPv4 core. In doing so, IPv6 reachability can be achieved without any impact on existing IPv4 core networks.

20.1.7   ADRENALINE

The ADRENALINE testbed is a hybrid platform whose transport plane consists of real ROADMs and fiber links as well as emulated optical nodes/links to study various topologies covering the region of Catalunya, Spain (Munoz et al., 2005). Beside a GMPLS-based control plane, ADRENALINE deploys a distributed management plane by combining SNMP and user-friendly XML-based tools. ADRENALINE provides not only soft-permanent connections, as done in LION, but also switched connections (lightpaths). Furthermore, soft-permanent connections under user initiative are supported by combining management and control plane technologies. Finally, ADRENALINE allows both the topology and major characteristics of the data communications network (DCN) to be modified for experimental studies.

20.1.8   ODIN

The Optical Dynamic Intelligent Network (ODIN) services architecture extends the ASON reference architecture of Fig. 5.1 through a *service layer* on top of the control plane (Mambretti et al., 2006). ODIN utilizes a GMPLS-based control plane for the dynamic provisioning of dedicated end-to-end lightpaths. ODIN is an intermediary optical network service layer between high-performance distributed processes (e.g., Grid applications) and lower-level optical network services and resources. The service layer is designed to provide a higher degree of virtualization (abstraction) of network services and resources and to allow edge processes to directly address and control network resources. A prototype of the ODIN architecture was implemented and tested on distributed Grid environments interconnected by several U.S. optical testbeds.

20.1.9   NetherLight/StarLight

Similar to ODIN, a *service plane* layered above legacy networks was introduced in Gommans et al. (2006) as a secure lightpath provisioning architecture that has the ability to integrate different approaches for AAA and different types of control planes (e.g., GMPLS). The service plane encompasses software agents for AAA and Grid network services and is able to bridge domains, establish trust, and expose control to credited users/applications via an API. The service plane was demonstrated in a transatlantic testbed between the two hubs NetherLight in Amsterdam and StarLight in Chicago, with further network clouds attached to them.

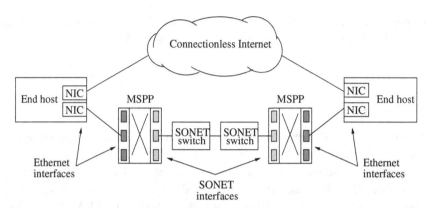

Figure 20.2 CHEETAH circuit-switched add-on service to the connectionless Internet. After Zheng et al. (2005). © 2005 IEEE.

20.1.10   CHEETAH

The Circuit-switched High-speed End-to-End Transport Architecture (CHEETAH) testbed, covering North Carolina, Georgia, and Tennessee, is designed as a circuit-switched add-on service to the connectionless Internet service (Zheng et al., 2005). As shown in Fig. 20.2, it consists of end hosts equipped with two Ethernet network interface cards (NICs), whereby the primary NIC is connected to the Internet while the secondary NIC is attached to an Ethernet port of the enterprise multiservice provisioning platform (MSPP). The MSPPs map GbE frames onto Ethernet-over-SONET (EoS) signals on SONET circuits that are dynamically set up using GMPLS. End hosts execute a routing decision algorithm that determines whether data is sent over the Internet or CHEE-TAH circuit based on the data transfer size and the loading conditions on both paths. CHEETAH provides a so-called *fallback* option that allows end hosts to fall back on the Internet path if the CHEETAH circuit set-up fails. Note that the fallback option enables the gradual upgrade of end hosts with CHEETAH capability, whereby a CHEETAH host may communicate with a non-CHEETAH host via the Internet.

20.1.11   USN

The U.S.-wide UltraScience Net (USN) provides on-demand dedicated lightpaths for e-science applications carrying initially 10Gb/s SONET and 10GbE signals between MSPPs located at the USN edges (Rao et al., 2005). In USN, a centralized scheduler is deployed which accepts user requests for setting up dedicated channels in future time slots subject to bandwidth availability and feasibility constraints. A signaling daemon on the central server uses GMPLS (or TL1 commands) to establish/release lightpaths by configuring the USN switches accordingly. The control plane is implemented using out-of-band VPN tunnels to encrypt control traffic and provide authenticated user and application access to USN.

20.2   Waveband switching

20.2.1   ATDnet testbed

Besides reducing port count, complexity, and costs of OXCs, wavebanding may also be used to upgrade the capacity of existing transparent optical networks without requiring any changes to the core network elements. On the Advanced Technology Demonstration network (ATDnet), an optical network testbed that links a number of government agency laboratories in the Washington, DC, metropolitan area, 4-channel 25-GHz-spaced waveband transmission in an all-optical ring consisting of ROADMs with a 200-GHz passband was demonstrated, whereby the four wavelengths generated at the network edge fit within the passband of the ROADMs (Toliver et al., 2003).

20.3   Photonic slot routing

20.3.1   AT&T Laboratories testbed

A PSR hubbed ring network in which *composite packets* are generated locally at each node with a single tunable transmitter was experimentally demonstrated in an AT&T Laboratories testbed (Boroditsky et al., 2003). A composite packet consists of multiple fixed-size packets that are generated serially at different wavelengths. These packets are sent to a wavelength stacker consisting of an optical circulator followed by fiber Bragg gratings (FBGs), each reflecting a different wavelength. The resultant composite packet is sent in a single photonic slot. The same set of gratings in conjunction with an additional circulator is used to unstack incoming composite packets. Besides stacking and unstacking, the switching of composite packets (photonic slots) was demonstrated by deploying a fast electro-optic 2×2 electro-optic cross-bar switch at each PSR node.

20.4   Optical flow switching

20.4.1   NGI ONRAMP

The Next Generation Internet Optical Network for Regional Access using Multi-wavelength Protocols (NGI ONRAMP) testbed in the Boston metropolitan area consists of a bidirectional feeder WDM ring connecting via ROADM-based access nodes to passive optical distribution networks on which the end users reside (Froberg et al., 2000). Optical flow switching was experimentally demonstrated over the ONRAMP testbed achieving Gigabit-per-second throughput of TCP data between end-user workstations.

20.4.2   CTVR

Optical IP switching (OIS), a switching paradigm similar to router-initiated OFS, was demonstrated in the Centre for Telecommunication Value Chain Research (CTVR) laboratory at Trinity College, Ireland (Ruffini et al., 2006). In OIS, a router monitors IP traffic and, when a flow with specific characteristics is detected, the router establishes an optical cut-through (bypassing) path between its upstream and downstream neighboring routers. The optical cut-through path initially involves only three adjacent routers, but can subsequently be extended by each router located at either end of the optical path. The CTVR testbed consists of several IP routers attached to MEMS-based photonic switches.

20.5   Optical burst switching

20.5.1   ATDnet

An OBS network overlaying three sites of the ATDnet (see Section 20.2.1) using *just-in-time* (JIT) signaling was demonstrated in Baldine et al. (2003). In JIT signaling, an OBS node configures its optical switches for the incoming burst immediately after receiving and processing the corresponding control packet.

20.5.2   JumpStart

The JIT-based JumpStart architecture and protocols were investigated in several U.S. testbeds (Baldine et al., 2005). JumpStart provides QoS-aware constraint-based routing and multicast services. Several applications were successfully tested (e.g., low-latency zero-jitter interactive visualization).

20.5.3   Optical Communication Center

The Optical Communication Center at Beijing University demonstrated a four-node OBS star network using *just-enough-time* (JET) signaling together with the *latest available unused channel with void filling* (LAUC-VF) scheduling algorithm (Guo et al., 2005a). JET signaling enables OBS nodes to make the so-called *delayed reservation* for incoming bursts by using the offset carried in each control packet. With delayed reservation, the optical switches at a given OBS node are configured right before the expected arrival time of the burst, resulting in an improved bandwidth efficiency. LAUC-VF keeps track of all void wavelength intervals and assigns an arriving burst a large enough void interval whose starting time is the latest but still earlier than the burst arrival time. Void filling helps improve the wavelength utilization and burst loss probability of OBS networks, at the expense of increased computational complexity.

20.5.4   University of Tokyo

A JET-based OBS network using *priority-based wavelength assignment* (PWA) and *deflection routing* for contention resolution was examined in Sun et al. (2005). Deflection routing provides a detour path for blocked bursts. With PWA, an edge OBS user increases the priority of a given wavelength if bursts have been successfully sent on this wavelength. Otherwise, the priority of the wavelength is decreased. An edge OBS user assigns the wavelength with the highest priority to an assembled burst in order to reduce burst loss and mitigate contention. The obtained results show that deflection routing helps reduce the burst blocking probability more than PWA.

20.5.5   JGN II

To avoid contention at core OBS nodes, edge OBS users may deploy *two-way* signaling to set up an end-to-end connection prior to burst transmission, as experimentally demonstrated in the Japan Gigabit Network (JGN) II testbed (Sahara et al., 2005). Two-way signaling is suitable for metro area OBS networks with acceptable round-trip propagation delays.

20.5.6   Key Laboratory

The Key Laboratory of Optical Communication & Lightwave Technologies at the Beijing University of Posts and Telecommunications consists of two edge OBS nodes connected via a single core OBS node which deploys LAUC-VF scheduling. Experimental studies showed that TCP performance degrades significantly for burst loss probabilities above 1% (W. Zhang et al., 2006). Various applications with different QoS requirements were demonstrated using an offset-time-based QoS scheme that provides service differentiation by assigning a different offset time to different service classes (Guo et al., 2005b).

20.6   Optical packet switching

20.6.1   RINGO

The ring optical network (RINGO) project, carried out by a consortium of Italian universities, experimentally investigates a unidirectional synchronous WDM ring network (Carena et al., 2004). Time is divided into equally sized slots, each being able to carry one fixed-size packet. Each node (or a subset of nodes) is assigned a different home wavelength channel with a receiver fixed tuned to it. Nodes deploy an array of fixed-tuned transmitters, one for each home channel. To avoid channel collisions, nodes check the status of all wavelengths on a slot-by-slot basis and may use empty slots to transmit locally generated packets.

20.6.2   HORNET

The Hybrid Optoelectronic Ring Network (HORNET) of Stanford University is a bidirectional synchronous time-slotted WDM ring network (White et al., 2003). Each node is equipped with a fast-tunable transmitter and a receiver fixed tuned to its home wavelength channel which may be shared by other nodes. HORNET uses a control wavelength channel to convey availability information of all data wavelength channels. The control channel undergoes OEO conversion at every node for processing and modifying the wavelength availability information. The so-called *segmentation and reassembly on demand* (SAR-OD) access protocol allows nodes to use consecutive empty slots for transmitting variable-size packets.

21 Summary

We considered various major optical switching techniques (GMPLS, WBS, PSR, OFS, OBS, and OPS) and provided a global overview of related testbed activities on next-generation optical switching networks. Table 21.1 summarizes the considered testbeds and their respective objectives. Many interesting and important features were successfully demonstrated which are expected to lead to standardization, commercial adoption, and revenue generation in the near term. Among others, DRAGON's interdomain routing and signaling extensions to GMPLS as well as integration of features for authentication, authorization, and accounting (AAA), ODIN's and NetherLight/StarLight's service plane, or CHEETAH's circuit-switched add-on service are promising examples for building the next generation of high-performance optical switching networks. According to Table 21.1 it appears that GMPLS and to some extent OBS network testbeds have received considerable attention worldwide, while WBS, PSR, OFS, and OPS may be currently viewed as concepts of less practical interest.

Table 21.1. Optical switching networks testbeds

Switching	Testbed	Objectives
GMPLS	LION (Europe)	Interoperability issues of soft-permanent connection set-up/tear-down and multilayer resilience
	GSN/GSN+ (Germany)	UNI and E-NNI interoperability tests in a multidomain multivendor environment for IP, Ethernet, and storage area client networks
	MUPBED (Europe)	Implementation of interaction models (GUNI, API) between Grid applications and control plane
	DRAGON (USA)	Deterministic allocation of interdomain network resources integrating features for authentication, authorization, and accounting (AAA); path computation based on traffic engineering, AAA, and scheduling routing information
	ONFIG (Singapore)	SAN over GMPLS controlled WDM ring using lightweight HyperSCSI transport protocol
	KDDI (Japan)	Seamless integration of IPv6 services via IPv6 provider edge routers using multiprotocol BGP extensions

(cont.)

Table 21.1. (*Cont.*)

Switching	Testbed	Objectives
	ADRENALINE (Spain)	Distributed management and web services for user-driven soft-permanent and switched connection set-up/tear-down; configurable data communications network
	ODIN (USA)	Introduction of service layer for virtualization of network services and resources
	NetherLight (Netherlands) StarLight (USA)	Introduction of service plane for integration of different AAA approaches and different control planes; application/user-driven provisioning of secure end-to-end lightpaths
	CHEETAH (USA)	Ethernet-over-SONET (EoS) circuits with fallback option on connectionless Internet
	USN (USA)	Centralized scheduler for setting up dedicated lightpaths in future time slots; VPN based control plane
WBS	ATDnet (USA)	Wavebanding for increasing the capacity of existing transparent wavelength-switching networks
PSR	AT&T Labs (USA)	Stacking, switching, and unstacking of composite packets in PSR hubbed ring network
OFS	NGI ONRAMP (USA)	ROADM-based feeder WDM ring connecting passive optical distribution networks with Gb/s TCP throughput
	CTVR (Ireland)	Optical IP switching (OIS) over MEMS-based photonic switched network
OBS	ATDnet (USA)	OBS network overlay using just-in-time (JIT) signaling
	JumpStart (USA)	JIT signaling and QoS-aware constraint-based routing
	Optical Communication Center (China)	Just-enough-time (JET) signaling; latest available unused channel with void filling (LAUC-VF) scheduling
	University of Tokyo (Japan)	JET signaling; contention resolution using priority-based wavelength assignment (PWA) and deflection routing
	JGN II (Japan)	Two-way signaling to avoid burst contention in metro area OBS networks
	Key Laboratory (China)	TCP performance and offset-time-based service differentiation using LAUC-VF scheduling
OPS	RINGO (Italy)	Synchronous unidirectional WDM ring using empty-slot access protocol to transmit fixed-size packets
	HORNET (USA)	Synchronous bidirectional WDM ring using control channel based SAR-OD access protocol to transmit variable-size packets

Bibliography

Abrams M., Becker P. C., Fujimoto Y., O'Byrne V., Piehler D. (2005). FTTP Deployments in the United States and Japan – Equipment Choices and Service Provider Imperatives. *IEEE/OSA Journal of Lightwave Technology 23*, 1 (Jan.), 236–246.

Adas A. (1997). Traffic Models in Broadband Networks. *IEEE Communications Magazine 35*, 7 (July), 82–89.

Aiello W., Bhatt S. N., Chung F. R. K., Rosenberg A. L., Sitaraman R. K. (2001). Augmented Ring Networks. *IEEE Transactions on Parallel and Distributed Systems 12*, 6 (July), 598–609.

Alexander S. B., Bondurant R. S., Byrne D. et al. (1993). A Precompetitive Consortium on Wide-Band All-Optical Networks. *IEEE/OSA Journal of Lightwave Technology 11*, 5/6 (May/June), 714–735.

Alharbi F., Ansari N. (2004a). A Novel Fairness Algorithm for Resilient Packet Ring Networks with Low Computational and Hardware Complexity. In *Proc., IEEE Workshop on Local and Metropolitan Area Networks (LANMAN)*. 11–16.

Alharbi F., Ansari N. (2004b). Low Complexity Distributed Bandwidth Allocation for Resilient Packet Ring Networks. In *Proc., IEEE Workshop on High Performance Switching and Routing (HPSR)*. 277–281.

An F., Gutierrez D., Kim K. S., Lee J. W., Kazovsky L. G. (2005). SUCCESS-HPON: A Next-Generation Optical Access Architecture for Smooth Migration from TDM-PON to WDM-PON. *IEEE Communications Magazine 43*, 11 (Nov.), S40–S47.

Arnaud B. S., Wu J., Kalali B. (2003). Customer-Controlled and -Managed Optical Networks. *IEEE/OSA Journal of Lightwave Technology 21*, 11 (Nov.), 2804–2810.

Ashwood-Smith P., Berger L. (2003). Generalized Multi-Protocol Label Switching (GMPLS) Signaling Contraint-Based Routed Label Distribution Protocol (CR-LDP) Extensions. IETF RFC 3472.

Assi C., Shami A., Ali M. A. (2001). Optical Networking and Real-Time Provisioning: An Integrated Vision for the Next-Generation Internet. *IEEE Network 15*, 4 (July/Aug.), 36–45.

Assi C. M., Ye Y., Dixit S., Ali M. A. (2003). Dynamic Bandwidth Allocation for Quality-of-Service over Ethernet PONs. *IEEE Journal on Selected Areas in Communications 21*, 9 (Nov.), 1467–1477.

Bakaul M., Nirmalathas A., Lim C., Novak D., Waterhouse R. (2005). Efficient Multiplexing Scheme for Wavelength-Interleaved DWDM Millimeter-Wave Fiber-Radio Systems. *IEEE Photonics Technology Letters 17*, 12 (Sept.), 2718–2720.

Baldi M., Ofek Y. (2002). Fractional Lambda Switching. In *Proc., IEEE International Conference on Communications (ICC)*. Vol. 5. 2692–2696.

Baldine I., Bragg A., Evans G. et al. (2005). JumpStart Deployments in Ultra-High-Performance Optical Networking Testbeds. *IEEE Communications Magazine 43*, 11 (Nov.), S18–S25.

Baldine I., Cassada M., Bragg A., Karmous-Edwards G., Stevenson D. (2003). Just-in-Time Optical Burst Switching Implementation in the ATDnet All-Optical Networking Testbed. In *Proc., IEEE GLOBECOM*. Vol. 5. 2777–2781.

Baldine I., Rouskas G. N., Perros H. G., Stevenson D. (2002). JumpStart: A Just-in-Time Signaling Architecture for WDM Burst-Switched Networks. *IEEE Communications Magazine 40*, 2 (Feb.), 82–89.

Ballart R., Ching Y.-C. (1989). SONET: Now It's the Standard Optical Network. *IEEE Communications Magazine 27*, 3 (Mar.), 8–15.

Banerjee A., Drake J., Lang J. et al. (2001a). Generalized Multiprotocol Label Switching: An Overview of Signaling Enhancements and Recovery Techniques. *IEEE Communications Magazine 39*, 7 (July), 144–151.

Banerjee A., Drake J., Lang J. P., Turner B., Kompella K., Rekhter Y. (2001b). Generalized Multiprotocol Label Switching: An Overview of Routing and Management Enhancements. *IEEE Communications Magazine 39*, 1 (Jan.), 144–150.

Barakat N., Sargent E. H. (2004). An Accurate Model for Evaluating Blocking Probabilities in Multi-Class OBS Systems. *IEEE Communications Letters 8*, 2 (Feb.), 119–121.

Barakat N., Sargent E. H. (2006). Separating Resource Reservations from Service Requests to Improve the Performance of Optical Burst-Switching Networks. *IEEE Journal on Selected Areas in Communications 24*, 4 (Apr.), 95–108.

Barry R. A., Chan V. W. S., Hall K. L. et al. (1996). All-Optical Network Consortium – Ultrafast TDM Networks. *IEEE Journal on Selected Areas in Communications 14*, 5 (June), 999–1013.

Battestilli T., Perros H. (2003). An Introduction to Optical Burst Switching. *IEEE Communications Magazine 41*, 8 (Aug.), S10–S15.

Battestilli T., Perros H. (2004). Optical Burst Switching for the Next Generation Internet. *IEEE Potentials 23*, 5 (Dec. 2004/Jan. 2005), 40–43.

Bengi K. (2002a). Access Protocols for an Efficient Optical Packet-Switched Metropolitan Area Ring Network Supporting IP Datagrams. In *Proc., Eleventh International Conference on Computer Communications and Networks*. 284–289.

Bengi K. (2002b). An Analytical Model for a Slotted WDM Metro Ring with A-Posteriori Access. In *Proc., Optical Network Design and Modelling (ONDM)*.

Bengi K. (2002c). An Optical Packet-Switched IP-over-WDM Metro Ring Network. In *Proc., 27th Annual IEEE Conference on Local Computer Networks (LCN)*. 43–52.

Bengi K., van As H. R. (2001). QoS Support and Fairness Control in a Slotted Packet-Switched WDM Metro Ring Network. In *Proc., IEEE GLOBECOM*. Vol. 3. 1494–1499.

Bengi K., van As H. R. (2002). Efficient QoS Support in a Slotted Multihop WDM Metro Ring. *IEEE Journal on Selected Areas in Communications 20*, 1 (Jan.), 216–227.

Benjamin D., Trudel R., Shew S., Kus E. (2001). Optical Services over the Intelligent Optical Network. *IEEE Communications Magazine 39*, 9 (Sept.), 73–78.

Berger L. (2003a). Generalized Multi-Protocol Label Switching (GMPLS) Signaling Functional Description. IETF RFC 3471.

Berger L. (2003b). Generalized Multi-Protocol Label Switching (GMPLS) Signaling Resource ReserVation Protocol-Traffic Engineering (RSVP-TE) Extensions. IETF RFC 3473.

Berthelon L., Audouin O., Bonno P. et al. (2000). Design of a Cross-Border Optical Core and Access Networking Field Trial: First Outcomes of the ACTS–PELICAN Project. *IEEE/OSA Journal of Lightwave Technology 18*, 12 (Dec.), 1939–1954.

Bianco A., Bonsignori M., Leonardi E., Neri F. (2002). Variable-Size Packets in Slotted WDM Ring Networks. In *Proc., Optical Network Design and Modelling (ONDM)*.

Bianco A., Distefano V., Fumagalli A., Leonardi E., Neri F. (1998). A-Posteriori Access Strategies in All-Optical Slotted WDM Rings. In *Proc., IEEE GLOBECOM.* Vol. 1. 300–306.

Boroditsky M., Frigo N. J., Lam C. F. et al. (2003). Experimental Demonstration of Composite-Packet-Switched WDM Network. *IEEE/OSA Journal of Lightwave Technology 21,* 8 (Aug.), 1717–1722.

Boroditsky M., Lam C. F., Smiljanić A. et al. (2001a). Experimental demonstration of composite packet switching on a WDM photonic slot routing network. In *Proc., Optical Fiber Communication (OFC).* Vol. 4. ThG6-1–ThG6-3.

Boroditsky M., Lam C. F., Woodward S. L. et al. (2001b). Composite packet switched WDM networks. In *Proc., The 14th Annual Meeting of the IEEE Lasers and Electro-Optics Society (LEOS).* Vol. 2. 738–739.

Brackett C. A., Acampora A. S., Sweitzer J. et al. (1993). A Scalable Multiwavelength Multihop Optical Network: A Proposal for Research on All-Optical Networks. *IEEE/OSA Journal of Lightwave Technology 11,* 5/6 (May/June), 736–753.

Byun H.-J., Nho J.-M., Lim J.-T. (2003). Dynamic Bandwidth Allocation Algorithm in Ethernet Passive Networks. *IEE Electronics Letters 39,* 13 (June), 1001–1002.

Caenegem R. V., Martínez J. M., Colle D. et al. (2006). From IP Over WDM to All-Optical Packet Switching: Economical View. *IEEE/OSA Journal of Lightwave Technology 24,* 4 (Apr.), 1638–1645.

Cai J., Fumagalli A., Chlamtac I. (2000). The Multitoken Interarrival Time (MTIT) Access Protocol for Supporting Variable Size Packets Over WDM Ring Network. *IEEE Journal on Selected Areas in Communications 18,* 10 (Oct.), 2094–2104.

Callegati F., Corazza G., Raffaelli C. (2002). Exploitation of DWDM for Optical Packet Switching with Quality of Service Guarantees. *IEEE Journal on Selected Areas in Communications 20,* 1 (Jan.), 190–201.

Cao X., Anand V., Li J., Xin C. (2005). Waveband Switching Networks with Limited Wavelength Conversion. *IEEE Communications Letters 9,* 7 (July), 646–648.

Cao X., Anand V., Qiao C. (2003a). A Waveband Switching Architecture and Algorithm for Dynamic Traffic. *IEEE Communications Letters 7,* 8 (Aug.), 397–399.

Cao X., Anand V., Qiao C. (2003b). Waveband Switching in Optical Networks. *IEEE Communications Magazine 41,* 4 (Apr.), 105–112.

Cao X., Anand V., Qiao C. (2004a). Multi-Layer versus Single-Layer Cross-Connect Architectures for Waveband Switching. In *Proc., IEEE INFOCOM.* Vol. 3. 1830–1840.

Cao X., Anand V., Xiong Y., Qiao C. (2003c). A Study of Waveband Switching with Multi-layer Multigranular Optical Cross-Connects. *IEEE Journal on Selected Areas in Communications 21,* 7 (Sept.), 1081–1095.

Cao X., Qiao C., Anand V., Li J. (2004b). Wavelength Assignment in Waveband Switching Networks with Wavelength Conversion. In *Proc., IEEE GLOBECOM.* Vol. 3. 1943–1947.

Carena A., Feo V. D., Finochietto J. M. et al. (2004). RingO: An Experimental WDM Optical Packet Network for Metro Applications. *IEEE Journal on Selected Areas in Communications 22,* 8 (Oct.), 1561–1571.

Carena A., Ferrero V., Gaudino R., Feo V. D., Neri F., Poggiolini P. (2002). RINGO: A Demonstrator of WDM Optical Packet Network on a Ring Topology. In *Proc., Optical Network Design and Modeling (ONDM).*

Cavazzoni C., Barosco V., D'Alessandro A. et al. (2003). The IP/MPLS Over ASON/GMPLS Test Bed of the IST Project LION. *IEEE/OSA Journal of Lightwave Technology 21,* 11 (Nov.), 2791–2803.

Chan V. W. S., Hall K. L., Modiano E., Rauschenbach K. A. (1998). Architectures and Technologies for High-Speed Optical Data Networks. *IEEE/OSA Journal of Lightwave Technology 16,* 12 (Dec.), 2146–2168.

Chang G.-K., Ellinas G., Gamelin J. K., Iqbal M. Z., Brackett C. A. (1996). Multiwavelength Reconfigurable WDM/ATM/SONET Network Testbed. *IEEE/OSA Journal of Lightwave Technology 14,* 6 (June), 1320–1340.

Chang G.-K., Yu J., Yeo Y.-K., Chowdhury A., Jia Z. (2006). Enabling Technologies for Next-Generation Optical Packet-Switching Networks. *Proceedings of the IEEE 94,* 5 (May), 892–910.

Chen J., Cidon I., Ofek Y. (1992). A Local Fairness Algorithm for the MetaRing, and its Performance Study. In *Proc., IEEE GLOBECOM.* Vol. 3. 1635–1641.

Chen W.-P., Hwang W.-S. (2002). A Packet Pre-Classification CSMA/CA MAC Protocol for IP over WDM Ring Networks. In *Proc., IEEE International Conference on Communication Systems.* Vol. 2. 1217–1221.

Chen Y., Qiao C., Yu X. (2004). Optical Burst Switching: A New Area in Optical Networking Research. *IEEE Network 18,* 3 (May/June), 16–23.

Chidgey P. J. (1994). Multi-Wavelength Transport Networks. *IEEE Communications Magazine 32,* 12 (Dec.), 28–35.

Chlamtac I., Elek V., Fumagalli A. (1997a). A Fair Slot Routing Solution for Scalability in All-Optical Packet Switched Networks. *Journal of High Speed Networks 6,* 3, 181–196.

Chlamtac I., Elek V., Fumagalli A., Szabó C. (1997b). Scalable WDM Network Architecture Based on Photonic Slot Routing and Switched Delay Lines. In *Proc., IEEE INFOCOM.* Vol. 2. 769–776.

Chlamtac I., Elek V., Fumagalli A., Szabó C. (1999a). Scalable WDM Access Network Architecture Based on Photonic Slot Routing. *IEEE/ACM Transactions on Networking 7,* 1 (Feb.), 1–9.

Chlamtac I., Fumagalli A., Wedzinga G. (1999b). Slot Routing as a Solution for Optically Transparent Scalable WDM Wide Area Networks. *Photonic Network Communications 1,* 1 (June), 9–21.

Chlamtac I., Ganz A., Karmi G. (1992). Lightpath Communications: A Novel Approach to High Bandwidth Optical WANs. *IEEE Transactions on Communications 40,* 7 (July), 1171–1182.

Cho W., Mukherjee B. (2001). Design of MAC Protocols for DWADM-Based Metropolitan-Area Optical Ring Networks. In *Proc., IEEE GLOBECOM.* Vol. 3. 1575–1579.

Choi S.-I., Huh J.-D. (2002). Dynamic Bandwidth Allocation Algorithm for Multimedia Services over Ethernet PONs. *ETRI Journal 24,* 6 (Dec.), 465–468.

Chu T.-S., Gans M. J. (1991). Fiber Optic Microcellular Radio. *IEEE Transactions on Vehicular Technology 40,* 3 (Aug.), 599–606.

Chuh Y., Kim J., Song Y., Park D. (2005). F-TCP: Light-Weight TCP for File Transfer in High Bandwidth-Delay Product Networks. In *Proc., IEEE International Conference on Parallel and Distributed Systems (ICPADS).* Vol. 1. 502–508.

Cidon I., Ofek Y. (1990). Metaring – A Full-Duplex Ring with Fairness and Spatial Reuse. In *Proc., IEEE INFOCOM.* 969–978.

Cidon I., Ofek Y. (1993). MetaRing – A Full-Duplex Ring with Fairness and Spatial Reuse. *IEEE Transactions on Communications 41,* 1 (Jan.), 110–120.

Cunningham D. G. (2001). The Status of the 10-Gigabit Ethernet Standard. In *Proc., European Conference on Optical Communication (ECOC).* Vol. 3. 364–367.

Danielsen S. L., Hansen P. B., Stubkjaer K. E. (1998). Wavelength Conversion in Optical Packet Switching. *IEEE/OSA Journal of Lightwave Technology 16,* 12 (Dec.), 2095–2108.

Darema F. (2005). Grid Computing and Beyond: The Context of Dynamic Data Driven Applications Systems. *Proceedings of the IEEE 93,* 3 (Mar.), 692–697.

Davik F., Kvalbein A., Gjessing S. (2005). Improvement of Resilient Packet Ring Fairness. In *Proc., IEEE GLOBECOM.* Vol. 1. 581–586.

Davik F., Yilmaz M., Gjessing S., Uzun N. (2004). IEEE 802.17 Resilient Packet Ring Tutorial. *IEEE Communications Magazine 42,* 3 (Mar.), 112–118.

Derr F., Huber M. N., Kettler G., Thorweihe N. (1995). An Optical Infrastructure for Future Telecommunications Networks. *IEEE Communications Magazine 33,* 11 (Nov.), 84–88.

Detti A., Eramo V., Listanti M. (2002). Performance Evaluation of a New Technique for IP Support in a WDM Optical Network: Optical Composite Burst Switching (OCBS). *IEEE/OSA Journal of Lightwave Technology 20,* 2 (Feb.), 154–165.

Develder C., Stavdas A., Bianco A. et al. (2004). Benchmarking and Viability Assessment of Optical Packet Switching for Metro Networks. *IEEE/OSA Journal of Lightwave Technology 22,* 11 (Nov.), 2435–2451.

Dittmann L., Develder C., Chiaroni D. et al. (2003). The European IST Project DAVID: A Viable Approach Toward Optical Packet Switching. *IEEE Journal on Selected Areas in Communications 21,* 7, 1026–1040.

Dueser M., Bayvel P. (2002a). Analysis of a Dynamically Wavelength-Routed Optical Burst Switched Network Architecture. *IEEE/OSA Journal of Lightwave Technology 20,* 4 (Apr.), 574–585.

Dueser M., Bayvel P. (2002b). Performance of a Dynamically Wavelength-Routed Optical Burst Switched Network. *IEEE Photonics Technology Letters 14,* 2 (Feb.), 239–241.

Dutta R., Rouskas G. N. (2002). Traffic Grooming in WDM Networks: Past and Future. *IEEE Network 16,* 6 (Nov./Dec.), 46–56.

Dwivedi A., Wagner R. E. (2000). Traffic Model for USA Long-Distance Optical Network. In *Proc., OFC,* Paper TuK1.

El-Bawab T. S., Shin J.-D. (2002). Optical Packet Switching in Core Networks: Between Vision and Reality. *IEEE Communications Magazine 40,* 9, 60–65.

Elek V., Fumagalli A., Wedzinga G. (2001). Photonic Slot Routing: A Cost-Effective Approach to Designing All-Optical Access and Metro Networks. *IEEE Communications Magazine 39,* 11 (Nov.), 164–172.

Elwalid A., Mitra D., Saniee I., Widjaja I. (2003). Routing and Protection in GMPLS Networks: From Shortest Paths to Optimized Designs. *IEEE/OSA Journal of Lightwave Technology 21,* 11 (Nov.), 2828–2838.

Eramo V., Listanti M., Spaziani M. (2005). Resources Sharing in Optical Packet Switches with Limited-Range Wavelength Converters. *IEEE/OSA Journal of Lightwave Technology 23,* 2 (Feb.), 671–687.

Fabianek B. (2006). Optical Networking Testbeds in Europe. In *Proc., OFC/NFOEC.*

Fan C., Maier M., Reisslein M. (2004). The AWG||PSC Network: A Performance-Enhanced Single-Hop WDM Network with Heterogeneous Protection. *IEEE/OSA Journal of Lightwave Technology 22,* 5 (May), 1242–1262.

Feng W., Chang F., Feng W., Walpole J. (2005). A Traffic Characterization of Popular On-Line Games. *IEEE/ACM Transactions on Networking 13,* 3 (June), 488–500.

Finn S. G., Barry R. A. (1996). Optical Services in Future Broadband Networks. *IEEE Network 10,* 6 (Nov./Dec.), 7–13.

Foh C. H., Andrew L., Wong E., Zukerman M. (2004). FULL-RCMA: A High Utilization EPON. *IEEE Journal on Selected Areas in Communications 22,* 8 (Oct.), 1514–1524.

Foisel H.-M., Gerlach C., Gladisch A., Szuppa S., Weber A. (2005). Global Seamless Network Demonstrator: A Comprehensive ASON/GMPLS Testbed. *IEEE Communications Magazine 43*, 11 (Nov.), S34–S39.

Fransson J., Johansson M., Roughan M., Andrew L., Summerfield M. A. (1998). Design of a Medium Access Control Protocol for a WDMA/TDMA Photonic Ring Network. In *Proc., IEEE GLOBECOM*. Vol. 1. 307–312.

Frazier H. (1998). The 802.3z Gigabit Ethernet Standard. *IEEE Network 12*, 3 (May/June), 6–7.

Frazier H., Johnson H. (1999). Gigabit Ethernet: From 100 to 1,000 Mbps. *IEEE Internet Computing 3*, 1 (Jan./Feb.), 24–31.

Frigo N. J. (1997). A Survey of Fiber Optics in Local Access Architectures. *Optical Fiber Telecommunications IIIA*, 461–522.

Froberg N. M., Henion S. R., Rao H. G. et al. (2000). The NGI ONRAMP Test Bed: Reconfigurable WDM Technology for Next Generation Regional Access Networks. *IEEE/OSA Journal of Lightwave Technology 18*, 12 (Dec.), 1697–1708.

Frost V. S., Melamed B. (1994). Traffic Modeling for Telecommunications Networks. *IEEE Communications Magazine 32*, 3 (Mar.), 70–81.

Fukashiro Y., Shrikhande K., Avenarius M. et al. (2000). Fast and Fine Tuning of a GCSR Laser Using a Digitally Controlled Driver. In *Proc., OFC*, Paper WM43. Vol. 2. 338–340.

Fumagalli A., Cai J., Chlamtac I. (1998). A Token Based Protocol for Integrated Packet and Circuit Switching in WDM Rings. In *Proc., IEEE GLOBECOM*. Vol. 4. 2339–2344.

Fumagalli A., Cai J., Chlamtac I. (1999). The Multi-Token Inter-Arrival Time (MTIT) Access Protocol for Supporting IP over WDM Ring Network. In *Proc., IEEE International Conference on Communications (ICC)*. Vol. 1. 586–590.

Gambini P., Renaud M., Guillemot C. et al. (1998). Transparent Optical Packet Switching: Network Architecture and Demonstrators in the KEOPS Project. *IEEE Journal on Selected Areas in Communications 16*, 7, 1245–1259.

Gambiroza V., Yuan P., Balzano L., Liu Y., Sheafor S., Knightly E. (2004). Design, Analysis, and Implementation of DVSR: A Fair High-Performance Protocol for Packet Rings. *IEEE/ACM Transactions on Networking 12*, 1 (Feb.), 85–102.

Ganguly B., Chan V. (2002). A Scheduled Approach to Optical Flow Switching in the ONRAMP Optical Access Network Testbed. In *Proc., Optical Fiber Communication (OFC)*. 215–216.

Ganguly B., Modiano E. (2000). Distributed Algorithms and Architectures for Optical Flow Switching in WDM Networks. In *Proc., IEEE Symposium on Computers and Communications (ISCC)*. 134–139.

Garcia M., Garcia D. F., Garcia V. G., Bonis R. (2004). Analysis and Modeling of Traffic on a Hybrid Fiber-Coax Network. *IEEE Journal on Selected Areas in Communications 22*, 9 (Nov.), 1718–1730.

Ge A., Callegati F., Tamil L. S. (2000). On Optical Burst Switching and Self-Similar Traffic. *IEEE Communications Letters 4*, 3 (Mar.), 98–100.

Gemelos S. M., White I. M., Wonglumsom D., Shrikhande K., Ono T., Kazovsky L. G. (1999). WDM Metropolitan Area Network Based on CSMA/CA Packet Switching. *IEEE Photonics Technology Letters 11*, 11 (Nov.), 1512–1514.

Gerstel O., Ramaswami R. (2003). Optical Layer Survivability: A Post-Bubble Perspective. *IEEE Communications Magazine 41*, 9 (Sept.), 51–53.

Golab W., Boutaba R. (2004). Policy-Driven Automated Reconfiguration for Performance Management in WDM Optical Networks. *IEEE Communications Magazine 42*, 1 (Jan.), 44–51.

Gommans L., Dijkstra F., de Laat C. et al. (2006). Applications Drive Secure Lightpath Creation across Heterogeneous Domains. *IEEE Communications Magazine 44,* 3 (Mar.), 100–106.

Goralski W. J. (1997). *SONET – A Guide to Synchronous Optical Network.* New York: McGraw-Hill.

Gordon R. E., Chen L. R. (2006). New Control Algorithms in an Optical Packet Switch with Limited-Range Wavelength Converters. *IEEE Communications Letters 10,* 6 (June), 495–497.

Green P. (2001). Progress in Optical Networking. *IEEE Communications Magazine 39,* 1 (Jan.), 54–61.

Green P. E. (1993). *Fiber Optic Networks.* Upper Saddle River, NJ: Prentice Hall.

Green P. E. (1996). Optical Networking Update. *IEEE Journal on Selected Areas in Communications 14,* 5 (June), 764–779.

Green P. E. (2002). Paving the Last Mile with Glass. *IEEE Spectrum 39,* 12 (Dec.), 13–14.

Green P. E. (2006). *Fiber To The Home – The New Empowerment.* Hoboken, NJ: John Wiley & Sons.

Griffith D., Lee S. (2003). A 1 + 1 Protection Architecture for Optical Burst Switched Networks. *IEEE Journal on Selected Areas in Communications 21,* 9 (Nov.), 1384–1398.

Griffith D., Sriram K., Golmie N. (2005). Protection Switching for Optical Bursts Using Segmentation and Deflection Routing. *IEEE Communications Letters 9,* 10 (Oct.), 930–932.

Gumaste A., Zheng S. Q. (2005). Next-Generation Optical Storage Area Networks: The Light-Trails Approach. *IEEE Communications Magazine 43,* 3 (Mar.), 72–79.

Guo H., Lan Z., Wu J. et al. (2005a). A Testbed for Optical Burst Switching Network. In *Proc., OFC/NFOEC.*

Guo H., Wu J., Liu X., Lin J., Ji Y. (2005b). Multi-QoS Traffic Transmission Experiments on OBS Network Testbed. In *Proc., European Conference on Optical Communication (ECOC).* Vol. 3. 601–602.

Herzog M., Maier M. (2006). RINGOSTAR: An Evolutionary Performance-Enhancing WDM Upgrade of IEEE 802.17 Resilient Packet Ring. *IEEE Communications Magazine 44,* 2 (Feb.), S11–S17.

Herzog M., Adams S., Maier M. (2005). Proxy Stripping: A Performance-Enhancing Technique for Optical Metropolitan Area Ring Networks. *OSA Journal of Optical Networking 4,* 7 (July), 400–431.

Herzog M., Maier M., Wolisz A. (2005). RINGOSTAR: An Evolutionary AWG-Based WDM Upgrade of Optical Ring Networks. *IEEE/OSA Journal of Lightwave Technology 23,* 4 (Apr.), 1637–1651.

Hill G. R. (1990). Wavelength Domain Optical Network Techniques. *Proceedings of the IEEE 77,* 1 (Jan.), 121–132.

Hirosaki B., Emura K., Hayano S., Tsutsumi H. (2003). Next-Generation Optical Networks as a Value Creation Platform. *IEEE Communications Magazine 41,* 9 (Sept.), 65–71.

Ho P.-H., Mouftah H. T. (2001). Network Planning Algorithms for the Optical Internet Based on the Generalized MPLS Architecture. In *Proc., IEEE GLOBECOM.* Vol. 4. 2150–2154.

Ho P.-H., Mouftah H. T. (2002). Path Selection with Tunnel Allocation in the Optical Internet Based on Generalized MPLS Architecture. In *Proc., IEEE International Conference on Communications (ICC).* Vol. 5. 2697–2701.

Ho P.-H., Mouftah H. T., Wu J. (2003a). A Novel Design of Optical Cross-Connects with Multi-Granularity Provisioning Support for the Next-Generation Internet. In *Proc., IEEE International Conference on Communications (ICC).* Vol. 1. 582–587.

Ho P.-H., Mouftah H. T., Wu J. (2003b). A Scalable Design of Multigranularity Optical Cross-Connects for the Next-Generation Optical Internet. *IEEE Journal on Selected Areas in Communications 21,* 7 (Sept.), 1133–1142.

Hopper A., Williamson R. C. (1983). Design and Use of an Integrated Cambridge Ring. *IEEE Journal on Selected Areas in Communications SAC-1,* 5 (Nov.), 775–784.

Houghton A. (2003). Supporting the Rollout of Broadband in Europe: Optical Network Research in the IST Program. *IEEE Communications Magazine 41,* 9 (Sept.), 58–64.

Hsueh Y., Shaw W., Kazovsky L. G., Agata A., Yamamoto S. (2005a). SUCCESS PON Demonstrator: Experimental Exploration of Next-Generation Optical Access Networks. *IEEE Communications Magazine 43,* 8 (Aug.), S26–S33.

Hsueh Y.-L., Rogge M. S., Yamamoto S., Kazovsky L. G. (2005b). A Highly Flexible and Efficient Passive Optical Network Employing Dynamic Wavelength Allocation. *IEEE/OSA Journal of Lightwave Technology 23,* 1 (Jan.), 277–286.

Huber D. E., Steinlin W., Wild P. J. (1983). SILK: An Implementation of a Buffer Insertion Ring. *IEEE Journal on Selected Areas in Communications SAC-1,* 5 (Nov.), 766–774.

Hunter D. K., Andonovic I. (2000). Approaches to Optical Internet Packet Switching. *IEEE Communications Magazine 38,* 9, 116–122.

Hunter D. K., Chia M. C., Andonovic I. (1998). Buffering in Optical Packet Switches. *IEEE/OSA Journal of Lightwave Technology 16,* 12 (Dec.), 2081–2094.

Hurwitz J., Feng W. (2004). End-to-End Performance of 10-Gigabit Ethernet on Commodity Systems. *IEEE Micro 24,* 1 (Jan./Feb.), 10–22.

Iannone E., Sabella R., de Stefano L., Valeri F. (1996). All-Optical Wavelength Conversion in Optical Multicarrier Networks. *IEEE Transactions on Communications 44,* 6 (June), 716–724.

Iovanna P., Sabella R., Settembre M. (2003). A Traffic Engineering System for Multilayer Networks Based on the GMPLS Paradigm. *IEEE Network 17,* 2 (Mar./Apr.), 28–37.

Jain R. (1993). FDDI: Current Issues and Future Plans. *IEEE Communications Magazine 31,* 9 (Sept.), 98–105.

Jelger C. S., Elmirghani J. M. H. (2001). A Simple MAC Protocol for WDM Metropolitan Access Ring Networks. In *Proc., IEEE GLOBECOM.* Vol. 3. 1500–1504.

Jelger C. S., Elmirghani J. M. H. (2002a). Performance of a Slotted MAC Protocol for WDM Metropolitan Access Ring Networks under Self-Similar Traffic. In *Proc., IEEE International Conference on Communications (ICC).* Vol. 5. 2806–2811.

Jelger C. S., Elmirghani J. M. H. (2002b). Photonic Packet WDM Ring Networks Architecture and Performance. *IEEE Communications Magazine 40,* 11 (Nov.), 110–115.

Jeong M., Cankaya H. C., Qiao C. (2002). On a New Multicasting Approach in Optical Burst Switched Networks. *IEEE Communications Magazine 40,* 11 (Nov.), 96–103.

Jeong M., Qiao C., Xiong Y., Cankaya H. C., Vandenhoute M. (2003). Tree-Shared Multicast in Optical Burst-Switched WDM Networks. *IEEE/OSA Journal of Lightwave Technology 21,* 1 (Jan.), 13–24.

Jiang X., Safaei F., Boustead P. (2005). Latency and Scalability: A Survey of Issues and Techniques for Supporting Networked Games. In *Proc., IEEE International Conference on Networks.*

Johansson S., Manzalini A., Giannoccaro M. et al. (1998). A Cost-Effective Approach to Introduce an Optical WDM Network in the Metropolitan Environment. *IEEE Journal on Selected Areas in Communications 16,* 7 (Sept.), 1109–1122.

Jourdan A., Chiaroni D., Dotaro E., Eilenberger G. J., Masetti F., Renaud M. (2001). The Perspective of Optical Packet Switching in IP-Dominant Backbone and Metropolitan Networks. *IEEE Communications Magazine 39,* 3 (Mar.), 136–141.

Kani J.-I., Teshima M., Akimoto K. et al. (2003). A WDM-Based Optical Access Network for Wide-Area Gigabit Access Services. *IEEE Communications Magazine 41,* 2 (Feb.), S43–S48.

Khanvilkar S., Khokhar A. (2004). Virtual Private Networks: An Overview with Performance Evaluation. *IEEE Communications Magazine 42,* 10 (Oct.), 146–154.

Khorsandi S., Shokrani A., Lambadaris I. (2005). A New Fairness Model for Resilient Packet Rings. In *Proc., IEEE International Conference on Communications (ICC).* Vol. 1. 288–294.

Klonidis D., Politi C. T., Nejabati R., O'Mahony M. J., Simeonidou D. (2005). OPSnet: Design and Demonstration of an Asynchronous High-Speed Optical Packet Switch. *IEEE/OSA Journal of Lightwave Technology 23,* 10 (Oct.), 2914–2925.

Knight P., Lewis C. (2004). Layer 2 and 3 Virtual Private Networks: Taxonomy, Technology, and Standardization Efforts. *IEEE Communications Magazine 42,* 6 (June), 124–131.

Kobayashi H., Kaminow I. P. (1996). Duality Relationships Among "Space," "Time," and "Wavelength" in All-Optical Networks. *IEEE/OSA Journal of Lightwave Technology 14,* 3 (Mar.), 344–351.

Kobrinski H., Vecchi M. P., Goodman M. S. et al. (1990). Fast Wavelength-Switching of Laser Transmitters and Amplifiers. *IEEE Journal on Selected Areas in Communications 8,* 6 (Aug.), 1190–1202.

Kompella K., Rekhter Y. (2005a). Label Switched Paths (LSP) Hierarchy with Generalized Multi-Protocol Label Switching (GMPLS) Traffic Engineering (TE). IETF RFC 4206.

Kompella K., Rekhter Y. (2005b). OSPF Extensions in Support of Generalized Multi-Protocol Label Switching (GMPLS). IETF RFC 4203.

Kompella K., Rekhter Y. (2005c). Routing Extensions in Support of Generalized Multi-Protocol Label Switching (GMPLS). IETF RFC 4202.

Kompella K., Rekhter Y., Berger L. (2005). Link Bundling in MPLS Traffic Engineering (TE). IETF RFC 4201.

Kong H., Phillips C. (2006). Prebooking Reservation Mechanism for Next-Generation Optical Networks. *IEEE Journal of Selected Topics in Quantum Electronics 12,* 4 (Jul./Aug.), 645–652.

Koonen T. (2006). Fiber to the Home/Fiber to the Premises: What, Where, and When. *Proceedings of the IEEE 94,* 5 (May), 911–934.

Kramer G. (2005). *Ethernet Passive Optical Networks.* New York: McGraw-Hill.

Kramer G., Pesavento G. (2002). Ethernet Passive Optical Network (EPON): Building a Next-Generation Optical Access Network. *IEEE Communications Magazine 40,* 2 (Feb.), 66–73.

Kramer G., Mukherjee B., Maislos A. (2003). Chapter 8: Ethernet Passive Optical Networks. In *IP over WDM: Building the Next Generation Optical Internet,* S. Dixit, ed. Wiley.

Kramer G., Mukherjee B., Pesavento G. (2001). Ethernet PON (EPON): Design and Analysis of an Optical Access Network. *Photonic Network Communications 3,* 3 (July), 307–319.

Kramer G., Mukherjee B., Pesavento G. (2002a). IPACT: A Dynamic Protocol for an Ethernet PON (EPON). *IEEE Communications Magazine 40,* 2 (Feb.), 74–80.

Kramer G., Mukherjee B., Ye Y., Dixit S., Hirth R. (2002b). Supporting Differentiated Classes of Service in Ethernet Passive Optical Networks. *OSA Journal of Optical Networking 1,* 8 (Aug.), 280–298.

Kuznetsov M., Froberg N. M., Henion S. R. et al. (2000). A Next-Generation Optical Regional Access Network. *IEEE Communications Magazine 38,* 1 (Jan.), 66–72.

Kvalbein A., Gjessing S. (2004). Analsysis and Improved Performance of RPR Protection. In *Proc., IEEE International Conference on Networks (ICON).* Vol. 1. 119–124.

Kvalbein A., Gjessing S. (2005). Protection of RPR Strict Order Traffic. In *Proc., IEEE Workshop on Local and Metropolitan Area Networks (LANMAN)*.

Kwong K. H., Harle D., Andonovic I. (2004). Dynamic Bandwidth Allocation Algorithm for Differentiated Services over WDM EPONs. In *Proc., IEEE International Conference on Communications Systems (ICCS)*. 116–120.

Lang J. (2005). Link Management Protocol (LMP). IETF RFC 4204.

Lang J. P., Drake J. (2002). Mesh Network Resiliency Using GMPLS. *Proceedings of the IEEE 90*, 9 (Sept.), 1559–1564.

Lannoo B., Colle D., Pickavet M., Demeester P. (2007). Radio-over-Fiber-Based Solution to Provide Broadband Internet Access to Train Passengers. *IEEE Communications Magazine 45*, 2 (Feb.), 56–62.

Lavrova O. A., Rossi G., Blumenthal D. J. (2000). Rapid Tunable Transmitter with Large Number of ITU Channels Accessible in Less Than 5 ns. In *Proc., ECOC*. 169–170.

Lee K.-C., Li V. O. K. (1993). A Wavelength-Convertible Optical Network. *IEEE/OSA Journal of Lightwave Technology 11*, 5/6 (May/June), 962–970.

Lee S., Sriram K., Kim H., Song J. (2005). Contention-Based Limited Deflection Routing Protocol in Optical Burst-Switched Networks. *IEEE Journal on Selected Areas in Communications 23*, 8 (Aug.), 1596–1611.

Lee S. S. W., Yuang M. C., Tien P.-L. (2004). A Langragean Relaxation Approach to Routing and Wavelength Assignment for Multi-Granularity Optical WDM Networks. In *Proc., IEEE GLOBECOM*. Vol. 3. 1936–1942.

Leenheer M. D., Thysebaert P., Volckaert B. et al. (2006). A View on Enabling-Consumer Oriented Grids through Optical Burst Switching. *IEEE Communications Magazine 44*, 3 (Mar.), 124–131.

Lehman T., Sobieski J., Jabbari B. (2006). DRAGON: A Framework for Service Provisioning in Heterogeneous Grid Networks. *IEEE Communications Magazine 44*, 3 (Mar.), 84–90.

Lethien C., Loyez C., Vilcot J.-P. (2005). Potentials of Radio over Multimode Fiber Systems for the In-Buildings Coverage of Mobile and Wireless LAN Applications. *IEEE Photonics Technology Letters 17*, 12 (Dec.), 2793–2795.

Li C.-C., Kau S.-W., Hwang W.-S. (2002). A CSMA/CP MAC Protocols for IP over WDM Metropolitan Area Ring Networks. In *Proc., IEEE International Conference on Communication Systems*. Vol. 2. 1212–1216.

Li G., Yates J., Wang D., Kalmanek C. (2002). Control Plane Design for Reliable Optical Networks. *IEEE Communications Magazine 40*, 2 (Feb.), 90–96.

Li M., Ramamurthy B. (2006). Integrated Intermediate Waveband and Wavelength Switching for Optical WDM Mesh Networks. In *Proc., IEEE INFOCOM*.

Li M., Yao W., Ramamurthy B. (2005a). A Novel Cost-Efficient On-Line Intermediate Waveband-Switching Scheme in WDM Mesh Networks. In *Proc., IEEE GLOBECOM*. Vol. 4. 2019–2023.

Li M., Yao W., Ramamurthy B. (2005b). Same-Destination-Intermediate Grouping vs. End-to-End Grouping for Waveband Switching in WDM Mesh Networks. In *Proc., IEEE International Conference on Communications (ICC)*. Vol. 3. 1807–1812.

Liao W., Loi C.-H. (2004). Providing Service Differentiation for Optical-Burst-Switched Networks. *IEEE/OSA Journal of Lightwave Technology 22*, 7 (July), 1651–1660.

Lin C.-T., Chen J., Peng P.-C. et al. (2007). Hybrid Optical Access Network Integrating Fiber-to-the-Home and Radio-over-Fiber Systems. *IEEE Photonics Technology Letters 19*, 8 (Apr.), 610–612.

Lin J., Wu J. (2006). Optical Networking Testbeds in China. In *Proc., OFC/NFOEC*.

Lin W.-P. (2005). A Robust Fiber-Radio Architecture for Wavelength-Division-Multiplexing Ring-Access Networks. *IEEE/OSA Journal of Lightwave Technology 23*, 9 (Sept.), 2610–2620.

Lin W.-P., Peng W.-R., Chi S. (2004). A Robust Architecture for WDM Radio-over-Fiber Access Networks. In *Proc., OFC*. Paper FG3. Vol. 2.

Liu J., Ansari N., Ott T. J. (2003). FRR for Latency Reduction and QoS Provisioning in OBS Networks. *IEEE Journal on Selected Areas in Communications 21*, 7 (Sept.), 1210–1219.

Lu X., Mark B. L. (2004). Performance Modeling of Optical-Burst Switching with Fiber Delay Lines. *IEEE Transactions on Communications 52*, 12 (Dec.), 2175–2183.

Ma M., Zhu Y., Cheng T. H. (2003). A Bandwidth Guaranteed Polling MAC Protocol for Ethernet Passive Optical Networks. In *Proc., IEEE INFOCOM*. Vol. 1. 22–31.

Maeda M. W. (1998). Management and Control of Transparent Optical Networks. *IEEE Journal on Selected Areas in Communications 16*, 7 (Sept.), 1008–1023.

Maier M. (2006). On the Future of Optical Ring Networks. In *Proc., IEEE Sarnoff Symposium*.

Maier M., Herzog M. (2007). Long-Lifetime Capacity Upgrades of Ring Networks for Unpredictable Traffic. *IEEE Journal on Selected Areas in Communications 25*, 4 (Apr.), 44–54.

Maier M., Reisslein M. (2004). AWG-Based Metro WDM Networking. *IEEE Communications Magazine 42*, 11 (Nov.), S19–S26.

Maier M., Reisslein M. (2006). Ring in the New for WDM Resilient Packet Ring. *IEEE Potentials 25*, 1 (Jan./Feb.), 22–26.

Maier M., Herzog M., Reisslein M. (2007). STARGATE: The Next Evolutionary Step Towards Unleashing the Potential of WDM EPONs. *IEEE Communications Magazine 45*, 5 (May), 50–56.

Maier M., Herzog M., Scheutzow M., Reisslein M. (2005). PROTECTORATION: A Fast and Efficient Multiple-Failure Recovery Technique for Resilient Packet Ring Using Dark Fiber. *IEEE/OSA Journal of Lightwave Technology 23*, 10 (Oct.), 2816–2838.

Maier M., Scheutzow M., Herzog M., Reisslein M. (2006). Multicasting in IEEE 802.17 Resilient Packet Ring. *OSA Journal of Optical Networking 5*, 11 (Nov.), 841–857.

Mambretti J., Lillethun D., Lange J., Weinberger J. (2006). Optical Dynamic Intelligent Network Services (ODIN): An Experimental Control-Plane Architecture for High-Performance Distributed Environments Based on Dynamic Lightpath Provisioning. *IEEE Communications Magazine 44*, 3 (Mar.), 92–99.

Mannie E. (2004). Generalized Multi-Protocol Label Switching (GMPLS) Architecture. IETF RFC 3945.

Marsan M. A., Bianco A., Leonardi E., Meo M., Neri F. (1996a). MAC Protocols and Fairness Control in WDM Multirings with Tunable Transmitters and Fixed Receivers. *IEEE/OSA Journal of Lightwave Technology 14*, 6 (June), 1230–1244.

Marsan M. A., Bianco A., Leonardi E., Meo M., Neri F. (1996b). On the Capacity of MAC Protocols for All-Optical WDM Multi-Rings with Tunable Transmitters and Fixed Receivers. In *Proc., IEEE INFOCOM*. Vol. 3. 1206–1216.

Marsan M. A., Bianco A., Leonardi E., Morabito A., Neri F. (1997c). SR^3: A Bandwidth-Reservation MAC Protocol for Multimedia Applications over All-Optical WDM Multi-Rings. In *Proc., IEEE INFOCOM*. Vol. 2. 761–768.

Marsan M. A., Bianco A., Leonardi E., Morabito A., Neri F. (1999). All-Optical WDM Multi-Rings with Differentiated QoS. *IEEE Communications Magazine 37*, 2 (Feb.), 58–66.

Marsan M. A., Bianco A., Leonardi E., Neri F., Toniolo S. (1997a). An Almost Optimal MAC Protocol for All-Optical WDM Multi-Rings with Tunable Transmitters and Fixed Receivers. In *Proc., IEEE International Conference on Communications (ICC)*. Vol. 1. 437–442.

Marsan M. A., Bianco A., Leonardi E., Neri F., Toniolo S. (1997b). MetaRing Fairness Control Schemes in All-Optical WDM Rings. In *Proc., IEEE INFOCOM*. Vol. 2. 752–760.

Marsan M. A., Leonardi E., Meo M., Neri F. (2000). Modeling slotted WDM rings with discrete-time Markovian models. *Computer Networks 32*, 599–615.

Mason B. (2000). Widely Tunable Semiconductor Lasers. In *Proc., ECOC*. Vol. 2. 157–158.

Maxemchuk N. F., Ouveysi I., Zukerman M. (2005). A Quantitative Measure for Telecommunications Networks Topology Design. *IEEE/ACM Transactions on Networking 13*, 4 (Aug.), 731–742.

McGarry M. P., Maier M., Reisslein M. (2004). Ethernet PONs: A Survey of Dynamic Bandwidth Allocation (DBA) Algorithms. *IEEE Communications Magazine 42*, 8 (Aug.), S8–S15.

McGarry M. P., Maier M., Reisslein M. (2006). WDM Ethernet Passive Optical Networks. *IEEE Communications Magazine 44*, 2 (Feb.), S18–S25.

Médard M., Marquis D., Barry R. A., Finn S. G. (1997). Security Issues in All-Optical Networks. *IEEE Network 11*, 3 (May/June), 42–48.

Medina R. (2002). Photons vs Electrons. *IEEE Potentials 21*, 2 (April/May), 9–11.

Modiano E. (1999). WDM-Based Packet Networks. *IEEE Communications Magazine 37*, 3 (Mar.), 130–135.

Modiano E., Lin P. J. (2001). Traffic Grooming in WDM Networks. *IEEE Communications Magazine 39*, 7 (July), 124–129.

Mukherjee B. (1992). WDM-Based Local Lightwave Networks Part I: Single-Hop Systems. *IEEE Network 6*, 3 (May), 12–27.

Mukherjee B. (2000). WDM Optical Communication Networks: Progress and Challenges. *IEEE Journal on Selected Areas in Communications 18*, 10 (Oct.), 1810–1824.

Mukherjee B. (2006). *Optical WDM Networks*. Springer.

Munoz R., Pinart C., Martinez R. et al. (2005). The ADRENALINE Testbed: Integrating GMPLS, XML, and SNMP in Transparent DWDM Networks. *IEEE Communications Magazine 43*, 8 (Aug.), S40–S48.

Mynbaev D. K., Scheiner L. L. (2000). *Fiber-Optic Communications Technology*. Pearson Education.

Nadeau T. D., Rakotoranto H. (2005). GMPLS Operations and Management: Today's Challenges and Solutions for Tomorrow. *IEEE Communications Magazine 43*, 7 (July), 68–74.

Neuts M., Rosberg Z., Vu H. L., White J., Zukerman M. (2002). Performance Analysis of Optical Composite Burst Switching. *IEEE Communications Letters 6*, 8 (Aug.), 346–348.

Noirie L. (2003). The Road Towards All-Optical Networks. In *Proc., Optical Fiber Communication Conference (OFC)*. Vol. 2. 615–616.

Nuzman C., Widjaja I. (2006). Time-Domain Wavelength Interleaved Networking with Wavelength Reuse. In *Proc., IEEE INFOCOM*.

Ohmoto R., Ohtsuka H., Ichikawa H. (1993). Fiber-Optic Microcell Radio Systems with a Spectrum Delivery Scheme. *IEEE Journal on Selected Areas in Communications 11*, 7 (Sept.), 1108–1117.

Ohyama T., Yamada T., Akahori Y. (2001). Hybrid integrated multiwavelength photoreceivers consisting of photo-diodes and an arrayed-waveguide grating. In *Proc., IEEE Lasers and Electro-Optics Society (LEOS)*. 835–836.

Oki E., Shiomoto K., Shimazaki D., Yamanaka N., Imajuku W., Takigawa Y. (2005). Dynamic Multilayer Routing Schemes in GMPLS-Based IP+Optical Networks. *IEEE Communications Magazine 43,* 1 (Jan.), 108–114.

O'Mahony M. J., Simeonidou D., Hunter D. K., Tzanakaki A. (2001). The Application of Optical Packet Switching in Future Communication Networks. *IEEE Communications Magazine 39,* 3 (Mar.), 128–135.

O'Mahony M. J., Simeonidou D., Yu A., Zhou J. (1995). The Design of a European Optical Network. *IEEE/OSA Journal of Lightwave Technology 13,* 5 (May), 817–828.

Øverby H. (2005). Packet Loss Rate Differentiation in Slotted Optical Packet Switched Networks. *IEEE Photonics Technology Letters 17,* 11 (Nov.), 2469–2471.

Øverby H., Stol N., Nord M. (2006). Evaluation of QoS Differentiation Mechanisms in Asynchronous Bufferless Optical Packet-Switched Networks. *IEEE Communications Magazine 44,* 8 (Aug.), 52–57.

Papadimitriou D., Drake J., Ash J., Farrel A., Ong L. (2005). Requirements for Generalized MPLS (GMPLS) Signaling Usage and Extensions for Automatically Switched Optical Network (ASON). IETF RFC 4139.

Papagiannaki K., Taft N., Zhang Z.-L., Diot C. (2005). Long-Term Forecasting of Internet Backbone Traffic. *IEEE Transactions on Neural Networks 16,* 5 (Sept.), 1110–1124.

Park J. T. (2004). Resilience in GMPLS Path Management: Model and Mechanism. *IEEE Communications Magazine 42,* 7 (July), 128–135.

Park S.-J., Lee C.-H., Jeong K.-T., Park H.-J., Ahn J.-G., Song K.-H. (2004). Fiber-to-the-Home Services Based on Wavelength-Division-Multiplexing Passive Optical Network. *IEEE/OSA Journal of Lightwave Technology 22,* 11 (Nov.), 2582–2591.

Pasqualini S., Kirstaedter A., Iselt A. et al. (2005). Influence of GMPLS on Network Providers' Operational Expenditures: A Quantitative Study. *IEEE Communications Magazine 43,* 7 (July), 28–34.

Pattavina A. (2005). Architectures and Performance of Optical Packet Switching Nodes for IP Networks. *IEEE/OSA Journal of Lightwave Technology 23,* 3 (Mar.), 1023–1032.

Payne D. B., Stern J. R. (1986). Transparent Single-Mode Fiber Optical Networks. *IEEE/OSA Journal of Lightwave Technology LT-4,* 7 (July), 864–869.

Pew Internet & American Life Project. (2006). Browsing the Web for fun. http://www.pewinternet.org.

Phuritatkul J., Ji Y., Zhang Y. (2006). Blocking Probability of a Preemption-Based Bandwidth-Allocation Scheme for Service Differentiation in OBS Networks. *IEEE/OSA Journal of Lightwave Technology 24,* 8 (Aug.), 2986–2993.

Puype B., Vasseur J.-P., Groebbens A. et al. (2005). Benefits of GMPLS for Multilayer Recovery. *IEEE Communications Magazine 43,* 7 (July), 51–59.

Qiao C. (2000). Labeled Optical Burst Switching for IP-over-WDM Integration. *IEEE Communications Magazine 38,* 9 (Sept.), 104–114.

Rajagopalan B., Pendarakis D., Saha D., Ramamoorthy R. S., Bala K. (2000). IP over Optical Networks: Architectural Aspects. *IEEE Communications Magazine 38,* 9 (Sept.), 94–102.

Ramaswami R. (2002). Optical Fiber Communication: From Transmission to Networking. *IEEE Communications Magazine 40,* 5 (May), 138–147.

Ramaswami R. (2006). Optical Networking Technologies: What Worked and What Didn't. *IEEE Communications Magazine 44,* 9 (Sept.), 132–139.

Ramaswami R., Sivarajan K. N. (2001). *Optical Networks – A Practical Perspective*. San Mateo, CA: Morgan Kaufmann.

Rao N. S. V., Wing W. R., Carter S. M., Wu Q. (2005). UltraScience Net: Network Testbed for Large-Scale Science Applications. *IEEE Communications Magazine 43*, 11 (Nov.), S12–S17.

Renaud M., Masetti F., Guillemot C., Bostica B. (1997). Network and System Concepts for Optical Packet Switching. *IEEE Communications Magazine 35*, 4 (Apr.), 96–102.

Robichaud Y. F., Huang C. (2005). Improved Fairness Algorithm to Prevent Tail Node Induced Oscillations in RPR. In *Proc., IEEE International Conference on Communications (ICC)*. Vol. 1. 402–406.

Rosberg Z., Vu H. L., Zukerman M. (2003). Burst Segmentation Benefit in Optical Switching. *IEEE Communications Letters 7*, 3 (Mar.), 127–129.

Rosen E., Viswanathan A., Callon R. (2001). Multiprotocol Label Switching Architecture. IETF RFC 3031.

Ross F. E. (1986). FDDI – a Tutorial. *IEEE Communications Magazine 24*, 5 (May), 10–17.

Rubin I., Hua H.-K. (1995a). An All-Optical Wavelength-Division Meshed-Ring Packet-Switching Network. In *Proc., IEEE INFOCOM*. Vol. 3. 969–976.

Rubin I., Hua H.-K. (1995b). SMARTNet: An All-Optical Wavelength-Division Meshed-Ring Packet-Switching Network. In *Proc., IEEE GLOBECOM*. Vol. 3. 1756–1760.

Rubin I., Hua H.-K. H. (1997). Synthesis and Throughput Behavior of WDM Meshed-Ring Networks Under Nonuniform Traffic Loading. *IEEE/OSA Journal of Lightwave Technology 15*, 8 (Aug.), 1513–1521.

Rubin I., Ling J. (1999). All-Optical Cross-Connect Meshed-Ring Communications Networks using a Reduced Number of Wavelengths. In *Proc., IEEE INFOCOM*. Vol. 2. 924–931.

Ruffini M., O'Mahony D., Doyle L. (2006). A Testbed Demonstrating Optical IP Switching (OIS) in Disaggregated Network Architectures. In *Proc., TRIDENTCOM*.

Sadot D., Boimovich E. (1998). Tunable Optical Filters for Dense WDM Networks. *IEEE Communications Magazine 36*, 12 (Dec.), 50–55.

Sahara A., Kasahara R., Yamazaki E., Aisawa S., Koga M. (2005). The Demonstration of Congestion-Controlled Optical Burst Switching Network Utilizing Two-Way Signaling – Field Trial in JGN II Testbed. In *Proc., OFC/NFOEC*.

Sahasrabuddhe L. H., Mukherjee B. (1999). Light-Trees: Optical Multicasting for Improved Performance in Wavelength-Routed Networks. *IEEE Communications Magazine 37*, 2 (Feb.), 67–73.

Saleh A. A. M., Simmons J. M. (1999). Architectural Principles of Optical Regional and Metropolitan Area Networks. *IEEE/OSA Journal of Lightwave Technology 17*, 12 (Dec.), 2431–2448.

Sato K., Yamanaka N., Takigawa Y. et al. (2002). GMPLS-Based Photonic Multilayer Router (Hikari Router) Architecture: An Overview of Traffic Engineering and Signaling Technology. *IEEE Communications Magazine 40*, 3 (Mar.), 96–101.

Sen S., Wang J. (2004). Analyzing Peer-to-Peer Traffic Across Large Networks. *IEEE/ACM Transactions on Networking 12*, 2 (Apr.), 219–232.

Sengupta S., Kumar V., Saha D. (2003). Switched Optical Backbone for Cost-Effective Scalable Core IP Networks. *IEEE Communications Magazine 41*, 6 (June), 60–70.

Seo S.-W., Bergman K., Prucnal P. R. (1996). Transparent Optical Networks with Time-Division Multiplexing. *IEEE Journal on Selected Areas in Communications 14*, 5 (June), 1039–1051.

Shami A., Bai X., Ghani N., Assi C. M., Mouftah H. T. (2005). QoS Control Schemes for Two-Stage Ethernet Passive Optical Access Networks. *IEEE Journal on Selected Areas in Communications 23,* 8 (Aug.), 1467–1478.

Sherif S. R., Hadjiantonis A., Ellinas G., Assi C., Ali M. A. (2004). A Novel Decentralized Ethernet-Based PON Access Architecture for Provisioning Differentiated QoS. *IEEE/OSA Journal of Lightwave Technology 22,* 11 (Nov.), 2483–2497.

Shin D. J., Jung D. K., Shin H. S. et al. (2005). Hybrid WDM/TDM-PON with Wavelength-Selection-Free Transmitters. *IEEE/OSA Journal of Lightwave Technology 23,* 1 (Jan.), 187–195.

Shinohara H. (2005). Broadband Access in Japan: Rapidly Growing FTTH Market. *IEEE Communications Magazine 43,* 9 (Sept.), 72–78.

Shiomoto K., Oki E., Imajuku W., Okamoto S., Yamanaka N. (2003). Distributed Virtual Network Topology Control Mechanism in GMPLS-Based Multiregion Networks. *IEEE Journal on Selected Areas in Communications 21,* 8 (Oct.), 1254–1262.

Shokrani A., Khorsandi S., Lambadaris I., Khan L. (2005). Virtual Queuing: An Efficient Algorithm for Bandwidth Management in Resilient Packet Rings. In *Proc., IEEE International Conference on Communications (ICC).* Vol. 2. 982–988.

Shrikande K., Srivatsa A., White I. M. et al. (2000). CSMA/CA MAC Protocols for IP-HORNET: An IP over WDM Metropolitan Area Ring Network. In *Proc., IEEE GLOBECOM.* Vol. 2. 1303–1307.

Shrikhande K., White I. M., Rogge M. S. et al. (2001). Performance Demonstration of a Fast-Tunable Transmitter and Burst-Mode Packet Receiver for *HORNET.* In *Proc., OFC.* Paper ThG2.

Shrikhande K. V., White I. M., Wonglumsom D.-R. et al. (2000). HORNET: A Packet-over-WDM Multiple Access Metropolitan Area Ring Network. *IEEE Journal on Selected Areas in Communications 18,* 10 (Oct.), 2004–2016.

Simeonidou D., Nejabati R., Zervas G., Klonidis D., Tzanakaki A., O'Mahony M. J. (2005). Dynamic Optical-Network Architectures and Technologies for Existing and Emerging Grid Services. *IEEE/OSA Journal of Lightwave Technology 23,* 10 (Oct.), 3347–3357.

Smiljanic A., Boroditsky M., Frigo N. J. (2001). Optical Packet-Switched Ring Network with Flexible Bandwidth Allocation. In *Proc., IEEE Workshop on High Performance Switching and Routing.* 83–87.

Smiljanic A., Boroditsky M., Frigo N. J. (2002). High-Capacity Packet-Switched Optical Ring Network. *IEEE Communications Letters 6,* 3 (Mar.), 111–113.

Sohraby K., Zhang Z., Chu X., Li B. (2003). Resource Management in an Integrated Optical Network. *IEEE Journal on Selected Areas in Communications 21,* 7 (Sept.), 1052–1062.

Spadaro S., Solé-Pareta J., Careglio D., Wajda K., Szymański A. (2004). Positioning of the RPR Standard in Contemporary Operator Environments. *IEEE Network 18,* 2 (Mar./Apr.), 35–40.

Spencer M. J., Summerfield M. A. (2000). WRAP: A Medium Access Control Protocol for Wavelength-Routed Passive Optical Networks. *IEEE/OSA Journal of Lightwave Technology 18,* 12 (Dec.), 1657–1676.

Summerfield M. A. (1997). MAWSON: A Metropolitan Area Wavelength Switched Optical Network. In *Proc., APCC '97.* 327–331.

Sun Y., Hashiguchi T., Minh V. Q., Wang X., Morikawa H., Aoyama T. (2005). Design and Implementation of an Optical Burst-Switched Network Testbed. *IEEE Communications Magazine 43,* 11 (Nov.), S48–S55.

Suwala G., Swallow G. (2004). SONET/SDH-like Resilience for IP Networks: A Survey of Traffic Protection Mechanisms. *IEEE Network 18,* 2 (Mar./Apr.), 20–25.

Szegedi P., Lakatos Z., Spaeth J. (2006). Signaling Architectures and Recovery Time Scaling for Grid Applications in IST Project MUPBED. *IEEE Communications Magazine 44,* 3 (Mar.), 74–82.

Tachibana T., Kasahara S. (2006). Burst-Cluster Transmission: Service Differentiation Mechanism for Immediate Reservation in Optical Burst Switching Networks. *IEEE Communications Magazine 44,* 5 (May), 46–55.

Takada A., Park J. H. (2002). Architecture of Ultrafast Optical Packet Switching Ring Network. *IEEE/OSA Journal of Lightwave Technology 20,* 12 (Dec.), 2306–2315.

Tan T. C. (2000). Gigabit Ethernet and Structured Cabling. *Electronics & Communication Engineering Journal 12,* 4 (Aug.), 156–166.

Tang P. K., Ong L. C., Alphones A., Luo B., Fujise M. (2004). PER and EVM Measurements of a Radio-over-Fiber Network for Cellular and WLAN System Applications. *IEEE/OSA Journal of Lightwave Technology 22,* 11 (Nov.), 2370–2376.

Tatipamula M., Faucheur F. L., Otani T., Esaki H. (2005). Implementation of IPv6 Services over a GMPLS-Based IP/Optical Network. *IEEE Communications Magazine 43,* 5 (May), 114–122.

Teng J., Rouskas G. N. (2005). Wavelength Selection in OBS Networks Using Traffic Engineering and Priority-Based Concepts. *IEEE Journal on Selected Areas in Communications 23,* 8 (Aug.), 1658–1669.

Thompson G. O. (1997). Work Progress on Gigabit Ethernet. *Computer 30,* 5 (May), 95–96.

Toliver P., Runser R., Young J., Jackel J. (2003). Experimental Field Trial of Waveband Switching and Transmission in a Transparent Reconfigurable Optical Network. In *Proc., OFC.* Vol. 2. 783–784.

Tong F. (1998). Multiwavelength Receivers for WDM Systems. *IEEE Communications Magazine 36,* 12 (Dec.), 42–49.

Turner J. (1999). Terabit Burst Switching. *Journal of High Speed Networks 8,* 1, 3–16.

van de Voorde I., van der Plas G. (1997). Full Service Optical Access Networks: ATM Transport on Passive Optical Networks. *IEEE Communications Magazine 35,* 4 (Apr.), 70–75.

Vanderbauwhede W. A., Harle D. A. (2005). Architecture, Design, and Modeling of the OPSnet Asynchronous Optical Packet Switching Node. *IEEE/OSA Journal of Lightwave Technology 23,* 7 (July), 2215–2228.

Varma E. L., Sankaranarayanan S., Newsome G., Lin Z.-W., Epstein H. (2001). Architecting the Services Optical Network. *IEEE Communications Magazine 39,* 9 (Sept.), 80–87.

Vaughan-Nichols S. J. (2002). Will 10-Gigabit Ethernet Have a Bright Future? *Computer 35,* 6 (June), 22–24.

Veeraraghavan M., Karri R., Moors T., Karol M. (2001). Architectures and Protocols that Enable New Applications on Optical Networks. *IEEE Communications Magazine 39,* 3 (Mar.), 118–127.

Veeraraghavan M., Zheng X., Huang Z. (2006). On the Use of Connection-Oriented Networks to Support Grid Computing. *IEEE Communications Magazine 44,* 3 (Mar.), 118–123.

Venkateswaran R. (2001). Virtual Private Networks. *IEEE Potentials 20,* 1 (Feb./Mar.), 11–15.

Verma S., Chaskar H., Ravikanth R. (2000). Optical Burst Switching: A Viable Solution for Terabit IP Backbone. *IEEE Network 14,* 6 (Nov./Dec.), 48–53.

Vigoureux M., Berde B., Andersson L. et al. (2005). Multilayer Traffic Engineering for GMPLS-Enabled Networks. *IEEE Communications Magazine 43,* 7 (July), 44–50.

Vlachos K. G., Monroy I. T., Koonen A. M. J., Peucheret C., Jeppesen P. (2003). STOLAS: Switching Technologies for Optically Labeled Signals. *IEEE Communications Magazine 41*, 11 (Nov.), S9–S15.

Vokkarane V. M., Jue J. P. (2003). Prioritized Burst Segmentation and Composite Burst-Assembly Techniques for QoS Support in Optical Burst-Switched Networks. *IEEE Journal on Selected Areas in Communications 21*, 7 (Sept.), 1198–1209.

Vokkarane V. M., Jue J. P. (2005). Segmentation-Based Nonpreemptive Channel Scheduling Algorithms for Optical Burst-Switched Networks. *IEEE/OSA Journal of Lightwave Technology 23*, 10 (Oct.), 3125–3137.

Vu H. L., Zukerman M. (2002). Blocking Probability for Priority Classes in Optical Burst Switching Networks. *IEEE Communications Letters 6*, 5 (May), 214–216.

Wagner R. E., Alferness R. C., Saleh A. A. M., Goodman M. S. (1996). MONET: Multiwavelength Optical Networking. *IEEE/OSA Journal of Lightwave Technology 14*, 6 (June), 1349–1355.

Wang D., Ramakrishnan K. K., Kalmanek C., Doverspike R., Smiljanić A. (2004). Congestion Control in Resilient Packet Rings. In *Proc., IEEE International Conference on Network Protocols (ICNP)*. 108–117.

Wei J. Y., R. I. McFarland, Jr. (2000). Just-in-Time Signaling for WDM Optical Burst Switching Networks. *IEEE/OSA Journal of Lightwave Technology 18*, 12 (Dec.), 2019–2037.

Wei J. Y., Shen C.-C., Wilson B. J., Post M. J., Tsai Y. (1998). Connection Management for Multiwavelength Optical Networking. *IEEE Journal on Selected Areas in Communications 16*, 7 (Sept.), 1097–1108.

White I. M., Rogge M. S., Hsueh Y.-L., Shrikhande K., Kazovsky L. G. (2002a). Experimental Demonstration of the HORNET Survivable Bidirectional Ring Architecture. In *Proc., OFC*. Paper WW1.

White I. M., Rogge M. S., Shrikhande K., Kazovsky L. G. (2002b). Design of a Control-Channel-Based Media-Access-Control Protocol for HORNET. *Journal of Optical Networking 1*, 12 (Dec.), 460–473.

White I. M., Rogge M. S., Shrikhande K., Kazovsky L. G. (2003). A Summary of the HORNET Project: A Next-Generation Metropolitan Area Network. *IEEE Journal on Selected Areas in Communications 21*, 9 (Nov.), 1478–1494.

White I. M., Shrikhande K., Rogge M. S. et al. (2000). Architecture and Protocols for HORNET: A Novel Packet-over-WDM Multiple-Access MAN. In *Proc., IEEE GLOBECOM*. Vol. 2. 1298–1302.

Widjaja I., Saniee I., Giles R., Mitra D. (2003). Light Core and Intelligent Edge for a Flexible, Thin-Layered, and Cost-Effective Optical Transport Network. *IEEE Communications Magazine 41*, 5 (May), S30–S36.

Williams P. J., Robbins D. J., Robson F. O., Whitbread N. D. (2000). High Power and Wide Quasi-Continuous Tuning, Surface Ridge SG-DBR Lasers. In *Proc., ECOC*. Vol. 2. 163–164.

Wilson B. J., Stoffel N. G., Pastor J. L. et al. (2000). Multiwavelength Optical Networking Management and Control. *IEEE/OSA Journal of Lightwave Technology 18*, 12 (Dec.), 2008–2057.

Wong E. W. M., Fumagalli A., Chlamtac I. (1995). Performance Evaluation of CROWNs: WDM Multi-Ring Topologies. In *Proc., IEEE International Conference on Communications (ICC)*. Vol. 2. 1296–1301.

Wu J., Wu J.-S., Tsao H.-W. (1994). A Fiber Distribution System for Microcellular Radio. *IEEE Photonics Technology Letters 6*, 9 (Sept.), 1150–1152.

Wu J.-S., Wu J., Tsao H.-W. (1998). A Radio-over-Fiber Network for Microcellular System Application. *IEEE Transactions on Vehicular Technology 47,* 1 (Feb.), 84–94.

Xiao C., Bing B., Chang G. K. (2005). An Efficient Reservation MAC Protocol with Preallocation for High-Speed WDM Passive Optical Networks. In *Proc., IEEE INFOCOM.* Vol. 1. 444–454.

Xin C., Qiao C. (2001). A Comparative Study of OBS and OFS. In *Proc., Optical Fiber Communication (OFC).* Vol. 4. ThG7-1–ThG7-3.

Xin C., Qiao C., Dixit S. (2004). Traffic Grooming in Mesh WDM Optical Networks – Performance Analysis. *IEEE Journal on Selected Areas in Communications 22,* 9 (Nov.), 1658–1669.

Xin C., Ye Y., Wang T.-S., Dixit S., Qiao C., Yoo M. (2001). On an IP-Centric Optical Control Plane. *IEEE Communications Magazine 39,* 9 (Sept.), 88–93.

Xinwan L., Jianping C., Guiling W., Hui W., Ailun Y. (2005). An Experimental Study of an Optical Burst Switching Network Based on Wavelength-Selective Optical Switches. *IEEE Communications Magazine 43,* 5 (May), S3–S10.

Xiong Y., Vandenhoute M., Cankaya H. C. (2000). Control Architecture in Optical Burst-Switched WDM Networks. *IEEE Journal on Selected Areas in Communications 18,* 10 (Oct.), 1838–1851.

Xu J., Qiao C., Li J., Xu G. (2004). Efficient Burst Scheduling Algorithms in Optical Burst-Switched Networks Using Geometric Techniques. *IEEE Journal on Selected Areas in Communications 22,* 9 (Nov.), 1796–1811.

Xu L., Perros H. G., Rouskas G. (2001). Techniques for Optical Packet Switching and Optical Burst Switching. *IEEE Communications Magazine 39,* 1 (Jan.), 136–142.

Yao S., Mukherjee B. (2003). Design of Hybrid Waveband-Switched Networks with OEO Traffic Grooming. In *Proc., Optical Fiber Communication (OFC).* Vol. 1. 357–358.

Yao S., Mukherjee B., Yoo S. J. B., Dixit S. (2003a). A Unified Study of Contention-Resolution Schemes in Optical Packet-Switched Networks. *IEEE/OSA Journal of Lightwave Technology 21,* 3 (Mar.), 672–683.

Yao S., Ou C., Mukherjee B. (2003b). Design of Hybrid Optical Networks With Waveband and Electrical TDM Switching. In *Proc., IEEE GLOBECOM.* Vol. 5. 2803–2808.

Yao S., Yoo S. J. B., Mukherjee B., Dixit S. (2001). All-Optical Packet Switching for Metropolitan Area Networks: Opportunities and Challenges. *IEEE Communications Magazine 39,* 3 (Mar.), 142–148.

Yoo M., Qiao C., Dixit S. (2000). QoS Performance of Optical Burst Switching in IP-over-WDM Networks. *IEEE Journal on Selected Areas in Communications 18,* 10 (Oct.), 2062–2071.

Yoo M., Qiao C., Dixit S. (2001). Optical Burst Switching for Service Differentiation in the Next-Generation Optical Internet. *IEEE Communications Magazine 39,* 2 (Feb.), 98–104.

Yu J., Gu J., Liu X., Jia Z., Chang G.-K. (2005). Seamless Integration of an 8×2.5 Gb/s WDM-PON and Radio-over-Fiber Using All-Optical Up-Conversion Based on Raman-Assisted FWM. *IEEE Photonics Technology Letters 17,* 9 (Sept.), 1986–1988.

Yu X., Li J., Cao X., Chen Y., Qiao C. (2004). Traffic Statistics and Performance Evaluation in Optical Burst Switched Networks. *IEEE/OSA Journal of Lightwave Technology 22,* 12 (Dec.), 2722–2738.

Yuan P., Gambiroza V., Knightly E. (2004). The IEEE 802.17 Media Access Protocol for High-Speed Metropolitan-Area Resilient Packet Rings. *IEEE Network 18,* 3 (May/June), 8–15.

Yuan X. C., Li V. O. K., Li C. Y., Wai P. K. A. (2003). A Novel Self-Routing Address Scheme for All-Optical Packet-Switched Networks with Arbitrary Topologies. *IEEE/OSA Journal of Lightwave Technology 21,* 2 (Feb.), 329–339.

Zang H., Jue J. P., Mukherjee B. (1999). Photonic Slot Routing in All-Optical WDM Mesh Networks. In *Proc., IEEE GLOBECOM.* Vol. 2. 1449–1453.

Zang H., Jue J. P., Mukherjee B. (2000). Capacity Allocation and Contention Resolution in a Photonic Slot Routing All-Optical WDM Mesh Network. *IEEE/OSA Journal of Lightwave Technology 18,* 12 (Dec.), 1728–1741.

Zhang L., An E.-S., Youn C.-H., Yeo H.-G., Yang S. (2003). Dual DEB-GPS Scheduler for Delay-Constraint Applications in Ethernet Passive Optical Networks. *IEICE Transactions on Communications E86-B,* 5 (May), 1575–1584.

Zhang Q., Vokkarane V. M., Jue J. P., Chen B. (2004). Absolute QoS Differentiation in Optical Burst-Switched Networks. *IEEE Journal on Selected Areas in Communications 22,* 9 (Nov.), 1781–1795.

Zhang T., Lu K., Jue J. P. (2006). Shared Fiber Delay Line Buffers in Asynchronous Optical Packet Switches. *IEEE Journal on Selected Areas in Communications 24,* 4 (Apr.), 118–127.

Zhang W., Wu J., Xu K., Lin J. T. (2006). TCP Performance Experiment on OBS Network Testbed. In *Proc., OFC/NFOEC.*

Zhang Z., Fu J., Guo D., Zhang L. (2001). Lightpath Routing for Intelligent Optical Networks. *IEEE Network 15,* 4 (July/Aug.), 28–35.

Zheng J., Mouftah H. T. (2005). Media Access Control for Ethernet Passive Optical Networks: An Overview. *IEEE Communications Magazine 43,* 2 (Feb.), 145–150.

Zheng X., Veeraraghavan M., Rao N. S. V., Wu Q., Zhu M. (2005). CHEETAH: Circuit-Switched High-Speed End-to-End Transport Architecture Testbed. *IEEE Communications Magazine 43,* 8 (Aug.), S11–S17.

Zhou L., Chai T. Y., Saradhi C. V. et al. (2006). Development of a GMPLS-Capable WDM Optical Network Testbed and Distributed Storage Application. *IEEE Communications Magazine 44,* 2 (Feb.), S26–S32.

Zirngibl M. (1998). Multifrequency Lasers and Applications in WDM Networks. *IEEE Communications Magazine 36,* 12 (Dec.), 39–41.

Index